Introduction to
ENVIRONMENTAL TOXICOLOGY
Impacts of Chemicals Upon Ecological Systems

Introduction to
ENVIRONMENTAL TOXICOLOGY

Impacts of Chemicals Upon Ecological Systems

Wayne G. Landis
Ming-Ho Yu

LEWIS PUBLISHERS
Boca Raton   Ann Arbor   London   Tokyo

Peiwen Qiao   Biology

Library of Congress Cataloging-in-Publication Data

Landis, Wayne G.
  Introduction to environmental toxicology : impacts of chemicals upon ecological systems / authors,
Wayne G. Landis, Ming-Ho Yu.
      p.   cm.
  Includes bibliographical references and index.
  ISBN 0-87371-515-2
  1. Pollution--Environmental aspects. 2. Pollutants--Toxicology.
  I. Ming-Ho Yu. II. Title.
  QH545.A1L35   1995
  574.5′222--dc20   94-23419
                                                                                                                    CIP

© 1995 by CRC Press, Inc.
Lewis Publishers is an imprint of CRC Press

No claim to original U.S. Government works
International Standard Book Number 0-87371-515-2
Library of Congress Card Number 94-23419
Printed in the United States of America 1 2 3 4 5 6 7 8 9 0
Printed on acid-free paper

Preface

We have prepared this text because we had no suitable book for teaching courses introducing environmental toxicology and biochemistry. A lot of this was prepared from our lecture notes and graphics and has benefited greatly from outside peer review. The emphasis is on environmental toxicology as it pertains to ecological systems. We have purposefully not emphasized occupational and human health issues. In addition, we have not attempted to construct an enclusive compendium of effects and toxicity, rather we have concentrated on basic mechanisms and processes. Since we plan to teach using this manuscript, we have sized it according to the demands of a two quarter or one semester course in environmental toxicology.

In preparing this text, we assumed that the readers would have a basic background in organic chemistry, biochemistry, genetics, physiology, ecology, and statistics. The book concentrates on toxicological issues and their integration with ecology so that the impacts of chemicals on ecological systems can be addressed.

One of the major difficulties in preparing this book has been the rate of change seen in the field. The U.S. Environmental Protection Agency prepared a new Framework for Ecological Risk Assessment, non-linear dynamics has become a major part of ecological theory, and new methods of examining effects at the level of community and ecosystem have been developed during the writing of this book. In two years we are sure that major revisions will be necessary to keep pace with developments. Due to the pace of research, new sections were being added right to the deadline for manuscript submittal. Hopefully, new editions will incorporate expanded sections on metals and the molecular biology of dioxin. We do suggest the addition of a set of primary literature readings in order to supplement and update the material in this text.

Wayne G. Landis is the Director of the Institute of Environmental Toxicology and Chemistry of Huxley College, Western Washington University. He received his undergraduate degree from Wake Forest University and his M.A. and Ph.D. in Zoology from Indiana University. With a background in protozoan genetics and ecology, his research has since concentrated on environmental toxicology. In the last several years he has published over 80 papers, received two patents on microbial degradation. In 1991 he chaired the annual Environmental Toxicology and Risk Assessment Symposium sponsored by the American Society for Testing and Materials held in Atlantic City and served as Organizational Chair for the Annual Meeting of the Society for Environmental Toxicology and Chemistry held in Seattle. During 1992 he served as President of the Pacific Northwest Chapter of the Society for Environmental Toxicology and Chemistry (PNWSETAC). Dr. Landis also has served on the editorial boards of several journals. Since 1989, he has also edited three books on aquatic toxicology and risk assessment published by the American Society for Testing and Materials.

Dr. Landis teaches courses in environmental and aquatic toxicology, environmental risk assessment and population biology. His current research includes developing new methods of evaluating environmental toxicity using birds and marine organisms, establishing interspecies structure-activity models, evaluating multispecies toxicity tests, and the description of how ecosystems respond to stressors.

Perhaps the most intriguing avenue of research has been the implementation of chaos and complexity theory to describe the dynamics of ecological systems after toxicant stress. This research has cast a great deal of doubt as to the existence of ecosystem recovery or stability in regards to the dynamics after a stressor event.

Ming-Ho Yu, Ph.D., is a Professor at Huxley College of Environmental Studies, Western Washington University, where he teaches courses in Environmental Toxicology and Nutrition. Dr. Yu received his B.S. degree in Agricultural Chemistry from National Taiwan University in Taipei, Taiwan, and his M.S. and Ph.D. degrees in Plant Nutrition and Biochemistry at Utah State University in Logan, Utah. He did his postdoctoral work at the University of Alberta in Edmonton, Alberta, Canada, and Utah State University. He was Visiting Professor at the Department of Public Health and Hygiene, Iwate medical University in Morioka, Japan.

Dr. Yu's main research interests are in the physiological and biochemical effects of fluoride and heavy metals on living systems, expecially on plants. He is a member of Sigma Xi, the American Association for the Advancement of Science, the American Chemical Society, the American Institute of Nutrition, the International Society for Fluoride Research, the New York Academy of Sciences, and the Society of Environmental Toxicology and Chemistry.

Dr. Yu is Vice President (President-elect) of the International Society for Fluoride Research, and is co-editor of *Environmental Sciences*, and international journal on environmental physiology and toxicology, published in Tokyo, Japan.

Acknowledgments

A major part of this book was written based on the notes and other course materials I used in teaching environmental toxicology-related courses at Western Washington University over the last 20 plus years. I want to thank my former students who took those classes from me. Many of them made critical comments on the course materials I used, and their comments inspired me greatly. Special appreciation is due to my wife Ervena, for her moral support during the course of preparing the manuscript.

M.H.Y.

The students of my environmental toxicology courses during the last 5 years at Western Washington University have suffered through the notes and figures that make up my contribution to this text, and I thank them for participating in this undertaking. Traci Litwiller compiled the methods summaries and conducted numerous literature searches. Lisa Holmquist was instrumental in the editing, and in the checking of the numerous references. Ruth Noellgen let me modify several of the figures from her thesis for this text. Linda S. Landis prepared the study questions and provided her unrelenting support of this project and in spite of the evenings often spent alone raising two delightful daughters, Margaret and Eva.

W.G.L.

We would also like to thank the numerous reviewers that made comments on each of the chapters. They are listed below and their efforts greatly improved the quality of the manuscript.

Rick Bennett   Chris Ingersoll
Nigel Blakley   Mike Lewis
Peter Chapman   Greg Linder
C.J. Driver   Beth Power
Donald L. Fox   Tom Sibley
Chris Grue   Frieda Taub
Jack Hardy   Sandra Thomson
Walter W. Heck   P.D. Whanger
Jamie Hobson   Maurice Zeeman

Table of Contents

Chapter 1
Introduction to Environmental Toxicology 1
Environmental Toxicology as an Interdisciplinary Science 1
Legislation 3
Introduction to the Textbook 4

Chapter 2
A Framework for Environmental Toxicology 7
Chemical and Physicochemical Characteristics 8
Bioaccumulation/Biotransformation/Biodegradation 8
Receptor and the Mode of Action 9
Biochemical and Molecular Effects 10
Physiological and Behavioral Effects 10
Population Parameters 11
Community Effects 12
Ecosystem Effects 13
Spatial and Temporal Scales 13
References and Suggested Readings 15
Study Questions 15

Chapter 3
An Introduction to Toxicity Testing 17
The Dose-Response Curve 18
Standard Methods 24
Disadvantages of Standard Methods 27
Classification of Toxicity Tests 27
Design Parameters for Toxicity Tests 29
Exposure Scenarios 31
Test Organisms 32
Comparison of Test Species 33
Statistical Design Parameters 34
Overview of Available Statistical Methods for the Evaluation of
Toxicity Tests 35
  Commonly Used Methods for the Calculation of Endpoints 35
  Comparison of Calculations of Several Programs for Calculating Probit
  Analysis 36

Data Analysis for Chronic and Multispecies Toxicity Tests 37
Analysis of Multispecies Toxicity Tests (Microcosms, Mesocosms)..............40
Nonmetric Clustering and Association Analysis 42
References and Suggested Readings 43
Study Questions...44

Chapter 4
Survey and Review of Typical Toxicity Test Methods 45

Single-Species Toxicity Tests 46
Daphnia 48-H Acute Toxicity Test...46
Algal 96-H Growth Toxicity Test..48
Acute Toxicity Tests with Aquatic Vertebrates and Macroinvertebrates 51
Terrestrial Vertebrate Toxicity Tests 55
Animal Care and Use Considerations...58
Frog Embryo Teratogenesis Assay: FETAX 61
Multispecies Toxicity Tests 63
Standardized Aquatic Microcosm 65
Mixed Flask Culture 67
FEFRA Microcosm 68
Soil Core Microcosm 70
Summary...71
Appendix: The Natural History and Utilization of Selected Test Species 72
Aquatic Vertebrates 72
Coho Salmon (*Oncorhynchus kisutch*)..72
Rainbow Trout (*Oncorhynchus gairdneri*) 73
Brook Trout (*Salvelinus fontinalis*)...73
Boldfish (*Carassius auratus*) 73
Fathead Minnow (*Pimephales promelas*) 74
Channel Catfish (*Ictalurus punctatus*) 74
Bluegill (*Lepomis macrochirus*)..74
Green Sunfish (*Lepomis cyanellus*) 74
Invertebrates — Freshwater 75
Daphnids (*Daphnia magna, D. pulex, D. pulicaria,*
Ceriodaphnia dubia)...75
Amphipods (*Gammarus lacustris, G. fasciatus,*
G. pseudolimnaeus, Hyalella azteca) 75
Crayfish (*Orconectes* sp., *Combarus* sp., *Procambarus* sp.,
Pacifastacus leniusculus) 76
Stoneflies (*Pteronarcys* sp.) 76
Mayflies (*Baetis* sp., *Ephemerella* sp., *Hexagenia limbata,*
H. bilineata) 76
Medges (*Chironomus* sp.) 77
Snails (*Physa integra, P. heterostropha, Amnicola limosa*):
(Mollusca, Gastropoda) 77

Planaria (*Dugesia tigrina*): (Platyhelminthes, Turbellaria) 78
Invertebrates: Saltwater 78
Copepods (*Acartia clausi, Acartia tonsa*) 78
Algae 78
Chlamydomonas reinhardi 78
Ulothrix sp. 79
Microcystis aeruginosa 79
Anabaena flos-aquae 79
Avian Species 79
Mallard (*Anas platyrhynchos*) 79
Northern Bobwhite (*Colinus virginianus*) 79
Ring-Necked Pheasant (*Phasianus colchicus*) 80
References and Suggested Readings 80
Study Questions 83

Chapter 5
Routes of Exposure and Modes of Action 85
The Damage Process 85
Atmospheric Pollutants and Plants 85
Plant Injury 86
Vertebrates 87
Exposure 87
Uptake 87
Transport 88
Storage 88
Metabolism 89
Excretion 89
Mechanisms of Action 89
Disruption or Destruction of Cellular Structure 90
Direct Chemical Combination with a Cellular Constituent 90
Effect on Enzymes 90
Secondary Action as a Result of the Presence of a Pollutant 92
Metal Shift 94
Common Modes of Action in Detail 95
Narcosis 95
Organophosphates 96
Monohaloacetic Acids 99
Introduction to QSAR 102
Construction of QSAR Models 104
Typical QSAR Development 106
Estimation of Toxicity Using QSAR 110
References and Suggested Readings 111
Study Questions 113

Chapter 6

Factors Modifying the Activity of Toxicants 115
Physicochemical Properties of Toxicants 115
Time and Mode of Exposure 116
Environmental Factors 116
    Temperature 116
    Humidity 117
    Light Intensity 117
Interaction of Pollutants 117
    Synergism and Potentiation 117
    Antagonism 118
Toxicity of Mixtures 118
Mixture Estimation System 121
Biological Factors Affecting Toxicity 122
    Plants... 122
    Animals 123
        Genetic Factors 123
        Developmental Factors 123
        Diseases.. 123
        Lifestyle 123
        Sex Variation 124
    Nutritional Factors 124
        Fasting/Starvation 124
        Proteins 125
        Carbohydrates 125
        Lipids 126
        Vitamin A 127
        Vitamin D 128
        Vitamin E 128
        Vitamin C 128
        Minerals 130
References and Suggested Readings 131
Study Questions.. 132

Chapter 7

Inorganic Gaseous Pollutants 135
Sulfur Oxides.. 135
    Sources of SO_2 135
    Characteristics of SO_2 136
    Effects on Plants... 137
    Effect on Animals 139
    Effect on Humans... 139
Nitrogen Oxides 140
    Forms and Formation of Nitrogen Oxides... 140

Major Reactive N Species in the Troposphere 140
Effects on Plants 141
Effects on Humans and Animals 142
    Physiological Effects 142
    Biochemical Effects 143
Ozone 143
    Sources 143
        Photochemical Smog 144
    Effects on Plants 145
    Effects on Humans and Animals 146
    Biochemical Effects 147
Carbon Monoxide 148
    Formation of CO 149
    Toxicological Effects 150
    Mechanism of Action 150
    Human Exposure to CO 151
Fluoride 152
    Environmental Sources and Forms of Fluoride 152
    Fluoride Pollution 152
    Effects on Plants 153
    Effect on Animals 154
    Effects on Human Health 155
    Biochemical Effects of Fluoride 156
References and Suggested Readings 156
Study Questions 159

Chapter 8
Biotransformation, Detoxification, and Biodegradation 161
Introduction 161
Metabolism of Environmental Chemicals: Biotransformation 161
    Types of Biotransformation 162
    Mechanisms of Biotransformation 162
    Consequence of Biotransformation 166
Microbial Degradation 167
Bioremediation 175
Isolation and Engineering of Degradative Organisms 177
The Genetics of Degradative Elements 179
An Example of a Detoxification Enzyme — The OPA Anhydrolases 180
Characteristics of the *OPD* Gene Product and Other Bacterial
OPA Anhydrolases 182
Eucaryotic OPA Anhydrolases 184
Characteristics of Other Invertebrate Metazoan Activities 186
Characteristics of the Fish Activities 186
Comparison of the OPA Anhydrases 186

Natural Role of the OPA Anhydrases 188
References and Suggested Readings 190
Study Questions 195

Chapter 9
Measurement and Evaluation of the Ecological Effects of
Toxicants 197
Introduction 197
Measurement of Ecological Effects at Various Levels of Biological
Organization 197
Bioaccumulation/Biotransformation/Biodegradation 200
Molecular and Physiological Indicators of Chemical Stress — Biomarkers 200
    Enzymatic and Biochemical Processes 201
Physiological and Histological Indicators 202
Toxicity Tests and Population Level Indicators 204
Sentinel Organisms or *In Situ* Biomonitoring 206
Population Parameters 207
Assemblage and Community Parameters 208
Interpretation of Effects at the Population, Community, and Ecosystem
Levels of Organization 211
Resource Competition as a Model of the Direct and Indirect Effects
of Pollutants 211
    Case 1 215
    Case 2 215
Modeling of Populations Using Age Structure and Survivorship Models 220
Population Biology, Nonlinear Systems, and Chaos 220
Community and Ecosystem Effects 227
Application of Multivariate Techniques 228
    Normalized Ecosystem Strain 228
    State Space of Ecosystems 230
    Nonmetric Clustering 231
Interpretation of Ecosystem Level Impacts — Stability and Ecosystem
Dynamics 237
Appendix: Multivariate Techniques — Nonmetric Clustering 243
References and Suggested Readings 244
Study Questions 249

Chapter 10
Ecological Risk Assessment and Environmental Toxicology 251
Introcution 251
Basics of Risk Assessment 252
Ecological Risk Assessment 253
Ecological Risk Assessment Framework 254
    Problem Formulation 254

Analysis 259
    Exposure Analysis 260
    Characterization of Ecological Effects..261
    Ecological Response Analyses 262
    Stressor-Response Profile 263
    Data Acquisition, Verification, and Monitoring 264
  Risk Characterization 264
    Integration 265
    Risk Description 266
    Interpretation of Ecological Significance...267
    Discussion Between the Risk Assessor and Risk Manager 267
    Data Acquisition, Verification, and Monitoring 268
References and Suggested Readings 268
Study Questions..268

Appendix A — U.S. EPA Document "A Framework for Ecological Risk Assessment" 271

Index 317

CHAPTER 1

Introduction to Environmental Toxicology

Environmental toxicology is the study of the impacts of pollutants upon the structure and function of ecological systems. For the purposes of this text, the emphasis will be upon ecological systems, at every level of biological organization, from molecular to ecosystem. The broad scope of environmental toxicology requires a multidisciplinary approach.

ENVIRONMENTAL TOXICOLOGY AS AN INTERDISCIPLINARY SCIENCE

Environmental toxicology takes and assimilates material from a variety of disciplines. Terrestrial and aquatic ecologists, chemists, molecular biologists, geneticists, and mathematicians are all important in the evaluation of the impacts of chemicals on biological systems (Figure 1.1). Ecology provides the basis of our ability to interpret the interactions of species in ecosystems and the impacts that toxicants may have upon the function and structure of a particular ecosystem. Molecular biology and pharmacokinetics operate at the opposite end of the biological hierarchy, describing the interactions of an organism with a toxicant at the molecular level. Analytical chemistry provides data on the environmental concentration of a compound and can also be used to estimate dose to an organism when tissues are analyzed. Organic chemistry provides the basic language and the foundation of both the abiotic and biotic interactions within an ecosystem. Biometrics, the application of statistics to biological problems, provides the tools for data analysis and hypothesis testing. Mathematical and computer modeling enables the researcher to predict effects and to increase the rigor of a hypothesis. Evolutionary biology provides the data for establishing comparisons from species to species and describes the adaptation of species to environmental change. Microbiology and molecular genetics may not only help the environmental toxicologist understand the fate and transformation of environmental pollutants, but may also provide the science and the efficient tools to clean up and restore an ecosystem. Finally, the science of risk assessment as applied to

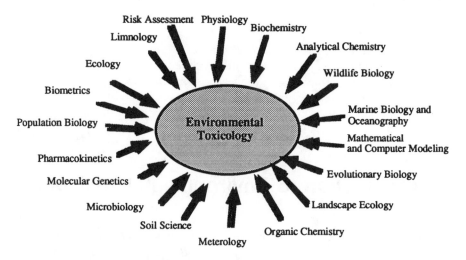

Figure 1.1 The components of environmental toxicology. Environmental toxicology borrows heavily from a variety of scientific disciplines. The very nature of the field is multidisciplinary, making a basic knowledge of biology, chemistry, mathematics, and physics essential.

environmental toxicology may form the framework to guide research and develop specific testable hypotheses.

As a discipline, environmental toxicology is relatively new. In 1991 the 15th Annual Symposium sponsored by the American Society for Testing and Materials and the 12th Annual Meeting sponsored by the Society of Environmental Toxicology and Chemistry on environmental toxicology were held. Of a rapidly evolving field, this text is only a snapshot of the directions and research of the late 1980s and early 1990s. The science evolved from the efficacy testing of pesticides in the 1940s to the cleanup of burning rivers, polluted lakes, and wildlife kills of the 1960s. The passage of the National Environmental Policy Act and the establishment of the U.S. Environmental Protection Agency forced the rapid development of the field. The Clean Air and Clean Water standards were required by law to be protective of human health and the environment. The Pellston workshops of the early 1970s provided a focal point for the discussion and consolidation of environmental toxicology. As standards development became important a relationship with the American Society for Testing and Materials evolved, which has resulted in Committee E-47 — Environmental Fate and Effects. This committee is responsible for the writing of many of the important methods used by environmental toxicologists worldwide. The Organization for Economic Cooperation and Development serves a similar role in Europe. In 1979 the Society for Environmental Toxicology and Chemistry was founded as a scientific society to support the growing needs of the field. In 1980, 85 persons attended the first SETAC Annual Meeting in Washington, D.C. In 1991, 2230 scientists and policy makers attended in Seattle.

As the field of environmental toxicology has grown, so has its sophistication and excitement. Environmental contamination is a fact of life and scientists are continually called upon to give expert advice, often with little data or time to develop the necessary information. Public outcry can lead to short-term funding and yet a myopic view. Often the concentration of the funding and research is upon the immediate care of dying and sick animals, usually warm-blooded vertebrates, without an appreciation of the damage done to the normal development of the structure and function of an ecosystem. Solutions are required, yet the development of the scientific knowledge and management expertise does not always occur. Once the dying animals are buried and the smell goes away, the long-term and irreversible changes within the ecosystem are often ignored. Likewise, overreaction and the implementation of treatment techniques that are extraordinarily expensive and do not provide a reasonable return can drain funds and other resources from important societal needs.

LEGISLATION

Unlike much of basic research, environmental toxicology has been often defined by and instigated by public policy as defined in legislation. Many of these laws in the United States, Canada, and Europe mandate toxicity testing, or require an assessment of toxicity. In the United States, federal law can often be supplemented by but not weakened by the states. For example, in the State of Washington, there are state and federal responsibilities for the assessment of damage due to a spill of oil or other hazardous substance. The State of Washington also has its own regulations for the control of toxic materials and also administers the National Pollution Discharge Elimination System (NPDES) permits. There are several pieces of legislation that are particularly relevant to the development of environmental toxicology.

The Federal Water Pollution Control Act of 1972, amended in 1976 (33 USC Sections 1251 to 1376), is commonly known as the Clean Water Act. The stated purpose is to restore and maintain the integrity of the nations waters. The regulations set by this legislation set maximum allowable concentrations of toxicants in discharges and receiving waters. The results of toxicity testing are commonly used to set these limits. In addition, NPDES permits now commonly require the use of toxicity tests performed on effluents from a variety of manufacturing sites to establish criteria for compliance.

The legislation that controls the registration of pesticides in the United States is the Federal Insecticide, Fungicide and Rodenticide Act, commonly referred to as FIFRA. Originally passed in 1947, the act has been amended by the Federal Environmental Pesticide Control Act of 1972, amendments to FIFRA in 1975, and the Federal Pesticide Act of 1978 (7 USC Section 135 et seq.). Pesticides by definition are toxic materials that are intentionally released to the environment. Many of these compounds provide a measurable economic benefit that is weighed against impact. Essential to the registration of pesticides has been a tiered testing scheme. In a tiered approach, there are specific tests to be performed at each level of the tier. If a compound exhibits particular characteristics it has the option of passing to the next level of testing. Typically, these tiers ranged from basic mechanistic data to field tests. In the approach commonly used before the fall of 1992, the top tier included field studies using large man-made ponds or investigations of terrestrial systems dosed with known quantities of pesticide. Field studies and other ecosystem level approaches are not currently routinely included. A great deal of toxicological data at every level of biological organization has been acquired as part of the registration process.

The Toxic Substance Control Act (1976, 42 USC Sections 2601 to 2629), referred to as TSCA, is an extremely ambitious program. TSCA attempts to charac-terize both human health and environmental impacts of every chemical manufactured in the United States. During the Premanufacturing Review Program, the EPA has but 90 days to assess the potential risk of a material to human health and the environment. Given the limited period of notification and the volume of compounds submitted, many of the evaluations use models that relate the structure of a compound to its potential toxicity. Structure activity models have proven useful in screening com-pounds for toxicity to aquatic and terrestrial organisms as well as mutagenicity and other endpoints. In addition to the toxicity estimation methods, there is a recom-mended but not binding series of measurements and toxicity tests that may be performed by the manufacturer. The toxicity tests typically involve a single-species approach.

Toxicity testing or the utilization of such data is routinely performed in support of the Comprehensive Environmental Response, Compensation and Liability Act of 1980 (42 USC Section 9601 et seq.), abbreviated as CERCLA, but more commonly referred to as superfund. This legislation requires that some assessment of the damage to ecological systems be considered. Research has been conducted that attempts to use a variety of toxicity tests to evaluate the potential damage of the chemical contaminants within a site to the environment. This need has given rise to interesting *in situ* methods of detecting toxicity. In the past, this program has generally been driven by human health considerations, but ecological impacts are now becoming important at several sites.

Although the federal legislation discussed above has provided the principal regulatory force in environmental toxicology, other mandates at the federal and state levels apply. These requirements will likely persist providing a continuing need for data acquisition in environmental toxicology.

INTRODUCTION TO THE TEXTBOOK

The purpose of this volume is to provide the background knowledge so that the short- and long-term effects of chemical pollution can be evaluated and the risks understood. There are nine more chapters, each with a specific building block towards the understanding of the status of the field of environmental toxicology.

Chapter 2 provides an overview of the field of environmental toxicology and introduces the progression from the initial introduction of the toxicant to the environment, its effect upon the site of action, and finally the impacts upon an ecosystem. Many of the terms used throughout this text are introduced in this section. After an introduction to toxicity testing, the remainder of the book is organized from the molecular chemistry of receptors to the ecological effects seen at the system level.

Chapter 3 is an introduction to toxicity testing. In this chapter the basics of designing a toxicity test and some of the basics of analysis are presented. The ability to understand and critique toxicity tests and bioassays is critical. Much of our understanding of the impacts of toxicants and the regulations governing acceptable levels are based on toxicity tests. Comparability and accuracy of toxicity tests are also crucial since these data are routinely used to derive structure activity relationships. Structure activity relationships are derived that relate the chemical structure of a material to its biological property, be it toxicity or biodegradation. These relationships are particularly useful when decisions are required with limited toxicological data.

After a chapter introducing the design parameters for toxicity tests, Chapter 4 presents a variety of methods that are used in environmental toxicology to assess the potential hazard of a material. A variety of tests are presented, from single species to ponds, and involving a wide variety of organisms. Tables are included that act as quick summaries of each of the tests described in the chapter. Perhaps not as exciting as contemplating the impacts of toxicants on ecosystems, the tests are the basis of our knowledge of toxicity. The setting of safe levels of chemicals in regulations, the measurement of impacts due to industry and residential outflows, and the estimate of risks are all based on the data derived from these tests. Included in this chapter are brief descriptions of many of the test organisms: freshwater, marine, and terrestrial.

Chapter 5 is an analysis of the routes of exposure allowing a toxicant to enter an organism and the modes of action at the molecular level that cause effects to reverberate throughout an ecological system. The crucial nature of understanding the routes of exposure and their importance in understanding the course of action of the toxicant is brought to light. As the compound reaches the cell, a number of interferences with the normal functioning of the organism take place, from acetylcholinesterase inhibition to the binding of common cellular receptors with disastrous outcomes. In addition to the biochemistry introduced in this chapter, a great deal of emphasis is placed on the determination of the activity of a compound by an analysis of its structure. Quantitative structure activity relationships (QSAR), used judiciously, have the ability to help set testing priorities and identify potentially toxic materials in mixtures. Heavily reliant upon the quality of the toxicity data discussed in Chapter 4, these methods use sophisticated statistical techniques for analysis of interaction of a toxicant with the receptor to estimate toxicity.

Even as the route of exposure and the molecular interactions that cause the toxic effects are delineated, that is not the entire story. Chapter 6 describes the myriad of physiological and environmental factors that can alter the exposure of the organism to the toxicant and also the response to the compound. Nutritional status, complexing elements in the environment as well as the organism and reproductive status can all drastically affect the response of an organism to an environmental exposure.

Many of the examples used in the preceding chapters emphasize organic pollutants, however, inorganic materials comprise an important class of contaminants. Chapter 7 describes the mode of action and the creation of a variety of inorganic gaseous pollutants, an increasingly important aspect of environmental toxicology. Major emphasis is placed on the atmospheric chemistry of each pollutant and the effects on a variety of organisms. The chemistry and toxicology of sulfur oxides, ozone, nitrogen oxides, carbon monoxide, and fluoride are reviewed in this chapter.

As a material enters an ecosystem, a variety of physical and biological transformations can take place, dramatically altering the property of the compound to cause toxicity. Chapter 8 reviews the mechanisms that alter the toxicity of a compound. This section is important in understanding and determining the exposure of the environment to a chemical toxicant. In addition, a knowledge of biodegradation and microbial ecology may also yield strategies for the reduction or elimination of xenobiotics.

One of the major sections of the textbook is the chapter dealing with the response of various ecological systems to the stress of toxicants. Chapter 9 deals with broad categories of responses to toxicants as well as specific examples. Biomonitoring and biomonitoring strategies are also discussed. Even as this text is being written, several new, exciting, and controversial ideas about the nature of complex systems, chaos, and the interactions with communities may drastically change our view of ecological systems and their management.

The discipline that ties together environmental toxicology is that of risk assessment. Chapter 10 provides a framework for the integration of the classical toxicology at the molecular and organismal levels and the prediction of events at the level of the community and ecosystem. Exciting research is currently underway examining the importance of indirect effects, landscape and global changes, and management of these risks. In Chapter 10 we review the current paradigm of the United States Environmental Protection Agency and apply it to the application of ecological risk assessment.

We hope that the reader finds the journey as exciting as we have.

CHAPTER 2

A Framework for Environmental Toxicology

Environmental toxicology can be simplified to the understanding of only three functions. These functions are presented in Figure 2.1. First, there is the interaction of the introduced chemical, xenobiotic, with the environment. This interaction controls the amount of toxicant or the dose available to the biota. Second, the xenobiotic interacts with its site of action. The site of action is the particular protein or other biological molecule that interacts with the toxicant. Third, the interaction of the xenobiotic with a site of action at the molecular level produces effects at higher levels of biological organization. If environmental toxicologists could write appropriate functions that would describe the transfer of an effect from its interaction with a specific receptor molecule to the effects seen at the community level, it would be possible to accurately predict the effects of pollutants in the environment. We are far from a suitable understanding of these functions. The remainder of the chapter introduces the critical factors for each of these functions. Unfortunately, we do not clearly understand how the impacts seen at the population and community levels are propagated from molecular interactions.

Techniques have been derived to evaluate effects at each step from the introduction of a xenobiotic to the biosphere to the final series of effects. These techniques are not uniform for each class of toxicant, and mixtures are even more difficult to evaluate. Given this background however, it is possible to outline the current levels of biological interaction with a xenobiotic:

Chemical and physicochemical characteristics
Bioaccumulation/biotransformation/biodegradation
Site of action
Biochemical monitoring
Physiological and behavioral
Population parameters
Community parameters
Ecosystem effects

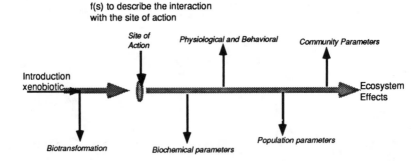

Figure 2.1 The three functions of environmental toxicology. Only three basic functions need to be described after the introduction of a xenobiotic into the environment. The first describes the fate and distribution of the material in the biosphere and the organism after the initial release to the environment (f(f)). The second function describes the interaction of the material with the site of action (f(s)). The last function describes the impact of this molecular interaction upon the function of an ecosystem (f(e)).

Each level of organization can be observed and examined at various degrees of resolution. The factors falling under each level are illustrated in Figure 2.2. Examples of these factors at each level of biological organization are given below.

CHEMICAL AND PHYSICOCHEMICAL CHARACTERISTICS

The interaction of the atoms and electrons within a specific molecule determines the impact of the compound at the molecular level. The contribution of the physico-chemical characteristics of a compound to the observed toxicity is called quantitative structure activity relationships (QSAR). QSAR has the potential of enabling environmental toxicologists to predict the environmental consequences of toxicants using only structure as a guide. The response of a chemical to ultraviolet radiation and its reactivity with the abiotic constituents of the environment determines a fate of a compound.

It must be remembered that in most cases the interaction at a molecular level with a xenobiotic is happenstance. Often this interaction is a by-product of the usual physiological function of the particular biological site with some other low molecular weight compound that occurs in the normal metabolism of the organism. Xenobiotics often mimic these naturally occurring organisms, causing degradation and detoxification in some cases, and in others toxicity.

BIOACCUMULATION/BIOTRANSFORMATION/BIODEGRADATION

A great deal can occur to a xenobiotic from its introduction to the environment to its interaction at the site of action. Many materials are altered in specific ways

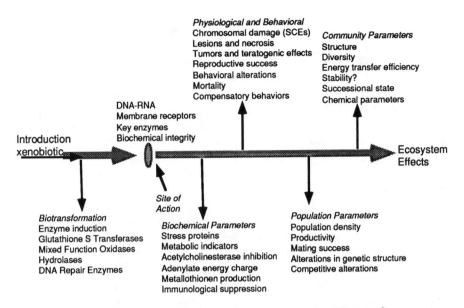

Figure 2.2 Parameters and indications of the interaction of a xenobiotic with the ecosystem. The examples listed are only a selection of the parameters that need to be understood for the explanation of the effects of a xenobiotic upon an ecosystem. However, biological systems appear to be organized within a hierarchy and that is how environmental toxicology must frame its outlook upon environmental problems.

depending upon the particular chemical characteristics of the environment. Bioaccumulation, the increase in concentration of a chemical in tissue compared to the environment, often occurs with materials that are more soluble in lipid and organics (lipophilic) than in water (hydrophilic). Compounds are often transformed into other materials by the various metabolic systems that reduce or alter the toxicity of materials introduced to the body. This process is biotransformation. Biodegradation is the process that breaks down a xenobiotic into a simpler form. Ultimately, the biodegradation of organics results in the release of CO_2 and H_2O to the environment.

RECEPTOR AND THE MODE OF ACTION

The site at which the xenobiotic interacts with the organism at the molecular level is particularly important. This receptor molecule or site of action may be the nucleic acids, specific proteins within nerve synapses or present within the cellular membrane, or it can be very nonspecific. Narcosis may affect the organism not by interaction with a particular key molecule, but by changing the characteristics of the cell membrane. The particular kind of interaction determines whether the effect is broad or more specific within the organism and phylogenetically.

BIOCHEMICAL AND MOLECULAR EFFECTS

There are broad ranges of effects at this level. We will use as an example, at the most basic and fundamental of changes, alterations to DNA.

DNA adducts and strand breakages are indicators of genotoxic materials, compounds that affect or alter the transmission of genetic material. One advantage to these methods is that the active site can be examined for a variety of organisms. The methodologies are proven and can be used virtually regardless of species. However, damage to the DNA only provides a broad classification as to the type of toxicant. The study of the normal variation and damage to DNA in unpolluted environments has just begun.

Cytogenetic examination of meiotic and mitotic cells can reveal damage to genetic components of the organism. Chromosomal breakage, micronuclei, and various trisomys can be detected microscopically. Few organisms, however, have the requisite chromosomal maps to score accurately more subtle types of damage. Properly developed, cytogenetic examinations may prove to be powerful and sensitive indicators of environmental contamination for certain classes of material.

A more complex system, directly affected by damage to certain regions of DNA and to cellular proteins, is the inhibition of the immunological system of an organism — immunological suppression. Immunological suppression by xenobiotics could have subtle but important impacts on natural populations. Invertebrates and other organisms have a variety of immunological responses that can be examined in the laboratory setting from field collections. The immunological responses of bivalves in some ways are similar to vertebrate systems and can be suppressed or activated by various toxicants. Mammals and birds have well documented immunological responses although the impacts of pollutants are not well understood. Considering the importance to the organism, immunological responses could be very valuable at assessing the health of an ecosystem at the population level.

PHYSIOLOGICAL AND BEHAVIORAL EFFECTS

Physiological and behavioral indicators of impact within a population are the classical means by which the health of populations is assessed. The major drawback has been the extrapolation of these factors based upon the health of an individual organism, attributing the damage to a particular pollutant, and extrapolating this to the population level.

Lesions and necrosis in tissues have been the cornerstone of much environmental pathology. Gills are sensitive tissues and often reflect the presence of irritant materials. In addition, damage to the gills has an obvious and direct impact upon the health of the organism. Related to the detection of lesions are those that are tumoragenic. Tumors in fish, especially flatfish, have been extensively studied as indicators of oncogenic materials in marine sediments. Oncogenesis has also been extensively studied in Medaka and trout as means of determining the pathways responsible for tumor development. Development of tumors in fish more commonly found in natural communities should follow similar mechanisms. As with many indicators of toxicant impact, relating the effect of tumor development to the health and reproduction of a wild population has not been as closely examined as the endpoint.

Reproductive success is certainly another measure of the health of an organism and is the principal indicator of the Darwinian fitness of an organism. In a laboratory situation it certainly is possible to measure fecundity and the success of offspring in their maturation. In nature these parameters may be very difficult to measure accurately. Many factors other than pollution can lead to poor reproductive success. Secondary effects, such as the impact of habitat loss on zooplankton populations essential for fry feeding will be seen in the depression or elimination of the young age classes.

Mortality is certainly easy to assay on the individual organism. Macroinvertebrates, such as bivalves and cnideria, can be examined and since they are relatively sessile, the mortality can be attributed to a factor in the immediate environment. Fish, being mobile, can die due to exposure kilometers away or because of multiple intoxications during their migrations. By the time the fish are dying, the other levels of the ecosystem are in a sad state.

The use of the cough response and ventilatory rate of fish has been a promising system for the determination and prevention of environmental contamination. Pioneered at Virginia Polytechnic Institute and State University, the measurement of the ventilatory rate of fish using electrodes to pick up the muscular contraction of the operculum has been brought to a very high stage of refinement. It is now possible to continually monitor the water quality as perceived by the test organisms with a desktop computer analysis system at a relatively low cost.

POPULATION PARAMETERS

A variety of endpoints have been used, including number and structure of a population, to indicate stress. Population numbers or density have been widely used for plant, animal, and microbial populations in spite of the problems in mark recapture and other sampling strategies. Since younger life stages are considered to be more sensitive to a variety of pollutants, shifts in age structure to an older population may indicate stress. In addition, cycles in age structure and population size occur due to the inherent properties of the age structure of the population and predator-prey interactions. Crashes in populations such as those of the stripped bass in the Chesapeake Bay do occur and certainly are observed. A crash often does not lend itself to an easy cause-effect relationship, making mitigation strategies difficult to create.

The determination of alterations in genetic structure, i.e., the frequency of certain marker alleles, has become increasingly popular. The technology of gel electrophoresis has made this a seemingly easy procedure. Population geneticists have long used this method to observe alterations in gene frequencies in populations of bacteria, protozoans, plants, various vertebrates and the famous Drosophilla. The largest drawback to this method is ascribing differential sensitivities to the genotypes in question. Usually, a marker is used that demonstrates heterogeneity within a particular species. Toxicity tests can be performed to provide relative sensitivities. However, the genes that have been looked at to date are not genes controlling the xenobiotic metabolism. These genes have some other physiological function and act as a marker for the remainder of the genes within a particular linkage group. Although with some problems, this method promises to provide both populational and biochemical data that may prove useful in certain circumstances.

Alterations in the competitive abilities of organisms can indicate pollution. Obviously, bacteria that can use a xenobiotic as a carbon or other nutrient source or that can detoxify a material have a competitive advantage, with all other factors being equal. Xenobiotics may also enhance species diversity if a particularly competitive species is more sensitive to a particular toxicant. These effects may lead to an increase in plant or algal diversity after the application of a toxicant.

COMMUNITY EFFECTS

The structure of biological communities has always been a commonly used indicator of stress in a biological community. Early studies on cultural eutrophication emphasized the impacts of pollution as they altered the species composition and energy flow of aquatic ecosystems. Various biological indices have been developed to judge the health of ecosystems by measuring aspects of the invertebrate, fish, or plant populations. Perhaps the largest drawback is the effort necessary to determine the structure of ecosystems and to understand pollution-induced effects from normal successional changes. There is also the temptation to reduce the data to a single index or other parameter that eliminates the dynamics and stochastic properties of the community.

One of the most widely used indexes of community structure have been species diversity. Many measures for diversity are used, from such elementary forms as species number to measures based on information theory. A decrease in species diversity is usually taken as an indication of stress or impact upon a particular ecosystem. Diversity indexes, however, hide the dynamic nature of the system and the effects of island biogeography and seasonal state. As demonstrated in microcosm experiments, diversity is often insensitive to toxicant impacts.

Related to diversity is the notion of static and dynamic stability in ecosystems. Traditional dogma stated that diverse ecosystems were more stable and therefore healthier than less rich ecosystems. May's work in the early 1970s did much to question these almost unquestionable assumptions about properties of ecosystems. We certainly do not doubt the importance of biological diversity, but diversity itself may indicate the longevity and size of the habitat rather than the inherent properties of the ecosystem. Rarely are basic principles such as island biogeography incorporated into comparisons of species diversity when assessments of community health are made. Diversity should be examined closely as to its worth in determining xenobiotic impacts upon biological communities.

Currently it is difficult to pick a parameter that describes the health of a biological community and have that form a basis of prediction. A single variable or magic number may not even be possible. In addition, what are often termed biological communities are based upon human constructs. The members of the marine benthic invertebrate community interact with many other types of organisms, microorganisms, vertebrates, and protists that in many ways determine the diversity and persistence of an organism. Communities can also be defined as functional groups, such as the intertidal community or alpine forest community, that may more accurately describe functional groupings of organisms.

ECOSYSTEM EFFECTS

Alterations in the species composition and metabolism of an ecosystem are the most dramatic impacts that can be observed. Acid precipitation has been documented to cause dramatic alterations in both aquatic and terrestrial ecosystems. Introduction of nutrients certainly increases the rate of eutrophication.

Effects can occur that alter the landscape pattern of the ecosystem. Changes in global temperatures have had dramatic effects upon species distributions. Combinations of nutrient inputs, utilization, and toxicants have dramatically altered the Chesapeake Bay system.

SPATIAL AND TEMPORAL SCALES

Not only are there scales in organization, but scales over space and time exist. It is crucial to note that all of the functions described in previous sections act at a variety of spatial and temporal scales (Suter and Barnthouse, 1993). Although in many instances these scales appear disconnected, they are in fact intimately intertwined. Effects at the molecular level have ecosystem-level effects. Conversely, impacts on a broad scale affect the very sequence of the genetic material as evolution occurs in response to the changes in toxicant concentrations or interspecies interactions.

The scales important in environmental toxicology range from the few angstroms of molecular interactions to the hundreds of thousands of square kilometers affected by large-scale events. Figure 2.3 presents some of the organizational aspects of ecological systems with their corresponding temporal and spatial scale. The diagram is only a general guide. Molecular activities and degradation may exist over short periods and volumes, but their ultimate impact may be global.

Perhaps the most important example of a new biochemical pathway generating a global impact was the development of photosynthesis. The atmosphere of Earth originally was reducing. Photosynthesis produces, as a by-product, oxygen. Oxygen, which is quite toxic, became a major constituent of the atmosphere. This change

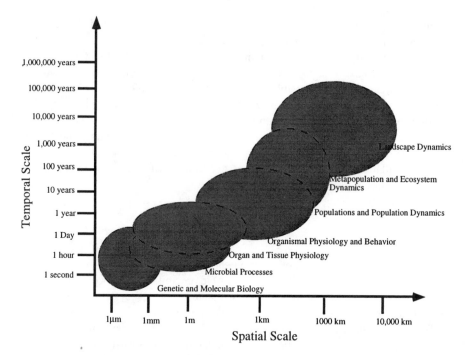

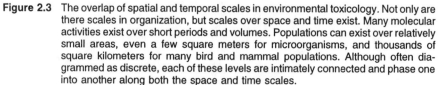

Figure 2.3   The overlap of spatial and temporal scales in environmental toxicology. Not only are there scales in organization, but scales over space and time exist. Many molecular activities exist over short periods and volumes. Populations can exist over relatively small areas, even a few square meters for microorganisms, and thousands of square kilometers for many bird and mammal populations. Although often diagrammed as discrete, each of these levels are intimately connected and phase one into another along both the space and time scales.

produced a mass extinction event, yet also provided for the evolution of much more efficient metabolisms.

Effects at the community and ecosystem level conversely have effects upon lower levels of organization. The structure of the ecological system may allow some individuals of populations to migrate to areas where the species are below a sustainable level or are at extinction. If the pathways to the depleted areas are not too long, the source population may rescue the population that is below a sustainable level. Instead of extinction, a population may be sustainable or even increase due to its rescue from a neighboring population. If the structure of the ecological landscape provides few opportunities for rescue, localized extinctions would be more likely.

As the effects of a toxicant can range over a variety of temporal scales, so can the nature of the input of the toxicant to the system (Figure 2.4). Household or garden use of a pesticide may be an event with a scale of a few minutes and a square meter. The addition of nutrients to ecological systems due to industrialization and agriculture may cover thousands of square kilometers and persist for hundreds or thousands of years. The duration and scale of anthropogenic inputs does vary a great deal.

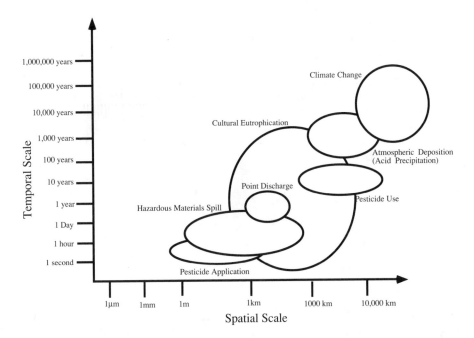

Figure 2.4 The overlap of spatial and temporal scales in chemical contamination. Just as there are scales of ecological processes, contamination events also range in scale. Pesticide applications can range from small-scale household use to large-scale agricultural applications. The addition of surplus nutrients and other materials due to agriculture or human habitation is generally large scale and long lived. Acid precipitation generated by the tall stacks of the midwestern United States is a fairly recent phenomena, but the effects will likely be long term. However, each of these events have molecular scale interactions.

However, it is crucial to realize that the interactions of the toxicant with the organism are still at the molecular level. Small effects can have global implications.

REFERENCES AND SUGGESTED READINGS

Suter, G.W., II and L.W. Barnthouse. 1993. Assessment Concepts. In *Ecological Risk Assessment*. G.W. Suter, II, Ed., Lewis Publishers, Boca Raton, pp. 21-47.

STUDY QUESTIONS

1. Define the three functions to be understood to simplify environmental toxicology.
2. Define QSAR.
3. Define bioaccumulation, biotransformation, and biodegradation.
4. What is "site of action"?
5. Describe limits to the use of DNA alteration as an indicator of genotoxic materials.

6. Describe immunological suppression.
7. Name three major physiological indicators of impact by a xenobiotic on a population.
8. Describe a problem using population parameters to indicate xenobiotic challenge.
9. Name two means by which a xenobiotic can alter competitive abilities of organisms.
10. What are the most dramatic impacts observable on ecosystems by xenobiotics?
11. Is the arrow describing the interactions of the ecological system with a chemical pollutant unidirectional?
12. Characterize ecological functions and processes by temporal and spatial scale.
13. What are the interactions between the scale of a chemical contamination and that of the affected ecological system?

CHAPTER 3

An Introduction to Toxicity Testing

Toxicity is the property or properties of a material that produces a harmful effect upon a biological system. A toxicant is the material that produces this biological effect. The majority of the chemicals discussed in this text are man-made or of anthropogenic origin. This is not to deny that extremely toxic materials are produced by biological systems; venom, botulinum endotoxin, and some of the fungal aflatoxins are extremely potent materials. However, compounds that are derived from natural sources are produced in low amounts. Anthropogenically derived compounds can be produced in the millions of pounds per year.

Materials introduced into the environment come from two basic types of sources. Point discharges are derived from such sources as sewage discharges, waste streams from industrial sources, hazardous waste disposal sites, and accidental spills. Point discharges are generally easy to characterize as to the types of materials released, rates of release, and total amounts. In contrast, nonpoint discharges are those materials released from agricultural runoffs, contaminated soils and aquatic sediments, atmospheric deposition, and urban runoff from such sources as parking lots and residential areas. Nonpoint discharges are much more difficult to characterize. In most situations, discharges from nonpoint sources are complex mixtures, amounts of toxicants are difficult to characterize, and rates and the timing of discharges are as difficult to predict as the rain. One of the most difficult aspects of nonpoint discharges is that the components can vary in their toxicological characteristics.

Many classes of compounds can exhibit environmental toxicity. One of the most commonly discussed and researched are the pesticides. Pesticide can refer to any compound that exhibits toxicity to an undesirable organism. Since the biochemistry and physiology of all organisms are linked by the stochastic processes of evolution, a compound toxic to a Norway rat is likely to be toxic to other small mammals. Industrial chemicals also are a major concern because of the large amounts transported and used. Metals from mining operations, manufacturing, and as contaminants in lubricants are also released to the environment. Crude oil and the petroleum products derived from the oil are a significant source of environmental toxicity because of their persistence and common usage in an industrialized society. Many of these compounds, especially metal salts and petroleum, can be found in normally uncontaminated environments. In many cases, metals such as copper and zinc are essential nutrients. However, it is not just the presence of a compound that poses a toxicological threat, but the relationships between its dose to an organism and its biological effects that determine what environmental concentrations are harmful.

Any chemical material can exhibit harmful effects when the amount introduced to an organism is high enough. Simple exposure to a chemical also does not mean that a harmful effect will result. Of critical importance is the dose, or actual amount of material that enters an organism, that determines the biological ramifications. At low doses no apparent harmful effects occur. In fact, many toxicity evaluations result in increased growth of the organisms at low doses. Higher doses may result in mortality. The relationship between dose and the biological effect is the dose-response relationship. In some instances, no effects can be observed until a certain threshold concentration is reached. In environmental toxicology, environmental concentration is often used as a substitute for knowing the actual amount or dose of a chemical entering an organism. Care must be taken to realize that dose may be only indirectly related to environmental concentration. The surface-to-volume ratio, shape, characteristics of the organisms external covering, and respiratory systems can all dramatically affect the rates of a chemical's absorption from the environment. Since it is common usage, concentration will be the variable from which mortality will be derived, but with the understanding that concentration and dose are not always directly proportional or comparable from species to species.

THE DOSE-RESPONSE CURVE

The graph describing the response of an enzyme, organism, population, or biological community to a range of concentrations of a xenobiotic is the dose-response curve. Enzyme inhibition, DNA damage, death, behavioral changes, and other responses can be described using this relationship.

Table 3.1 presents the data for a typical response over concentration or dose for a particular xenobiotic. At each concentration the percentage or actual number of organisms responding or the magnitude of effects is plotted (Figure 3.1). The distribution that results resembles a sigmoid curve. The origin of this distribution is straightforward. If only the additional mortalities seen at each concentration are plotted, the distribution that results is that of a normal distribution or a bell-shaped curve (Figure 3.2). This distribution is not surprising. Responses or traits from organisms that are controlled by numerous sets of genes follow bell-shaped curves. Length, coat color, and fecundity are examples of multigenic traits whose distribution results in a normal distribution.

The distribution of mortality vs. concentration or dose is drawn so that the cumulative mortality is plotted at each concentration. At each concentration the total numbers of organisms that have died by that concentration are plotted. The presentation

Table 3.1   Toxicity Data for Compound 1

	Dose								
	0.5	**1.0**	**2.0**	**3.0**	**4.0**	**5.0**	**6.0**	**7.0**	**8.0**
Compound 1									
Cumulative toxicity	0.0	2.0	7.0	23.0	78.0	92.0	97.0	100.0	100.0
Percent additional deaths at each concentration	0.0	2.0	5.0	15.0	55.0	15.0	5.0	3.0	0.0

Note: All of the toxicity data are given as a percentage of the total organisms of a particular treatment group. For example, if 7 out of 100 organisms died or expressed other endpoints at a concentration of 2 mg/kg, then the percentage responding would be 7%.

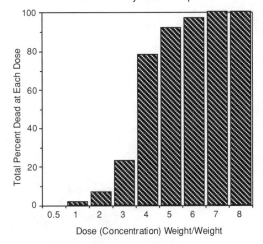

Plot of Toxicity Data Compound 1

Figure 3.1   Plot of cumulative mortality vs. environmental concentration or dose. The data are plotted as cumulative number of dead by each dose using the data presented in Table 3.1. The x-axis is in units of weight to volume (concentration) or weight of toxicant per unit weight of animal (dose).

in Figure 3.1 is usually referred to as a dose response curve. Data are plotted as continuous and a sigmoid curve usually results (Figure 3.3). Two parameters of this curve are used to describe it: (1) the concentration or dose that results in 50% of the measured effect and (2) the slope of the linear part of the curve that passes through the midpoint. Both parameters are necessary to accurately describe the relationship between chemical concentration and effect. The midpoint is commonly referred to as a LD_{50}, LC_{50}, EC_{50}, or IC_{50}. The definitions are relatively straightforward.

LD_{50} — The dose that causes mortality in 50% of the organisms tested estimated by graphical or computational means.

LC_{50} — The concentration that causes mortality in 50% of the organisms tested estimated by graphical or computational means.

Figure 3.2 Plot of mortality vs. environmental concentration or dose. Not surprisingly, the distribution that results is that of a normal distribution or a bell-shaped curve. Responses or traits from organisms that are controlled by numerous sets of genes follow bell-shaped curves. Length, coat color, and fecundity are examples of multigenic traits whose distribution results in a bell-shaped curve. The x-axis is in units of weight to volume (concentration) or weight of toxicant per unit weight of animal (dose).

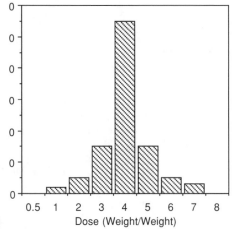

Percent Mortality at Each Dose

Dose (Weight/Weight)

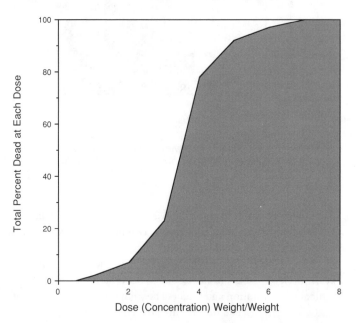

Dose-Response Sigmoid Curve

Total Percent Dead at Each Dose

Dose (Concentration) Weight/Weight

Figure 3.3 The sigmoid dose-response curve. Converted from the discontinuous bar graph of Figure 3.2 to a line graph. If mortality is a continuous function of the toxicant, the result is the typical sigmoid dose-response curve. The x-axis is in units of weight to volume (concentration) or weight of toxicant per unit weight of animal (dose).

EC$_{50}$ — The concentration that has an effect on 50% of the organisms tested estimated by graphical or computational means. Often this parameter is used for effects that are not death.

IC$_{50}$ — Inhibitory concentration that reduces the normal response of an organism by 50% estimated by graphical or computational means. Growth rates of algae, bacteria, and other organisms are often measured as an IC$_{50}$.

One of the primary reasons for conducting any type of toxicity test is to rank chemicals as to their toxicity. Table 3.2 provides data on toxicity for two different compounds. It is readily apparent that the midpoint for compound 2 will likely be higher than that of compound 1. A plot of the cumulative toxicity (Figure 3.4) confirms that the concentration that causes mortality to half of the population for compound 2 is higher than compound 1. Linear plots of the data points are superimposed upon the curve (Figure 3.5) confirming that the midpoints are different. Notice, however, that the slopes of the lines are similar.

In most cases the toxicity of a compound is usually reported using only the midpoint reported in a mass per unit mass (mg/kg) or volume (mg/l). This practice is misleading and can lead to a misunderstanding or the true hazard of a compound to a particular xenobiotic. Figure 3.6 provides an example of two compounds with the same LC$_{50}$s. Plotting the cumulative toxicity and superimposing the linear graph the concurrence of the points is confirmed (Figure 3.7). However the slopes of the lines are different with compound 3 having twice the toxicity of compound 1 at a concentration of 2. At low concentrations, those that are often found in the environment, compound 3 has the greater effect.

Conversely, compounds may have different LC$_{50}$s, but the slopes may be the same. Similar slopes may imply a similar mode of action. In addition, toxicity is not generated by the unit mass of xenobiotic but by the molecule. Molar concentrations or dosages provide a more accurate assessment of the toxicity of a particular compound. This relationship will be explored further in our discussion of quantitative structure activity relationships. Another weakness of the LC$_{50}$, EC$_{50}$, and IC$_{50}$ is that they reflect the environmental concentration of the toxicant over the specified time of the test. Compounds that move into tissues slowly may have a lower toxicity in

Table 3.2 Toxicity Data for Compounds 2 and 3

	Dose								
	0.5	1.0	2.0	3.0	4.0	5.0	6.0	7.0	8.0
Compound 2									
Cumulative toxicity	1.0	3.0	6.0	11.0	21.0	36.0	86.0	96.0	100.0
Percent additional deaths at each concentration	1.0	2.0	3.0	5.0	10.0	15.0	50.0	10.0	4.0
Compound 3									
Cumulative toxicity	0.0	5.0	15.0	30.0	70.0	85.0	95.0	100.0	100.0
Percent additional deaths at each concentration	0.0	5.0	10.0	15.0	40.0	15.0	10.0	5.0	0.0

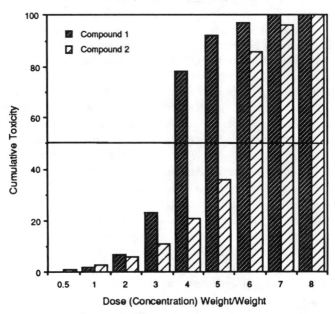

Figure 3.4 Comparison of dose-response curves — 1. One of the primary goals of toxicity testing is the comparison or ranking of toxicity. The cumulative plots comparing compound 1 and compound 2 demonstrate the distinct nature of the two different toxicity curves.

a 96-h test simply because the concentration in the tissue has not reached toxic levels within the specified testing time. L. McCarty has written extensively on this topic and suggests that a "Lethal Body Burden" or some other measurement be used to reflect tissue concentrations. These ideas are discussed in a later chapter.

Often other terminology is used to describe the concentrations that have a minimal or nonexistent effect. Those that are currently common are NOEC, NOEL, NOAEC, NOAEL, LOEC, LOEL, MTC, and MATC.

NOEC — No observed effects concentration, determined by graphical or statistical methods.

NOEL — No observed effects level, determined by graphical or statistical methods. This parameter is reported as a dose.

NOAEC — No observed adverse effects concentration, determined by graphical or statistical methods. The effect is usually chosen for its impact upon the species tested.

NOAEL — No observed adverse effects level, determined by graphical or statistical methods.

LOEC — Lowest observed effects concentration, determined by graphical or statistical methods.

LOEL — Lowest observed effects level, determined by graphical or statistical methods.

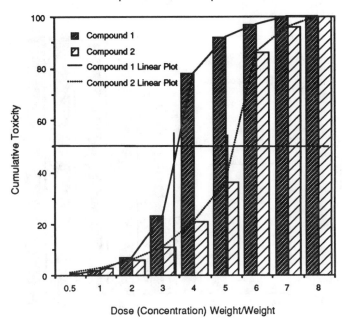

Comparison of Dose Response Curves-2

Figure 3.5 Comparison of dose-response curves — 2. Plotting the dose-response curve demonstrates that the concentrations that cause mortality to 50% of the population are distinctly different. However, the slopes of the two curves appear to be the same. In many cases this may indicate that the compounds interact similarly at the molecular level.

MTC — Minimum threshold concentration, determined by graphical or statistical methods.

MATC — Minimum allowable toxicant concentration, determined by graphical or statistical methods.

These concentrations and doses usually refer to the concentration or dose that does not produce a statistically significant effect. The ability to accurately determine a threshold level or no effect level is dependent upon a number of criteria including:

Sample size and replication.
Number of endpoints observed.
Number of dosages or concentration.
The ability to measure the endpoints.
Intrinsic variability of the endpoints within the experimental population.
Statistical methodology.

Given the difficulty of determining these endpoints cautions should be taken when using these parameters. An implicit assumption of these endpoints is that there is a threshold concentration or dose. That is, the organism, through compensatory

Comparison of Dose Response Curves-4

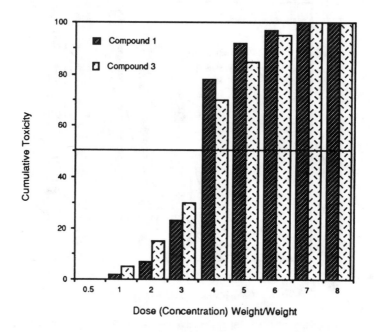

Dose (Concentration) Weight/Weight

Figure 3.6   Comparison of dose-response curves — 3. Cumulative toxicity plots for compounds 1 and 3. Notice that the plots intersect at roughly 50% mortality.

mechanisms or the inherent mode of the toxicity of the chemical, can buffer the effects of the toxicant at certain levels of intoxication (Figure 3.8). In some cases biological effects occur at succeeding lower concentrations until the chemical is removed from the environment. There is much debate about which model of dose vs. effects is more accurate and useful.

STANDARD METHODS

Over the years a variety of test methods have been standardized. These protocols are available from the American Society for Testing and Materials (ASTM), the Organization for Economic Cooperation (OECD), and the National Toxicology Program (NTP), and are available as United States Environmental Protection Agency publications, the *Federal Register*, and often from the researchers that developed the standard methodology.

There are distinct advantages to the use of a standard method or guideline in the evaluation of the toxicity of chemicals or mixtures:

Uniformity and comparability of test results.
Allows replication of the result by other laboratories.
Provides criteria as to the suitability of the test data for decision making.

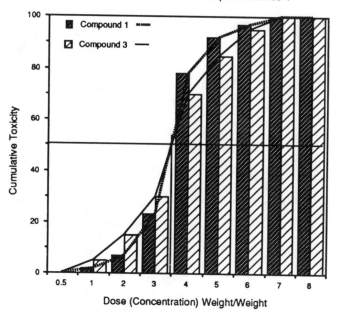

Figure 3.7 Comparison of dose-response curves — 4. Although the midpoints of the curves for compounds 1 and 3 are the same, at low concentrations more typical of exposure in the environment compound 3 is more toxic.

Logistics are simplified, little or no developmental work.

Data can be compiled with that of other laboratories for use when large data sets are required. Examples are quantitative structure activity research and risk assessment.

The method establishes a defined baseline from which modifications can be made to answer specific research questions.

Over the years numerous protocols have been published. Usually, a standard method or guide has the following format for the conduct of a toxicity test using the ASTM methods and guides as an example:

The scope of the method or guide is identified.

Reference documents, terminology specific to the standards organization, a summary, and the utility of the methodology are listed and discussed.

Hazards and recommended safeguards are now routinely listed.

Apparatuses to be used are listed and specified. In aquatic toxicity tests the specifications of the dilution water are given a separate listing, reflecting its importance.

Specifications for the material undergoing test are provided.

Test organisms are listed along with criteria for health, size, and sources.

Experimental procedure is detailed. This listing includes overall design, physical and chemical conditions of the test chambers or other containers, range of concentrations, and measurements to be made.

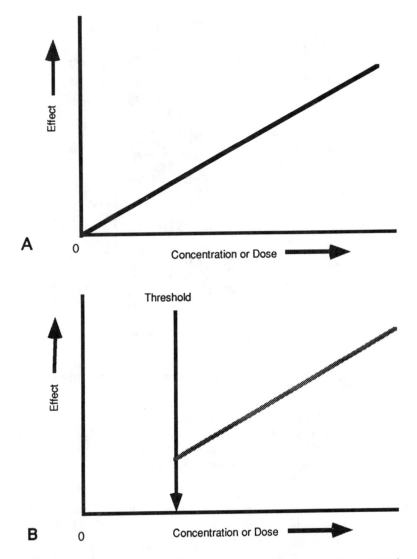

Figure 3.8 Threshold concentration. There are two prevailing ideas on the toxicity of compounds at low concentrations. Often it is presumed that a compound has a toxic effect as long as any amount of the compound is available to the organism (A). Only at zero concentration will the effect disappear. The other prevailing idea is that a threshold dose exists below which the compound is present but no effects can be discerned (B). There is a great deal of debate about which model is accurate.

Analytical methodologies for making the measurements during the experiment are often given a separate listing.

Acceptability criteria are listed by which to judge the reliability of the toxicity test.

Methods for the calculation of results are listed. Often several methods of determining the EC_{50}, LD_{50}, or NOEL are referenced.

Specifications are listed for the documentation of the results.

Appendixes are often added to provide specifics for particular species of strains of animals and the alterations to the basic protocol to accommodate these organisms.

DISADVANTAGES OF STANDARD METHODS

Standard methods do have a disadvantage: the methods are generally designed to answer very specific questions that are commonly presented. As in the case of acute and chronic toxicity tests the question is the ranking of the toxicity of a chemical in comparison to other compounds. When the questions are more detailed or the compound has unusual properties, deviations from the standard method should be undertaken. The trap of standard methods is that they may be used blindly — first ask the question, then find or invent the most appropriate method.

CLASSIFICATION OF TOXICITY TESTS

There are a large number of toxicity tests that have been developed in environmental toxicology because of the large variety of species and ecosystems that have been investigated. However, it is possible to classify the tests using the length of the experiments relative to the life span of the organism and the complexity of the biological community. Figure 3.9 provides a summary of this classification.

Acute toxicity tests cover a relatively short period of an organism's life span. In the case of fish, daphnia, rats, and birds periods of 24 to 48 h have been used. Even in the case of the short-lived *Daphnia magna*, a 48-h period is just barely long enough for it to undergo its first molting. Vertebrates with generally longer life spans undergo an even smaller portion of their life during these toxicity tests. A common misconception is that toxicity tests of similar periods of time using bacteria, protists, and algae also constitute acute toxicity tests. Many bacteria can divide in less than 1 h under optimal conditions. Most protists and algae are capable of undergoing binary fission is less than a 24-h period. A 24-h period to an algal cell may be an entire generation. The tests with unicellular organisms are probably better classified as chronic or growth toxicity tests.

Generally, chronic and sublethal toxicity tests last for a significant portion of an organism's life expectancy. There are many types of toxicity tests that do this. Reproductive tests often examine the reproductive capabilities of an organism. By their nature, these tests must include: (1) the gestational period for females and (2) for males a significant portion of the time for spermatogenesis. Growth assays may include an accounting of biomass produced by protists and algae or the development of newly hatched chicks. Chronic tests are not usually multigenerational.

Multispecies toxicity tests, as their name implies, involve the inclusion of two or more organisms and are usually designed so that the organisms interact. The effects of a toxicant upon various aspects of population dynamics such as predator-prey

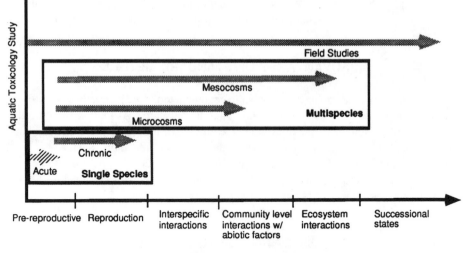

Figure 3.9 Classification of toxicity tests in environmental toxicology. Generally, the two parameters that are involved are the length of the test relative to the test organism and the species composition of the test system.

interactions and competition are a goal of these tests. Usually these tests are called microcosm or small cosmos toxicity tests. There is no clear definition of what volume, acreage, or other measure of size constitute a microcosm. A larger microcosm is a mesocosm. Mesocosms usually, but not always, have more trophic levels and generally a greater complexity than a microcosm toxicity test. Often mesocosms are outside and subject to the natural variations in rainfall, solar intensity, and atmospheric deposition. Microcosms are commonly thought of as creatures of the laboratory. Mesocosms are generally large enough to enable looking at structural and functional dynamics that are generally thought of as ecosystem level. Unfortunately, one man's mesocosm is another person's microcosm, making classification difficult. The types of multispecies comparisons are detailed in their own section.

The most difficult, costly, and controversial level of toxicity testing is the field study. Field studies can be observational or experimental. Field studies can include all levels of biological organization and are also affected by the temporal, spatial, and evolutionary heterogeneities that exist in natural systems. One of the major challenges in environmental toxicology is the ability to translate the toxicity tests performed under controlled conditions in the laboratory or test site to the structure and function of real ecosystems. This inability to translate the generally reproducible and repeatable laboratory data to effects upon the systems that environmental toxicology tries to protect is often called the lab-to-field dilemma. Comparisons of laboratory data to field results are an ongoing and important part of research in environmental toxicology.

DESIGN PARAMETERS FOR TOXICITY TESTS

Besides the complexity of the biological system and the length of the test, there are more practical aspects to toxicity tests. In aquatic test systems the tests may be classified as static, static renewal, recirculating, or flow through.

In a static test the test solution is not replaced during the test. This has the advantages of being simpler and cost-effective. The amount of chemical solution required is small and so is the toxic waste generation. No special equipment is required. However, oxygen content and toxicant concentration generally decrease through time while metabolic waste products increase. This method of toxicant application is generally used for short-term tests using small organisms or surprisingly, the large multispecies microcosm and mesocosm type tests.

The next step in complexity is the static-renewal test. In this exposure scheme a toxicant solution is replaced after a specified time period by a new test solution. This method has the advantage of replacing the toxicant solution so that metabolic waste can be removed and toxicant and oxygen concentrations can be returned to the target concentrations. Still, a relatively small amount of material is required to prepare test solutions and only small amounts of toxic waste are generated. More handling of the test vessels and the test organisms is required, increasing the chances of accidents or stress to the test organisms. This method of toxicant application is generally used for longer-term tests such as daphnid chronic and fish early life history tests.

A recirculating methodology is an attempt to maintain the water quality of the test solution without altering the toxicant concentration. A filter may be used to remove waste products or some form of aeration may be used to maintain dissolved oxygen concentration at a specified level. The advantages to this system are the maintenance of the water quality of the test solution. Disadvantages include an increase in complexity, an uncertainty that the methods of water treatment do not alter the toxicant concentration, and the increased likelihood of mechanical failure.

Technically, the best method in ensuring a precise exposure and water quality is the use of a flow-through test methodology. A continuous-flow methodology usually involves the application of peristaltic pumps, flow meters, and mixing chambers to ensure an accurate concentration. Continuous-flow methods are rarely used. The usual method is an intermittent flow using a proportional diluter (Figure 3.10) to mix the stock solution with diluent to obtain the desired test solutions.

There are two basic types of proportional diluters used to ensure accurate delivery of various toxicant concentrations to the test chambers: the venturi and the solenoid systems. The venturi system has the advantage of few moving parts and these systems can be fashioned at minimal cost. Unfortunately, some height is required to produce enough vacuum to ensure accurate flow and mixing of stock solution of toxicant and the dilution water. A solenoid system consists of a series of valves controlled by sensors in the tanks that open the solenoid valves at the appropriate times to ensure proper mixing. The solenoid system has the advantages of being easy to set up and transport and often they are extremely durable. Often the tubing can be stainless steel or

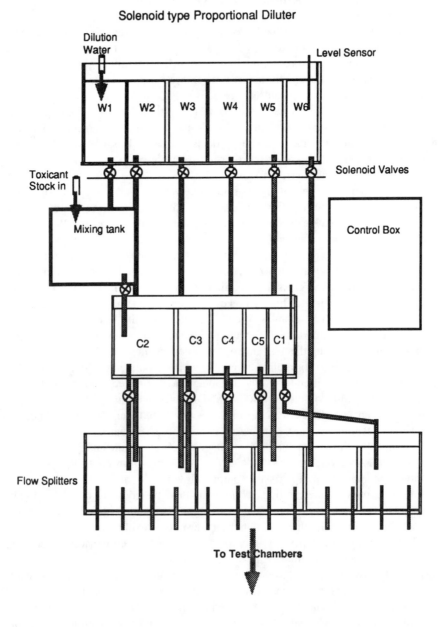

Figure 3.10   Schematic of a proportional diluter with flow controlled by solenoid valves. This
mechanism ensures that an accurate concentration of the test material is reliably
introduced to the test organisms at a specified rate.

polypropylene instead of glass. The disadvantages to the solenoid system are an increase in moving parts, expense, and when the electricity stops so does the diluter. Both of these systems use gravity to move the solutions through the diluter.

EXPOSURE SCENARIOS

In aquatic test systems exposure is usually a whole-body exposure. That means that the toxicant can enter the organism through the skin, cell wall, and respiratory system (gills, stomata), and by ingestion. Occasionally a toxicant is injected into an aquatic organism but that is not usually the case in toxicity tests to screen for effects. Whole-body exposures are less common when dealing with terrestrial species. Often an amount of xenobiotic is injected into the musculature (intramuscular), peritoneum (intraperitoneal), or into a vein (intravenous) on a weight of toxicant per unit weight of the animal basis. Other toxicity tests place a specified amount into the stomach by a tube (gavage) so that the amount of material entering the organism can be carefully quantified. However, feeding studies are conducted so that a specific concentration of toxicant is mixed with a food or water to ensure toxicant delivery. Unfortunately, many compounds are not palatable and the test organisms quickly cease to eat.

Other routes of exposure include inhalation exposure for atmospherically borne pollutants. In many cases of an originally atmospheric exposure, dermal exposure may occur. An alternative method of ensuring an inhalation exposure is to provide an air- or water-tight seal limiting exposure to the respiratory apparatus. In the case of rodents, nose-only exposures can be used to limit coat and feet contamination. Dermal exposures are important in the uptake of substances from contaminated soils or from atmospheric deposition.

Plants and soil- and sediment-dwelling organisms have other potential routes of exposure that may be used in toxicity testing. Plants are often exposed through the soil or to an atmospheric deposition. Soil invertebrates are often placed in a standardized soil laced with a particular concentration of the test substance. Sediment tests are usually with contaminated sediments or with a material added to a standardized sediment.

Often overlooked in toxicity testing can be the multiple routes of exposure that may be inadvertently available during the toxicity test. An inhalation study that exposes the animal to a toxicant in the atmosphere must also take into account deposition of the material on the feathers or fur and the subsequent self-cleaning causing an oral exposure. Likewise, exposure is available dermally through the bare feet, face, or eyes of the animal. In field pesticide experiments where the exposure might be assumed to be through the ingestion of dead pests, contaminated foliage, soil, and airborne particulates can increase the available routes of exposure, thereby increasing the actual dose to the organism. Soil organisms often consume the soil for nutrition, adding ingestion to a dermal route of exposure.

TEST ORGANISMS

One of the most crucial aspects of a toxicity test is the suitability and health of the test organisms, or in the case of multispecies toxicity tests, the introduced community. It is also important to clearly define the goals of the toxicity test. If the protection of a particular economic resource such as a salmon fishery is of overriding importance, it may be important to use a salmonid and its food sources as test species. Toxicity tests are performed to gain an overall picture of the toxicity of a compound to a variety of species. Therefore, the laboratory test species is taken only as representative of a particular class or in many cases, phyla.

Some of the criteria for choosing a test species for use in a toxicity test are listed and discussed below.

1. *The test organism should be widely available through laboratory culture, procurement from a hatchery or other culture facility, or collection from the field.* In many cases marine organisms are difficult to culture successfully in the laboratory environment requiring field collection.

2. *The organism should be successfully maintained in the laboratory environment and available in sufficient quantities.* Many species do not fare well in the laboratory; our lack of knowledge of the exact nutritional requirements, overcrowding, and stress induced by the mere presence of laboratory personnel often make certain species unsuitable for toxicity testing.

3. *The genetics, genetic composition, and history of the culture should be known.* Perhaps the best documented organisms in laboratory culture are *Escherichia coli* and the laboratory strains of the Norway Rat. *E. coli* has been widely used in molecular genetics and biology as the organism of choice. Laboratory rats have long been used as test organisms for the evaluation of human health effects and research and are usually identified by a series of numbers. Often, each strain has a defined genealogy. Often strains of algae and protozoans are identified by strain and information is available as to their collection site. The American Type Culture Collection is a large repository of numerous procaryotic and eucaryotic organisms. The Star Culture Collection at the University of Texas is a repository for many unicellular algae. However, the majority of toxicity tests in environmental toxicology are conducted with organisms of unknown origin or field collection. Indeed, often the cultures originated from collections and the genetic relationships to the organisms used by other laboratories is poorly known.

4. *The relative sensitivities to various classes of toxicants of the test species should be known relative to the endpoints to be measured.* This criterion is not often realized in environmental toxicology. The invertebrate *Daphnia magna* is one of the most commonly used organisms in aquatic toxicology, yet only the results for approximately 500 compounds are listed in the published literature. The fathead minnow has been the subject of a concerted test program at the U.S. EPA Environmental Research Laboratory-Duluth conducted by G. Vieth over the last 10 years, yet fewer than a thousand compounds have been examined. In contrast, the acute toxicity of over 2000 compounds has been examined using the Norway Rat as the test species.

5. *The sensitivity of the test species should be representative of the particular class or phyla that the species represents.* Again this is an ideal criterion, not often met in the case of most test species. The limiting factor here is often the lack of information on the sensitivity of the organisms not routinely used for toxicity testing. In the case of teleost fish, a fish is a fish, as demonstrated by Suter (1993) in a recent review. What this means is that most of the time the toxicity of a compound to a fathead minnow is comparable to the toxicity of the compound to a salmonid. This fact is not surprising given the relative evolutionary distance of the vertebrates compared to the invertebrate classes. There is the myth of the "most sensitive species" and that is the organism that should be tested. Cairns (1986) has discussed the impossibility of such an organism, yet it is still held as a criterion to the selection of a test organism. In most cases it is not known what organisms and what endpoints are the most sensitive to a particular toxicant. The effects of toxicants to fungi, nonvascular plants, and mosses are poorly understood, yet these are major components of terrestrial ecosystems. Also, our knowledge of what species exist in a particular type of ecosystem over time and space is still limited. Often the dilemma has to be faced where it is a goal to protect an endangered species from extinction, yet no toxicological data are or can be made available.

6. *In multispecies toxicity tests the interactions among the component species should be understood.* The formulation of a successful multispecies toxicity test has been a major challenge. A continuing discussion has been focused upon the composition of the species of the test system. Many of the multispecies systems rely upon the inoculation of the test vessels from a natural source. These systems depend upon the chance establishment of a viable community within the test chambers. Other designers of multispecies systems inoculate organisms derived from standard laboratory cultures in a way to establish the functional components of the ecosystem. Laboratory-derived multispecies systems often lack the species richness that natural inocula provide. However, natural inocula lack the repeatability in species composition that the other design school can provide. Adherents to the use of natural inocula also point to the coevolution of the species contained within the test culture. This argument supposes (1) that the laboratory derived species did not originate from comparable ecosystems and (2) that the coevolution is important in the reaction of the community to the introduction of a toxicant. The tradeoffs in view of providing predictions of impacts on biological communities are not really understood at this point.

COMPARISON OF TEST SPECIES

Often the question of the best test species for screening for environmental toxicity has been debated. A wide variety is currently available representing a number of phyla and families, although a wide swath of biological categories is not represented by any test species. In the aquatic arena, an interesting paper by Doherty (1983) compared four test species for sensitivity to a variety of compounds. The test species were rainbow trout, bluegill sunfish (*Lepomis macrochirus*), fathead minnow, and *D. magna*. A particular strength of the study was the reliance upon data from Betz

Laboratories in addition to literature values. Having data from one laboratory reduces the interlaboratory error that is often a part of toxicity testing.

The results were very interesting. There was a high level of correlation ($r > 88\%$) among the four species in all combinations. Of course, three of the species are teleost fish. However, the Daphnia also fit the pattern. The exceptions about the correlations were compounds that contained chromium. *D. magna* was much more sensitive than the fish species.

Many other comparisons such as these have been made and are discussed in more detail in Chapter 9. However, in the selection of a test species for screening purposes, there seem to be high correlations between species for a broad number of toxicants. However, due to evolutionary events and happenstance, some organisms may be much more sensitive to a particular class of compound. So far, there is no *a priori* means of detecting such sensitivities without substantial biochemical data.

STATISTICAL DESIGN PARAMETERS

In the design of a toxicity test there is often a compromise between the statistical power of the toxicity test and the practical considerations of personnel and logistics. In order to make these choices in an efficient and informed manner several parameters are considered:

What is the specific question (questions) to be answered by this toxicity test?
What are the available statistical tools?
What power, in a statistical sense, is necessary to answer the specific questions?
What are the logistical constraints of a particular toxicity test?

The most important parameter is a clear identification of the specific question that the toxicity test is supposed to answer. The determination of the LC_{50} within a tight confidence interval will often require many fewer organisms than the determination of an effect at the low end of the dose-response curve. In multispecies toxicity tests and field studies the inherent variability or noise of these systems requires massive data collection and reduction efforts. It is also important to determine ahead of time whether a hypothesis-testing or regression approach to data analysis should be attempted.

Over the last several years a variety of statistical tests and other tools have become widely available as computer programs. This increase in statistical tools available can increase the sophistication of the data analysis and in some cases reduce the required work load. Unfortunately, the proliferation of these packages has led to *post hoc* analysis and the misapplication of the methods.

The power of the statistical test is a quantitative measure of the ability to differentiate accurately differences in populations. The usual case in toxicity testing is the comparison of a treatment group to control group. Depending on the expected variability of the data and the confidence level chosen, an enormous sample size or number of replicates may be required to achieve the necessary discrimination. If the sample size or replication is too large then the experimental design may have to be altered.

The logistical aspects of an experimental design should intimately interact with the statistical design. In some cases the toxicity evaluation may be untenable because of the numbers of test vessels or field samples required. Upon full consideration it may be necessary to rephrase the question or use another test methodology.

OVERVIEW OF AVAILABLE STATISTICAL METHODS FOR THE EVALUATION OF TOXICITY TESTS

A number of programs exist for the calculation of the chemical concentration that produces an effect in a certain percentage of the test population. The next few paragraphs review some of the advantages and disadvantages of the various techniques. The goal is to provide an overview, not a statistical text.

COMMONLY USED METHODS FOR THE CALCULATION OF ENDPOINTS

As reviewed by Stephan (1977) and Bartell et al. (1992), there are several methods available for the estimation of toxic endpoints. The next few paragraphs discuss some of the advantages and disadvantages of the popular methods.

Graphical interpolation essentially is the plotting of the dose response curve and reading the concentration that corresponds to the LC_{50} or the LC_{10}. This technique does not require concentrations that give a partial kill, say 7 out of 20 test organisms. In addition, data that provide atypical dose response curves can be analyzed since no previous assumptions are necessary. Another feature that is important is that the raw data must be observed by the researcher, illuminating any outliers or other features that would classify the dose-response curve as atypical. The disadvantage to using a graphical technique is that confidence intervals cannot be calculated and the interpretation is left to human interpolation. Graphing and graphical interpolation would generally be recommended as an exploratory analysis no matter which computational method is finally used. Graphing the data allows a determination of the properties of the data and often highlights points of interest or violations of the assumptions involved in the other methods of endpoint calculation.

The probit method is perhaps the most widely used method for calculating toxicity vs. concentration. As its name implies, the method used a probit transformation of the data. A disadvantage of the method is that it requires two sets of partial kills. However, a confidence interval is easily calculated and can then be used to compare toxicity results. There are several programs available for the calculation, and as discussed below, provide comparable results.

If only one or no partial kills are observed in the data, the Litchfield and Wilcoxin method can be employed. This method can provide confidence intervals, but is partially graphical in nature and employs judgment by the investigator. The probit method is generally preferred but the Litchfield and Wilcoxin method can be used when the partial kill criteria for the probit are not met.

Another transformation of the data is used in the logit method. A logit transformation of the data can be used, and the curve fitted by a maximum likelihood method. As with some of the other methods, a dearth of partial kill concentrations requires assumptions by the investigator to calculate an EC or LC value.

The Spearmen-Karber method must have toxicant concentrations that cover 0 to 100% mortality. Derived values are often comparable to the probit.

Perhaps the most widely applicable method, other than the graphical interpolation, is the moving average. The method can be used only to calculate the LC_{50} and there is the assumption that the dose response curve has been correctly linearized. As with the other methods, a partial kill is required to establish a confidence interval.

COMPARISON OF CALCULATIONS OF SEVERAL PROGRAMS FOR CALCULATING PROBIT ANALYSIS

Each of the methods for the estimation of an LC_{50} or other toxicological endpoint is available as a computer program. Examples of commonly available programs are: TOXSTAT, SAS-PROBIT, SPSS-PROBIT, DULUTH-TOX, and a program written by C. Stephan, ASTM-PROBIT. Bromaghin and Engeman (1989) and in a separate paper Roberts (1989) compared several of these programs using model data sets.

Bromaghin and Engeman considered the proposed ASTM-PROBIT to be a subset of the SAS Institute program, the SAS log 10 option. Two different data sets were used. The first data set was constructed using a normal distribution with a mean (LD_{50}) of 4.0 and a standard deviation of 1.25. Eleven dosage levels, quite a few compared to a typical aquatic toxicity test, ranging from 1.5 to 6.5 in increments of 0.5 were selected. The second set of test data was normally distributed with a mean equal to 8 and a standard deviation equal to 10. Five dosage levels, more typical of a toxicity test, ranging from 2 to 32 by multiples of 2, were used. In other words the concentrations were 2, 4, 8, 16, and 32. One hundred organisms were assumed to have been used at each test concentration in each data set. The response curves were generated based on two different criteria. First, that the response was normal with regard to the dosage. Second, the response is assumed to be normal with respect to either the base 10 or natural logarithm.

As shown in Table 3.3, the resulting estimated value was dependent on the method and the underlying assumptions used to calculate the LC_{50}. SAS log 10 and the ASTM-PROBIT were consistently identical in the calculated values of the LD_{50}s and the accompanying fiducial limits. Interestingly, the assumption of the normality being based on dose or the log 10 was important. In the first data set, when the normality of the data is based on the log 10 of the dose, the SAS default overestimated the LD_{50} in such a manner that the value was outside the limits given by the SAS log 10 and the ASTM method. In the second data set, the use of the appropriate calculation option was even more crucial. The inappropriate computational method missed the mark in each case and was accompanied by large fiducial limits. Bromaghin

Table 3.3   Estimates of LD_{50} Using Probit Analysis and SAS-PROBIT and ASTM-PROBIT

Data set (true LD_{50})	Normality with respect to:	Calculation method with estimate (95% fiducial limits)		
		SAS default	SAS log 10	ASTM
1 (4.0)	Dose	4.00(3.88–4.12)	3.80(3.59–4.02)	3.80(3.58–4.02)
	Log 10 dose	4.11(4.01–4.21)	3.99(3.90–4.10)	3.99(3.90–4.10)
2 (8.0)	Dose	8.02(5.35–10.36)	5.37(1.46–10.91)	5.37(1.46–10.91)
	Log 10 dose	12.28(8.04–16.57)	8.00(5.61–11.42)	8.00(5.61–11.42)

and Engeman (1989) conclude that these methods are not robust to departures from the underlying assumptions about the response distributions.

Roberts (1989) made a comparison between several commonly available programs used to calculate probit estimates of LD_{50}s. These programs were:

DULUTH-TOX, written by C. Stephan of the Duluth Environmental Protection Agency's Environmental Research Laboratory, was used to calculate toxicity endpoints.

ASTM-PROBIT was also written by C. Stephan as part of an ASTM Committe E-47 effort to produce a standard method of calculating toxicity estimates.

UG-PROBIT was developed by the Department of Mathematics and Statistics and the University of Guelph, Canada.

SPSS-PROBIT is a part of the SPSS statistical program available commercially and on many mainframes of universities and industry.

SAS-PROBIT is analogous to the SPSS-PROBIT in that it is part of widely available SAS statistical package.

After an extensive analysis, Roberts concluded that most of the programs provided useful and comparable LC_{50} estimates. The exception to this was the UG-PROBIT. The commercially available packages in SAS and SPSS had the advantages of graphical output and a method for dealing with control mortality. DULUTH-TOX and ASTM-TOX incorporated statistical tests to examine the data to assure that the assumptions of the probit calculations were met.

DATA ANALYSIS FOR CHRONIC AND MULTISPECIES TOXICITY TESTS

Analysis of variance (ANOVA) is the standard means of evaluating toxicity data to determine the concentrations that are significantly different in effects from the control or not dosed treatment. The usual procedure is (Gelber et al. 1985)

1. Transformation of the data
2. Testing for equivalence of the control or not dosed treatment with the carrier control
3. Analysis of variance performed on the treatment groups
4. Multiple comparisons between treatment groups to determine which groups are different from the control or not dosed treatment

Now we will examine each step.

In chronic studies, the data often are expressed as a percentage of control, although this is certainly not necessary. Hatchability, percentage weight gain, survival, and deformities are often expressed as percentage of the control series. The arcsine square root transformation is commonly used for this type of data before any analysis takes place. Many other types of transformations can be used depending upon the circumstances and types of data. The overall goal is to present the data in a normal distribution so that the parametric ANOVA procedure can be used.

Data such as weight and length and other growth parameters should not be included in the analysis if mortality occurred. Smaller organisms, because they are likely to absorb more of the toxicant on a per mass basis, are generally more sensitive, biasing the results.

If a carrier solvent has been used it is critical to compare the solvent control to the control treatment to ensure comparability. The common Student's t-test can be used to compare the two groups. If any differences exist, then the solvent control must be used as the basis of comparison. Unfortunately, a t-test is not particularly powerful with typical data sets. In addition, multiple endpoints are usually assessed in a chronic toxicity test. The change of a Type II error, stating that a difference exists when it does not, is a real possibility with multiple endpoints under consideration.

ANOVA has been the standby for detecting differences between groups in environmental toxicology. Essentially, the ANOVA uses variance within and between the groups to examine the distance of one group or treatment to another. An F-score is calculated on the transformed data with the null hypothesis since the effects upon all of the groups are the same. The test is powerful with the assumption met. If the F-score is not statistically significant, the treatments all have the same effect and the tested material has no effect. With a nonsignificant F-score (generally $P > 0.05$) the analysis stops. If the F-score is significant ($P < 0.05$) then the data are examined to determine which groups are different from the controls.

Multiple comparison tests are designed to select the groups that are significantly different from the control or each other. The most commonly used test is Dunnett's procedure. This test is designed to make multiple comparisons simultaneously. However, given the number of comparisons made in a typical chronic test, there is a significant chance that a statistically significant result will be found even if there are no treatment differences. The usual probability level is set at 0.05. Another way of looking at this is that comparisons with a statistically significant result will appear 5 times out of 100 even if no treatment differences exist. Beware of spurious statistical significance.

The overall purpose of the multiple comparisons is a determination of the MATC. The lowest concentration at which an effect is detected is the statistically determined lowest observed effect concentration (LOEC). The concentration that demonstrates no difference from the control is the no observed effects concentration (NOEC). The maximum allowable toxicant concentration is generally reported as LOEC > MATC > NOEC. The most sensitive endpoint is generally used for this estimation. Perhaps the greatest difficulty in estimating endpoints such as the NOEC and LOEC are their dependence upon the statistical power of the test. Often treatment numbers are determined by parameters other than statistical power, cost, safety, and other logistical factors. A greater statistical power would likely improve the ability to detect significant differences at subsequently lower concentrations. Along with statistical power, the placement of the test concentrations relative to the generally unknown dose-response curve can also alter the interpretation of the NOEC, LOEC, and the derived MATC. The closer the spacing and the more concentrations used, the more accurate are these derived parameters.

Gelber et al. (1985) suggest that a major improvement can be made in the analysis of chronic toxicity tests. They suggest that Williams' test (Williams 1971, 1972) is more powerful than Dunnett's since it is designed to detect increasing concentration (dose) response relationships. A removal of the preliminary ANOVA is also recommended, since performing the ANOVA and the multiple comparison tests both have a 5% error rate. They suggest performing multiple Williams' test to arrive at the concentration that is not significantly different from the control set.

The above methods are generally used to calculate a midpoint in the dose-response curve that results in 50% mortality or to test the null hypothesis that there is no effect. In ranking compounds in their acute or chronic toxicity this may be an appropriate approach. However, in the estimation of mortality at low concentrations, concentrations that are probably more realistic in a field situation, LC_{10}s or even LC_1s may be more appropriate. As proposed by C. E. Stephan, a regression or curve-fitting approach to the evaluation of laboratory toxicity data may be more appropriate for estimating environmental effects. In this instance a regression is used to calculate the best fit line through the data. Linear regression after a log transformation can be used along with other regression models. Confidence intervals of the LC_{10} or LC_1 estimation derived from a regression technique can be quite large; however, an estimate of effects at low concentrations can be derived.

Figure 3.11 plots the data in example 3 with the data transformed to a base 10 logarithm. The relationship for this data set is rather linear, and the toxicity at low concentrations can easily be estimated. In this instance, 100% mortality has a log of 2.0, the LC_{50} is 1.7, and the LC_{10} is equal to 1.0.

Hypothesis testing in the determination of NOELs and LOELs also has drawbacks largely related to the assumptions necessary for the computations. These characteristics have been listed by Stephan and Rodgers (1985) and compared to curve-fitting models for the estimation of endpoints.

First, use of typical hypothesis-testing procedures that clearly state the α value (typically 0.05) leaves the β value unconstrained and skews the importance of reporting the toxic result. In other words, the typical test will be conservative on the side of saying there is no toxicity even when toxicity is present.

Second, the threshold for statistical significance does not innately correspond to a biological response. In other words, hypothesis testing may produce a NOEL that is largely a statistical and experimental design artifact and not biological reality. As discussed earlier in the chapter, there is debate about the existence of a response threshold.

Third, a large variance in the response, due to poor experimental design or innate organismal variability in the response, will reduce the apparent toxicity of the compound using hypothesis testing.

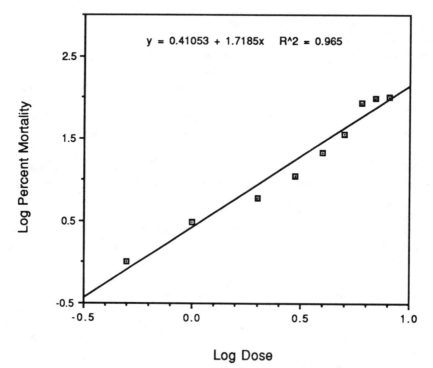

Figure 3.11   Plot of a log-log regression for toxicity data set 3.

Fourth, the results are sensitive to the factors of experimental design that determine the statistical power and resolution of the analysis methods. These design parameters are typically the number of replicates for each test concentration, and the number and spacing of the test concentrations.

Fifth, no dose-response relationship is derived using hypothesis testing methods. The lack of dose-response information means that the investigator has no means of evaluating the reasonableness of the test results. Conversely, a specific type of dose-response relationship is not required to conduct the analysis.

Given these disadvantages, hypothesis testing is still the prevalent method of determining NOELs and is the preferred method stated in publications such as that by Weber et al. 1989. Currently, there appear to be no alternative methods in widespread use. Hypothesis testing and the use of analysis of variance are still the predominant method of analysis even for tests incorporating greater levels of biological and chemical complexity.

ANALYSIS OF MULTISPECIES TOXICITY TESTS (MICROCOSMS, MESOCOSMS)

Typically, the goals of a multispecies toxicity test are:

- to detect changes in the population dynamics of the individual taxa that would not be apparent in single species tests,
- to detect community-level differences that are correlated with treatment groups thereby representing a deviation from the control group, and
- to observe the distribution and degradation of a compound in a physical model of a ecosystem.

A number of methods have been developed to attempt to satisfy the goals of multispecies toxicity testing. Analysis of variance (ANOVA) is the classical method to examine single variable differences from the control group. However, because multispecies toxicity tests generally run for weeks or even months, there are problems with using conventional ANOVA. These include the increasing likelihood of introducing a Type II error (accepting a false null-hypothesis), temporal dependence of the variables, and the difficulty of graphically representing the data set. Conquest and Taub (1989) developed a method to overcome some of the problems by using intervals of nonsignificant difference (IND). This method corrects for the likelihood of Type II errors and produces intervals that are easily graphed to ease examination. The method is routinely used to examine data from SAM toxicity tests, and it is applicable to other multivariate toxicity tests. The major drawback is the examination of a single variable at a time over the course of the experiment. While this addresses the first goal in multispecies toxicity testing listed above, it ignores the second. In many instances, community-level responses are not as straightforward as the classical predator-prey or nutrient-limitation dynamics usually picked as examples of single-species responses that represent complex interactions.

Multivariate methods have proved promising as a method of incorporating all the dimensions of an ecosystem. One of the first methods used in toxicity testing was the calculation of ecosystem strain developed by Kersting (1988) for a relatively simple (three-species) microcosm. This method has the advantage of using all the measured parameters of an ecosystem to look for treatment-related differences. At about the same time, Johnson (1988a, 1988b) developed a multivariate algorithm using the n-dimensional coordinates of a multivariate data set and the distances between these coordinates as a measure of divergence between treatment groups. Both of these methods have the advantage of examining the ecosystem as a whole rather than by single variables, and can track such processes as succession, recovery, and the deviation of a system due to an anthropogenic input.

However, a major disadvantage of both these methods, and of many conventional multivariate methods, is that all the data are often incorporated without regard to the units of measurement or the appropriateness of including all variables in the analysis. It can be difficult to combine variables such as pH, with units ranging from 0 to 14, with the numbers of bacterial cells per milliliter, where low numbers are in the 10^6 range, to say nothing of the conceptual difficulties of adding pH units to counts. Similarly, random variables (i.e., variables with no treatment-related response) indiscriminately incorporated into the analysis may contribute so much noise that they overshadow variables that do show treatment-related effects.

Ideally, a multivariate statistical test used for evaluating complex data sets will have the following characteristics:

- It will not combine counts from dissimilar taxa by means of sums of squares, or other mathematical techniques, as in the Euclidean and cosine distance measures.
- It will not require transformations of the data, such as normalizing the variance.
- It will work without modification on incomplete data sets.
- It will work without further assumptions on different data types (e.g., species counts or presence/absence data).
- Significance of a taxon to the analysis will not be dependent on the absolute size of its count, so that taxa having a small total variance, such as rare taxa, can compete in importance with common taxa, and taxa with a large, random variance will not automatically be selected, to the exclusion of others.
- It will provide an integral measure of "how good" the analysis is, i.e., whether the data set differs from a random collection of points.
- It will, in some cases, identify a subset of the taxa that serve as reliable indicators of the physical environment.

Recently developed for the analysis of ecological data (Matthews and Hearne 1991), nonmetric clustering is a multivariate derivative of artificial intelligence research that satisfies all these criteria, and has the potential of circumventing many of the problems of conventional multivariate analysis.

NONMETRIC CLUSTERING AND ASSOCIATION ANALYSIS

Unlike the more conventional multivariate statistics, nonmetric clustering is an outgrowth of artificial intelligence and a tradition of conceptual clustering. In this approach, an accurate description of the data is only part of the goal of the statistical analysis technique. Equally important is the intuitive clarity of the resulting statistics. For example, a linear discriminant function to distinguish between groups might be a complex function of dozens of variables, combined with delicately balanced factors. While the accuracy of the discriminant may be quite good, use of the discriminant for evaluation purposes is limited because humans cannot perceive hyperplanes in highly dimensional space. By contrast, conceptual clustering attempts to distinguish groups using as few variables as possible, and by making simple use of each one. Rather than combining variables in a linear function, for example, conjunctions of elementary "yes-no" questions could be combined: species A greater than 5, species B less than 2, and species C between 10 and 20. Numerous examples throughout the artificial intelligence literature have proved that this type of **conceptual** statistical analysis of the data provides much more useful insight into the patterns in the data, and is often more accurate and robust. Conceptual statistical analysis attempts to fit the data, but not at the expense of a simple, intuitive result. The use of nonmetric clustering and other methods have been compared in a number of field and laboratory tests (Matthews and Matthews 1991; Matthews, Matthews, and Hachmoller 1991a; Landis et al. 1993).

REFERENCES AND SUGGESTED READINGS

Bartell, S.M., R.H. Gardner, and R.V. O'Neill. 1992. *Ecological Risk Estimation*. Lewis Publishers, Boca Raton.

Bromaghin, J.F. and R.M. Engeman. 1989. Head to head comparison of SAS and ASTM-proposed probit computer programs. In *Aquatic Toxicology and Environmental Fate:* 11th Volume. ASTM STP 1007. G.W. Suter and M.A. Lewis, Eds., American Society for Testing and Materials, Philadelphia, PA, pp. 303-307.

Cairns, J.C., Jr. 1986. The myth of the most sensitive species. *Bioscience* 36:670-672.

Conquest, L.L. and F.B. Taub. 1989. Repeatability and reproducibility of the Standard Aquatic Microcosm: Statistical properties. In *Aquatic Toxicology and Hazard Assessment:* 12th Volume. ASTM STP 1027. U.M. Cowgill and L.R. Williams, Eds., American Society for Testing and Materials, Philadelphia, PA, pp. 159-177.

Doherty, F.G. 1983. Interspecies correlations of acute aquatic median lethal concentration for four standard testing species. *Environ. Sci. Technol.* 17:661-665.

Gelber, R.D., P.T. Levin, C.R. Mehta, and D.A. Schoenfeld. 1985. Statistical Analysis. In *Aquatic Toxicology*. G.M. Rand and S.R. Petrocelli, Eds., Hemisphere Publishing Corporation, Washington, D.C., pp. 110-123.

Johnson, A.R. 1988a. Evaluating ecosystem response to toxicant stress: a state space approach. In *Aquatic Toxicology and Hazard Assessment:* 10th Volume. ASTM STP 971. W.J. Adams, G.A. Chapman, and W.G. Landis, Eds., American Society for Testing and Materials, Philadelphia, PA, pp. 275-285.

Johnson, A.R. 1988b. Diagnostic variables as predictors of ecological risk. *Environ. Manage.* 12: 515-523.

Kersting, K. 1988. Normalized ecosystem strain in micro-ecosystems using different sets of state variables. *Verh. Int. Verein. Limnol.* 23: 1641-1646.

Kersting, K. and R. van Wungaarden. 1992. Effects of Chlorpyifos on a microecosystem. *Environ. Toxicol. Chem.* 11: 365-372.

Landis, W.G., R.A. Matthews, A.J. Markiewicz, and G.B. Matthews. 1993. Multivariate analysis of the impacts of the turbine fuel JP-4 in a microcosm toxicity test with implications for the evaluation of ecosystem dynamics and risk assessment. *Ecotoxicology* 2:271-300.

Matthews, G.B. and J. Hearne. 1991. Clustering without a metric. *IEEE Trans. Pattern Anal. Machine Intelligence* 13: 175-184.

Matthews, G.B. and R.A. Matthews. 1991. A model for describing community change. In *Pesticides in Natural Systems: How Can Their Effects Be Monitored? Proc. Conf.,* Environ. Res. Lab./ORD, Corvallis, OR, EPA 9109/9-91/011.

Matthews, G.B., R.A. Matthews, and B. Hachmoller. 1991a. Mathematical analysis of temporal and spatial trends in the benthic macroinvertebrate communities of a small stream. *Can. J. Fish. Aquat. Sci.* 48: 2184-2190.

Rand, G.M. and S.R. Petrocelli. 1985. Introduction. In *Aquatic Toxicology*. G.M. Rand and S.R. Petrocelli, Eds., Hemisphere Publishing Corporation, Washington, D.C., pp. 1-28.

Roberts, M.H. 1989. Comparison of several computer programs for probit analysis of dose related mortality data. In *Aquatic Toxicology and Environmental Fate:* 11th Volume. ASTM STP 1007. G.W. Suter and M.A. Lewis, Eds., American Society for Testing and Materials, Philadelphia, PA, pp. 308-320.

Stephan, C.E. 1977. Methods for calculating an LC50. In *Aquatic Toxicology and Hazard Evaluation*, ASTM STP 634, F.L. Mayer and J.L. Hamelink, Eds., American Society for Testing and Materials Philadelphia, PA, pp. 65-84.

Stephan, C.E. and J.R. Rodgers. 1985. Advantages of using regression analysis to calculate results of chronic toxicity tests. In *Aquatic Toxicology and Hazard Assessment: Eighth Symposium*. R.C. Bahner and D.J.H. Hansen, Eds., American Society for Testing and Materials, Philadelphia, PA, pp. 328-339.

Suter, G.W. 1993. Organism-level effects. In *Ecological Risk Assessment*. G.W. Suter, II, Ed., Lewis Publishers, Boca Raton, pp. 175-246.

Weber, C.I., W.H. Pettier, T.J. Norberg-King, W.E.B. Horning, II, F.A. Kessler, J.R. Menkedick, T.W. Neiheisel, P.A. Lewis, D.J. Klem, Q.H. Pickering, E.L. Robinson, J.M. Lazorchak, L.J. Wymer, and R.W. Freyberg. 1989. Short-Term Methods for Estimating the Chronic Toxicity of Effluents and Surface Waters to Freshwater Organisms. Second Edition. EPA-600/2-89/001. Environmental Monitoring Systems Laboratory, U.S. Environmental Protection Agency, Cincinnati, OH.

Williams, D.A. 1971. A test for differences between treatment means when survival dose levels are compared with a zero dose control. *Biometrics* 27:103-117.

Williams, D.A. 1972. A comparison of several dose levels with a zero dose control. *Biometrics* 28:519-531.

STUDY QUESTIONS

1. Anthropogenic toxicants introduced into the environment come from what types of sources?
2. What is a pesticide?
3. What determines a toxicant compound's environmentally harmful concentrations?
4. Define dose-response relationship. What is a dose-response curve?
5. Describe the two parameters of a dose-response curve which determine the curve.
6. Similar slopes of dose-response curves may imply what about the xenobiotics being compared?
7. Discuss the two prevailing concepts for studying toxicity of compounds at low concentrations.
8. What are the advantages to the use of a standard method in evaluation of the toxicity of chemicals or mixtures?
9. What are the two general parameters involved in the classification of toxicity tests in environmental toxicology?
10. Describe a microcosm and a mesocosm test.
11. Describe the lab-to-field dilemma.
12. What differences are there between a static and a static-renewal toxicity test?
13. What are the advantages and disadvantages of the recirculating methodology of toxicity testing?
14. Name and describe the best technical method for toxicity testing.
15. What is whole-body aquatic test systems exposure?
16. Discuss the six criteria for choosing a test species for use in a toxicity test.
17. Discuss the natural source vs. laboratory derived composition of species in multispecies toxicity tests.
18. What are the most important parameters when choosing statistical design parameters for a toxicity test?
19. Compare the various methods for calculating endpoints from an acute or chronic toxicity test.

CHAPTER 4

Survey and Review of Typical Toxicity Test Methods

The importance of understanding the test procedures that are crucial to environmental toxicology cannot be underestimated. The requirements of the tests dictate the design of the laboratory, logistics, and the required personnel. In every interpretation of an EC_{50} or an NOEL there should be a clear understanding of the test method used to obtain that estimate. The understanding should include the strengths and weaknesses of the test method and the vagaries of the test organism or organisms. Quite often it is the standard method that is modified by a researcher to answer more specific questions about the effects of xenobiotics. These standard tests form the basis of much of what we know about relative chemical toxicity in a laboratory setting.

Table 4.1 lists a number of toxicity tests currently available from a variety of standard sources. This table is not all-inclusive since there are more specialized tests for specific locations or situations. Many more methods exist, some of which are derivatives of basic toxicity tests. More important than memorization of each test procedure is a good understanding of the general thrust of the various toxicity tests, methods of data analysis, and experimental design.

The following survey starts with single species toxicity tests and concludes with field studies. These summaries are based on the standard methods published by the American Society for Testing and Materials, the U.S. Environmental Protection Agency, and other published sources. Many of these methods are listed in the reference section for this chapter. The survey is broken up into single species and multispecies tests. Although Chapter 3 discussed to some length the various types of toxicity: acute, chronic, partial life cycle, etc., it is in many ways logical to list them by the level of biological organization. That is what is done here. Since it is difficult to include every toxicity test in a volume of this size, representative tests have been chosen for summary. Inclusion here does not imply an endorsement by the author, but these tests serve as examples of the kinds of toxicity tests used to evaluate environmental hazards.

Table 4.1   Partial List of ASTM Standard Methods for Toxicity Evaluation or Testing

Biodegradation By a Shake-Flask Die-Away Method
Conducting a 90-Day Oral Toxicity Study in Rats
Conducting a Subchronic Inhalation Toxicity Study in Rats
Conducting Aqueous Direct Photolysis Tests
Determining the Anaerobic Biodegradation Potential of Organic Chemicals
Determining a Sorption Constant (K_{oc}) for an Organic Chemical in Soil and Sediments
Inhibition of Respiration in Microbial Cultures in the Activated Sludge Process
Algal Growth Potential Testing with *Selenastrum capricornutum*
Conducting Bioconcentration Tests with Fishes and Saltwater Bivalve Mollusks
Conducting Reproductive Studies with Avian Species
Conducting Subacute Dietary Toxicity Tests with Avian Species
Evaluating Environmental Fate Models of Chemicals
Measurement of Chlorophyll Content of Algae in Surface Waters
Standardized Aquatic Microcosm: Freshwater
Using Brine Shrimp Nauplii as Food for Test Animals in Aquatic Toxicology
Using Octanol-Water Partition Coefficient to Estimate Median Lethal Concentrations for Fish
    Due to Narcosis
Conduct of Micronucleus Assays in Mammalian Bone Marrow Erythrocytes
Conducting Acute Toxicity Tests on Aqueous Effluents with Fishes, Macroinvertebrates, and
    Amphibians
Conducting Acute Toxicity Tests with Fishes, Macroinvertebrates, and Amphibians
Conducting Early Life-Stage Toxicity Tests with Fishes
Conducting Life-Cycle Toxicity Tests with Saltwater Mysids
Conducting Renewal Life-Cycle Toxicity Tests with *Daphnia magna*
Conducting Sediment Toxicity Tests with Freshwater Invertebrates
Conducting 10-Day Static Sediment Toxicity Tests with Marine and Estuarine Amphipods
Conducting Static 96-h Toxicity Tests with Microalgae
Conducting Static Acute Aquatic toxicity Screening Tests with the Mosquito, *Wyeomyia
    smithii* (Coquillett)
Conducting Static Acute Toxicity Tests Starting with Embryos of Four Species of Saltwater
    Bivalve Mollusks
Conducting Static Toxicity Tests with the *Lemma gibba* G3
Conducting a Terrestrial Soil-Core Microcosm Test
Conducting Three-Brood, Renewal Toxicity Tests with *Ceriodaphnia dubia*
Hazard of a Material to Aquatic Organisms and Their Uses
Assessing the Performance of the Chinese Hamster Ovary Cell/Hypoxanthine Guanine
    Phosphoribosyl Transferase Gene Mutation Assay

SINGLE-SPECIES TOXICITY TESTS

DAPHNIA 48-H ACUTE TOXICITY TEST

This test along with the fish 96-h acute toxicity test is one of the standbys in aquatic toxicology. *Daphnia magna* and *D. pulex* are the common test species. *D. magna* require relatively hard water for its culture. *D. magna* are large, commonly available, and easy to culture. *D. pulex* are not quite as large as *D. magna* and tolerate softer water. It is recommended that the test organisms be derived from adults, three generations after introduction into the specific laboratory media.

Water quality is a major factor in the performance of any laboratory aquatic toxicity test. Care must be taken to eliminate other sources of mortality, such as chlorine of chlorinated organics, heavy metal contamination, and contamination by organics in the groundwater or reservoir supply. In some labs with access to a high-grade tap or well water only a minor purification system is required. However, in many cases a further filtration and distillation step may be required. Soft dilution water (40 to 48 mg/l as CaCO₃) is recommended for tests with *D. pulex,* and moderately hard water (80 to 100 mg/l as CaCO₃) is recommended for tests with *D. magna.* Water dilution is considered acceptable if *Daphnia* spp show adequate survival and reproduction when cultured in the water.

Sodium pentachlorophenate (NaPCP) is the reference toxicant that has been suggested for toxicity tests using daphnids. The use of a reference toxicant is important in confirming the health of the daphnia and the quality of the water and test methodology.

In general, ten neonates that are less than 24-h old are placed in 125-ml beakers containing 100 ml of test solution at five concentrations and a negative control. The tests are usually run in triplicate. Death is difficult to observe so immobility of the daphnia is used as the endpoint. An organism in considered immobile (nonmotile) if it does not resume swimming after prodding with a pipet or glass rod. Measurements are made at 24-h intervals. No feeding occurs during the course of this toxicity test.

The daphnia 48-h toxicity test is a useful screen for the toxicity of single compounds, mixtures, or effluents. In some cases the daphnid toxicity test has been used to evaluate the potential pathology or other potential problems with genetically engineered organisms. The advantages of the daphnid toxicity test are short time frame, small amounts of hazardous waste are generated, and the test is inexpensive. Often daphnids are more sensitive than vertebrates to a variety of toxicants. The disadvantages include the time-consuming maintenance of test stocks and the sensitivity of the organisms to water quality.

The chronic or partial life cycle toxicity test with *D. magna* is an attempt to look at growth and reproductive success of the test organisms. This test is contrasted to its acute counterpart in Table 4.2. The test follows a set of daphnia through the production of three broods with generally a measurement of growth (length or mass) of the original organisms along with the numbers of offspring derived from each animal.

One of the most controversial aspects of this test has been the food source during the study. A number of mixtures have been tried with interesting results. A mixture of trout chow and algae has been demonstrated to provided excellent growth, but there are concerns about the consistency of the ingredients. Many laboratories use a combination of algae, *Ankistrodesmus convolutus, A. falcatus, Chlamydomonas reinhardii,* and *Selenastrum capricornutum* as the food source.

This toxicity test is usually run as a static renewal but some researchers have used a continuous-flow setup with a proportional diluter. Handling the organisms during the transfer to new media is a potential problem for inexperienced technicians.

Occasionally it is difficult to set up concentrations for the test if the median values for the chronic endpoints are close to the values for a toxicant that induce mortality over the duration of the experiment. Loss of replicates can occur if the mortality rates are high enough. Use of the dose-response curve of the acute data should help in identifying useful boundary conditions for the higher concentrations of xenobiotic.

Table 4.2   Comparison of the *D. magna* 48-h Acute Toxicity Test with the Common *D. magna* Chronic Toxicity or Partial Life Cycle Test.

Test type	Chronic (partial life cycle)	Acute 48 h
Organisms	*D. magna*	*D. magna*
Age of test organisms	24-h old	24-h old
Number of organisms per chamber	10	10 (minimum)
Experimental design		
Test vessel type and size	100 ml beakers	250 ml
Test solution volume	80 ml	200 ml
Number of replicates per sample	2 (minimum)	3 (minimum)
Feeding regime	Various combinations of trout chow, yeast, alfalfa, green algae, and diatoms given in excess	Do not feed
Test duration	21 days	48 hr
Physical and chemical parameters		
Water temperature	20°C	20 ± 2°C
Light quality	Ambient laboratory levels	Ambient laboratory levels
Light intensity	Up to 600 lux	540 to 1080 lux
Photoperiod	16 h light and 8 h dark (with 15- to 30-min transition)	16 h light and 8 h dark
pH range	7.0–8.6	7.0–8.6
DO concentration	40–100%	60–100%
Aeration	Not necessary	none
Endpoint	Survival, growth, and reproduction	Immobilization

Closely related to the *D. magna* partial life cycle toxicity test is the three-brood renewal toxicity test with *Ceriodaphnia dubia* (Table 4.3). The test was developed in an attempt to shorten the amount of time, amount of toxicant, and the cost of performing chronic type toxicity tests. This methodology has proven useful in a variety of roles, especially in the testing of effluents. One of the drawbacks and, at the same time, advantages of the method is the small size of the test organism. Adult *C. dubia* are about the same size as first instar *D. magna*. Handling the first instars and even the adults often takes a dissecting microscope and a steady hand. Conversely, the small size enables the researcher to conduct the test in a minimum of space and the rapid reproduction rate makes the method one of the shortest life cycle type tests.

As with the *D. magna* tests, one of the problems has been in the successful formulation of a food to ensure the health and reproducible reproduction of the *C. dubia* during the course of the toxicity tests. A combination of trout chow, yeast, rye grass powder, and algae have been used. Nonetheless the *C. dubia* three-brood toxicity test has been proven to be useful and replicable.

ALGAL 96-H GROWTH TOXICITY TEST

The purpose of this toxicity test is to examine the toxicity of materials to a variety of freshwater and marine algae and it is summarized in Table 4.4. In aquatic systems

Table 4.3 Summary for Conducting Three-Brood, Renewal Toxicity Tests with *Ceriodaphnia dubia*

Test type	Static renewal/chronic
Organisms	*Ceriodaphnia dubia*
Age of test organisms	<12-h old
Experimental design	
Test vessel type and size	Test has been conducted with 30 ml beaker with 15 ml of test solution; can use any container made of glass, Type 316 stainless steel, or fluorocarbon plastic if a) each *C. dubia* is in a separate chamber or compartment and b) each chamber can maintain adequate DO levels for the organism; chambers should be covered with glass, stainless steel, nylon, or fluorocarbon plastic covers or Shimatsu closures
Number of replicates	10
Total number of organisms	At least 10
Number of organisms per chamber	1
Feeding regime	Various combinations of trout chow, yeast, rye grass powder, and algae have been used; types of algae include *Ankistrodesmus convolutus, A. falcatus, Chlamydomonas reinhardii,* and *Selenastrum capricornutum*
Test duration	7 days
Physical and chemical parameters	
Temperature	25 ± 1°C
Test solution pH	Not specified
DO concentration	40–100 %
Endpoint	Reproduction

algae are generally responsible for a large percentage of the primary production. Impacts upon the unicellular photosynthetic organisms could have long-lasting impacts to the community.

Numerous test organisms have been used in this toxicity test, but those currently recommended by the ASTM guidelines are

Freshwater
   Green algae: *Selenastrum capricornutum, Scenedesmus subspicatus, Chlorella vulgaris*
   Blue-green algae (bacteria): *Microcystus aeruginosa, Anabena flos-aquae*
   Diatom: *Navicula pelliculosa*
Saltwater
   Diatom: *Skeltonema costatum, Thalassiosira pseudonana*
   Flagellate: *Dunaliella tertiolecta*

Other test organisms can be used if necessary for a particular toxicity assessment or research. The methodology is very adaptable.

Depending upon the test organism, between 2×10^4 and 5×10^4 cells are used to inoculate the test vessel and the concentration of cells is determined daily. Cell counts are made daily by using a hemocytometer or an electronic particle counter

Table 4.4   Summary of Test Conditions for Conducting Static 96-h Toxicity Tests with Microalgae

Test type	Static
Organisms	Freshwater species: *Selenastrum capricornutum, Scenedesmus subspicatus, Chlorella vulgaris, Microcystis aeruginosa, Anabaena flos-aquae, Navicula pelliculosa;* Saltwater species: *Skeletonema costatum, Thalassiosira pseudonana,* and *Dunaliella tertiolecta*
Number of organisms per chamber (±10%)	*Selenastrum capricornutum* and other freshwater green algae 2×10^4 cells/ml *Navicula pelliculosa* 2×10^4 cells/ml *Microcystis aeruginosa* 5×10^4 cells/ml *Anabaena flos-aquae* 2×10^4 cells/ml Saltwater species 2×10^4 cells/ml
Experimental design	
Test vessel type and size	Sterile Erlenmeyer flasks of borosilicate glass, any size
Test solution volume	Not to exceed 50% of the flask volume for tests conducted on a shaker, and not more than 20% of the flask volume for tests not conducted on a shaker
Number of replicate chambers per sample	2 or more
Test duration	96 h
Physical and chemical parameters	
Water temperature	24 ± 2°C for freshwater green and blue-green algae 20 ± 2°C for *Navicula pelliculosa* and other saltwater algae
Light quality	Continuous "cool-white" fluorescent
Light intensity	Should not vary by more than ±15%: 60 μE m^{-2}/s^{-1} (4300 lm/m^2) for freshwater diatoms and green algae 30 μE m^{-2}/s^{-1} (2150 lm/m^2) for freshwater blue-green algae 82–90 μE m^{-2}/s^{-1} (5900 to 6500 lm/m^2) for *Thalassiosira* 60 μE m^{-2}/s^{-1} (4300 lm/m^2) for *Skeletonema*
Photoperiod	14 h light/10 h dark for *Skeletonema*
Test solution pH	7.5 ± 0.1 for freshwater 8.0 ± 0.1 for saltwater
Endpoint	Biomass, cell number, area underneath the growth curve

such as the Coulter Counter. Chlorophyll a can be measured spectrophotometrically or fluorometrically. The fluorometric determinations are more accurate at low concentrations of test organism. Other measurements that have been used include DNA content, ATP charge, and ^{14}C assimilation.

If only standing biomass is the endpoint to be measured then only cell concentration at the end of the exposure period has to be determined. However, measurements such as area under the curve and growth rate are important variables in determining the ecological impacts of a toxicant. These valuable endpoints require measurements of cell density each day for the duration of the toxicity test. Other measurements to ensure the replicability of the data include pH, temperature, and light intensity. Whenever possible, toxicant concentration should also be taken at the beginning and end of the test.

A good microbiological sterilization technique is required to ensure a minimum of cross-contamination with other algae and to prevent the introduction of bacteria. The degradation of the toxicant by introduced bacteria can alter the apparent toxicity, even to the point of eliminating the test compound from the media.

Another interesting aspect of this test is the enhancement of algal growth often found at low concentrations of toxicant. The spontaneous hydrolysis or other breakdown of the test compound may provide nutrients as well as nutrients contained in effluents. It is crucial that the data be appropriately plotted and analyzed.

ACUTE TOXICITY TESTS WITH AQUATIC VERTEBRATES AND MACROINVERTEBRATES

As with the daphnid toxicity tests, toxicity tests using a variety of fish species, amphibians, and macroinvertebrates have long been the standbys of aquatic toxicity evaluations. Table 4.5 summarizes the species and methods used in these tests.

One of the major problems in conducting these toxicity tests is the reliable supply of healthy test organisms. Many of the fish species used to stock ponds and lakes are available through hatcheries. Specialized suppliers also exist for the species that are routinely used for toxicity evaluations. In some cases it is required that wild organisms are collected and acclimated to the laboratory environment before conducting the toxicity test. Animals collected from wild have some advantages and some drawbacks. The major advantage is that if the organism is collected locally the sensitivity demonstrated in the toxicity test is representative of that particular native population. Care must be taken, however, to not unduly stress the collected organisms or the resultant stress may cause an overestimate of the toxicity of the compound being examined. The major difficulty of using organisms collected from wild populations is the variation among populations in sensitivity to the toxicant or to the laboratory culture collections. With mobile organisms it may be difficult to consistently collect organisms from the same breeding population. Also the act of collecting the organisms may seriously deplete their numbers, especially in areas near the testing facility. Care should be taken not to deplete local populations. Another solution is to maintain a habitat adjacent to the facility as a source of the test organisms under the control and regulation of the testing laboratory.

Another difficulty in conducting a broad series of toxicity tests is the assurance of adequate water quality and volume for a variety of species. For testing freshwater species the solution is often the investment in a well system with the water filtered and sterilized. Occasionally the testing facility may be adjacent to a body of water that can supply a consistent and uncontaminated source of water for the culture of the test organisms and also act as a source of dilution water. Laboratories on the Great Lakes or marine laboratories often have access to large volumes of relatively clean water. The least desirable but often the only option available is the use of distilled tap water for culture and dilution. At the least, the tap water should be doubly distilled and filtered before being used to make culture media. Systems that use distilled water supplied by a central system, filtered through an ion exchange system and then glass distilled have proven reliable. Unfortunately, the necessity of using distilled water

Table 4.5   Summary for Conducting Acute Toxicity Tests with Fishes, Macroinvertebrates, and Amphibians

Test type	Static, renewal, flow-through
Organisms	Freshwater
	Vertebrates
	Frog (*Rana* sp.), toad (*Bufo* sp.), coho salmon (*Oncorhynchus kisutch*), rainbow trout (*Salmo gairdneri*), brook trout (*Salvelinus fontinalis*), goldfish (*Carassius auratus*), fathead minnow (*Pimephales promelas*), channel catfish (*Ictalarus punctatus*), bluegill (*Lepomis macrochirus*), green sunfish (*Lepomis cyanellus*)
	Invertebrates
	Daphnids (*Daphnia magna, D. pulex, D. pulicaria*), amphipods (*Gammarus lacustris, G. fasciatus, G. pseudolimnaeus*), crayfish (*Orconectes* sp., *Combarus* sp., *Procambarus* sp., *Pacifastacus leniusculus*), stoneflies (*Pteronarcys* sp.), mayflies (*Baetis* sp., *Ephemerella* sp.), mayflies (*Hexagenia limbata, H. bilineata*), midges (*Chironomus* sp.), snails (*Physa integra, P. heterostropha, Amnicola limosa*), planaria (*Dugesia tigrina*)
	Saltwater
	Vertebrates
	Sheepshead minnow (*Cyprinodon variegatus*), mummichog (*Fundulus heteroclitus*), longnose killifish (*Fundulus similis*), silverside (*Menidia* sp.), threespine stickleback (*Gasterosteus aculeatus*), pinfish (*Lagodon rhomboides*), spot (*Leiostomus xanthurus*), shiner perch (*Cymatogaster aggregata*), tidepool sculpin (*Oligocottus maculosus*), sanddab (*Citharichthys stigmaeus*), flounder (*Paralichthys dentatus, P. lethostigma*), starry flounder (*Platichthys stellatus*), english sole (*Parophrys vetulus*), herring (*Clupea harengus*)
	Invertebrates
	Copepods (*Acartia clausi, A. tonsa*), shrimp (*Penaeus setiferus, P. duorarum, P. aztecus*), grass shrimp (*Palaemonetes pugio, P. intermedius, P. vulgaris*), sand shrimp (*Crangon septemspinosa*), shrimp (*Pandalus jordani, P. danae*), bay shrimp (*Crangon nigricauda*), mysid (*Mysidopsis bahia, M.bigelowi, M. almyra*), blue crab (*Callinectes sapidus*), shore crab (*Hemigrapsus* sp., *Pachygrapsus* sp.), green crab (*Carcinus maenas*), fiddler crab (*Uca* sp.), oyster (*Crassostrea virginica, C. gigas*), polychaete (*Capitella capitata*)
Age and size of test organisms:	All organisms should be as uniform as possible in age and size
	Fish: juvenile; weight between 0.1-5.0 g; total length of longest fish should be no more than twice that of the shortest fish
	Invertebrates: except for deposition tests with bivalve mollusks and tests with copepods, immature organisms should be used whenever possible
	Daphnids: less than 24-h old
	Amphipods, mayflies, and stone flies: early instar
	Midges: second or third instar
	Saltwater mysids: less than 24 h post-release from the brood sac
	Do not use ovigerous decapod crustaceans or polychaetes with visible developing eggs in coelom
	Amphibians: use young larvae whenever possible

Table 4.5 (continued)   Summary for Conducting Acute Toxicity Tests with Fishes, Macroinvertebrates, and Amphibians

Test type	Static, renewal, flow-through

Experimental design

   Test vessel type and size   Smallest horizontal dimension should be three times the largest horizontal dimension of the largest organism

   Depth should be at least three times the height of the largest organism

   Solution volume   At least 150 mm deep for organisms over 0.5 g each and at least 50 mm deep for smaller organisms

   Feeding regime   Feed at least once a day a food that will support normal function

   Test duration   Daphnids and midge larvae: 48 h

   All other species: 96 h in static tests, at least 96 h in renewal and flow-through test

Physical and chemical parameters

   Water temperature (°C)   Freshwater

        Vertebrates

            Frog, *Rana sp.* (22)

            Toad, *Bufo sp.* (22)

            Coho salmon, *Oncorhynchus kisutch* (12)

            Rainbow trout, *Salmo gairdneri* (12)

            Brook trout, *Salvelinus fontinalis* (12)

            Goldfish, *Carassius auratus* (17, 22)

            Fathead minnow, *Pimephales promelas* (25)

            Channel catfish, *Ictalurus punctatus* (17, 22)

            Bluegill, *Lepomis macrochirus* (17, 22)

            Green sunfish, *Lepomis cyanellus* (17, 22)

        Invertebrates

            Daphnids, *Daphnia magna, D. pulex, D. pulicaria* (20)

            Amphipods, *Gammarus lacustris, G. fasciatus, G. pseudolimnaeus* (17)

            Crayfish, *Orconectes sp. Combarus sp. Procambarus sp.* (17, 22)

            Pacifastacus leniusculus (17)

            Stoneflies, *Pteronarcys sp.* (12)

            Mayflies, *Baetis sp. Ephemerella sp.* (17)

            Mayflies, *Hexagenia limbata, H. bilineata* (22)

            Midges, *Chironomus sp.* (22)

            Snails, *Physa integra, P. heterostropha, Amnicola limosa* (22)

            Planaria, *Dugesia tigrina* (22)

        Saltwater

        Vertebrates

            Sheepshead minnow, *Cyprinodon variegatus* (22)

            Mummichog, *Fundulus heteroclitus* (22)

            Longnose killifish, *Fundulus similis* (22)

            Silverside, *Menidia sp.* (22)

            Threespine stickleback, *Gasterosteus aculeatus* (17)

            Pinfish, *Lagodon rhomboides* (22)

            Spot, *Leiostomus xanthurus* (22)

            Shiner perch, *Cymatogaster aggregata* (12)

            Tidepool sculpin, *Oligocottus maculosus* (12)

            Sanddab, *Citharichthys stigmaeus* (12)

            Flounder, *Paralichthys dentatus, lethostigma* (22)

            Starry flounder, *Platichthys stellatus* (12)

            English sole, *Parophrys vetulus* (12)

            Herring, *Clupea harengus* (12)

Table 4.5 (continued)   Summary for Conducting Acute Toxicity Tests with Fishes, Macroinvertebrates, and Amphibians

Test type	Static, renewal, flow-through
	Invertebrates
	Copepods, *Acartia clausi* (12)
	Acartia tonsa (22)
	Shrimp, *Penaeus setiferus, P. duorarum, P. aztecus* (22)
	Grass shrimp, *Palaemonetes pugio, P. intermedius, P. vulgaris* (22)
	Sand shrimp, *Crangon septemspinosa* (17)
	Shrimp, *Pandalus jordani, P. danae* (12)
	Bay shrimp, *Crangon nigricauda* (17)
	Mysid, *Mysidopsis bahia, M.bigelowi, M. almyra* (27)
	Blue crab, *Callinectes sapidus* (22)
	Shore crab, *Hemigrapsus sp., Pachygrapsus sp.* (12)
	Green crab, *Carcinus maenas* (22)
	Fiddler crab, *Uca sp.* (22)
	Oyster, *Crassostrea virginica, C. gigas* (22)
	Polychaete, *Capitella capitata* (22)
Light quality	Not specified
Light intensity	Not specified
Photoperiod	16 h light / 8 h dark with a 15–30 min transition period
Test solution pH	Very soft: 6.4-6.9
	Soft: 7.2-7.6
	Hard: 7.6-8.0
	Very hard: 8.0-8.4
DO concentration	60–100% for static test during first 48 h
	40–100% for static test after 48 h
	60–100% for renewal and flow-through tests (all times)
Endpoint	Death, immobilization

cuts down on the volumes available for large scale flow-through tests systems. Finally, it is important to constantly monitor the quality of the water source. The choice of deionizing or filtering units is also important. Apparently, some resins do leach out small amounts of materials toxic to fish and invertebrates. A positive control using a toxicant with well-known LC_{50} values should give an indication of the suitability of the test solutions. Measurement of variables such as hardness, pH, alkalinity, and in the case of marine systems salinity, can prevent disasters or unreliable test results.

The fish species used in these tests can be far ranging although the most popular are the fathead minnow (*Pimephales promelas*), bluegill (*Lepomis macrochirus*), the channel catfish (*Ictalarus punctatus*), and the rainbow trout (*Oncorhynchus gairdneri*). Andromonas fish are usually represented by the Coho salmon (*O. kisutch*). Marine species used are often the sheepshead minnow (*Cyprinodon variegatus*), mummichog (*Fundulus heteroclitus*), and silversides (*Menidia sp.*).

A variety of invertebrates are also used in these series of tests. Freshwater invertebrates are often represented by daphnids, insect larvae, crayfish, and mollusks. Various mysid, shrimp, and crab species are used to represent marine invertebrates.

Table 4.6 Summary of Test Conditions for Conducting a Subchronic Inhalation Toxicity Study in Rats

Test type	Subchronic
Organisms	Variety of rodent species may be used; rat is preferred
Age and size of organisms	Ideally before 6 weeks old, not more than 8 weeks old; weight variation not to exceed ±20% for each sex
Experimental design	
Test chamber size	Weight of rat (g) Floor area/rat (cm^2)

Weight of rat (g)	Floor area/rat (cm^2)
<100	109.68 (17.0 in.^{2})
100–200	148.40 (23.0 in.^{2})
200–300	187.11 (29.0 in.^{2})
300–400	258.08 (40.0 in.^{2})
400–500	387.15 (60.0 in.^{2})
>500	451.64 (70.0 in.^{2})

	Height should be at least 17.8 cm (7 in.)
Exposure to test substance	Ideally for 6 h/day on a 7-day/week basis; if necessary, exposure on a 5-day/week basis is considered acceptable; test substance is introduced into the chamber air supply; a suitable analytical control system should be used
Number of test groups	3
Number of organisms per group	20 rats (10 male, 10 female)
Number of organisms per chamber	1 individual
Feeding regime	Withhold food and water during exposure period
Test duration	90 days
Clinical examinations	Urinalysis, hematology, blood chemistry, and necropsy
Physical and chemical parameters	
Temperature	22 ± 2°C
Humidity	Ideally 40–60%
Oxygen content	19%
Dynamic airflow	12 to 15 air changes per hour
Endpoint	Death

TERRESTRIAL VERTEBRATE TOXICITY TESTS

In parallel to the short-term toxicity tests with aquatic species are the standard mammal and bird toxicity tests. The methodologies are typically classed as to period and mode of exposure. Two examples of mammalian tests are summarized in Tables 4.6 and 4.7. The small-mammal toxicity tests were originally and are still used primarily for the extrapolation of toxicity and hazard to humans. The advantage to this developmental process is that a great deal of toxicity data occurs for a variety of compounds, both in their structure and their mode of action. Often the only toxicity data available for a compound is a rat or mouse toxicity endpoint. An enormous amount of physiological and behavioral data are available due to the extensive testing, and much of what forms the foundation of traditional toxicology was formed using these methods. The strains of rodents used are often well

Table 4.7   Summary of Test Conditions for Conducting a 90-Day Oral Toxicity Study in Rats

Test type	Subchronic
Organisms	Rats; other rodents may be used with appropriate modifications and justifications
Age and size of organism	Ideally before rats are 6 weeks old and not more than 8 weeks old; weight variation should not exceed ±20% of the mean weight for each sex
Feeding regime	Any unmedicated commercial diet that meets the minimum nutritional standards of the test species
Experimental design	
Test chamber size	

Weight of rat (g)	Floor area/rat (cm^2)
<100	109.69 (17.0 in.^{2})
100–200	148.40 (23.0 in.^{2})
200–300	187.11 (29.0 in.^{2})
300–400	258.08 (40.0 in.^{2})
400–500	387.15 (60.0 in.^{2})
>500	451.64 (70.0 in.^{2})

	Height should be at least 17.8 cm (7 in.)
Test chamber type	All-metal cages with wire-mesh bottoms, suspended in racks
Number of test groups	At least 4
Number of test organisms per group	20 (10 male, 10 female)
Number of test organisms per chamber	1 individual
Dosage	Administer through the diet, the drinking water, by capsule, or by gavage; if by gavage, a 5-day/week dosing regimen is acceptable
Test duration	90 days
Clinical examinations	Urinalysis, hematology, blood chemistry, and necropsy
Physical and chemical parameters	
Temperature	22 ± 2°C
Endpoint	Death

characterized genetically with some having extensive pedigrees available. The drawback to environmental toxicology, however, is that the focus has traditionally been the extrapolation of the toxicity data to primates and not towards other classes of mammals. It is difficult to accurately extrapolate rodent oral toxicity data to cattle since cattle have drastically different digestive systems. It is possible to use other species of rodents and other small mammals with strains having originated from wild caught organisms and these tests may prove useful in assessing the interspecific variability of a toxic response.

In contrast, the avian toxicity tests have been developed over the last two decades in order to assess the effects of environmental contaminants, especially the effects of pesticides to nontarget species. The methods are similar in general to other short-term toxicity tests. A variety of species from different families of birds have been used, although standardization as to strain of each species has not been as extensive as with the mammalian toxicity tests. Examples of an acute feeding study and a reproductive test are presented in Tables 4.8 and 4.9.

Table 4.8 Summary for Conducting Subacute Dietary Toxicity Tests with Avian Species

Test type	Avian subacute dietary
Organisms	Test to be done primarily with Northern bobwhite (*Colinus virginianus*), Japanese quail (*Coturnix japonica*), mallard *(Anas platyrhynchos*), and ring-necked pheasant (*Phasianus colchicus*)
Age of organism	14, 14, 5, and 10 days, respectively
Experimental design	
Test chamber	Construction materials in contact with birds should not be toxic, nor be capable of adsorbing or absorbing test substances; materials that can be dissolved by water or loosened by pecking should not be used; stainless or galvanized steel, or materials coated with plastics are acceptable; any material or pen shape is acceptable provided the birds are able to move about freely and that pens can be kept clean
Test substance	One concentration should kill more than 0% but less than 50% and one concentration should kill more than 50% but less than 100%. These results can be obtained with four to six treatment levels
Number of organisms per group	Minimum of 10 birds for each test concentration
Number of organisms per replicate	Minimum of 5
Feeding	Test substance is mixed with feed. Birds shall be fed *ad libitum*
Test duration	Treated diets are available for 5 days then replaced with untreated feed. Birds are held for a minimum of 3 days following treatment
Clinical examinations	Body weight (record at beginning and end) and feed consumption
Physical and chemical parameters	
Temperature	A temperature gradient from ca. 38°C to ca. 22°C should be established in brooders
Photoperiod	Minimum of 14 h of light
Humidity	45–70% (higher relative humidities may be appropriate for waterfowl)
Ventilation	Sufficient to supply 10–15 air changes per hour
Endpoint	Mortality

It should not be assumed that one method exists for each of these tests. In many cases subtle differences exist between protocols that are acceptable. Table 4.10 compares two methods, one being the ASTM consensus method and the other from the U.S. EPA. The ASTM method is broader and includes species that the U.S. EPA method does not. This allows the U.S. EPA method to be more specific since fewer species are involved. Both tests are for a maximum of 14 days. Other differences are in the experimental chambers. The ASTM standard includes a general description of the test chamber, while the U.S. EPA standard includes the size and specifications for the materials. Although both standards are used, differences do exist and it is important to understand the specifications and the potential differences when comparing toxicity results.

Table 4.9 Summary for Conducting Reproductive Studies with Avian Species

Test type	Avian reproduction
Organisms	Ring-necked pheasant (*Phasianus colchicus*), bobwhite (*Colinus virginianus*), Japanese quail (*Coturnix japonica*), chicken (*Tympanuchus cupido*), mallard (*Anas platyrhynchos*), black duck (*Anas rubripes*), screech owl (*Otus asio*), American kestrel, ring dove (*Streptopelia risoria*), gray partridge, crowned guinea-fowl
Age of organism	Should be within ±10% of the mean age of the group
Feeding	Feed and water should be available *ad libitum*. Feed consumption should be measured for 7-day periods throughout the study
Experimental design	
Test chamber type and size	Materials that can be dissolved by water or loosened by pecking should not be used; stainless steel, galvanized steel, or materials coated with perfluorocarbon plastics are acceptable; any design is acceptable such that the birds are able to move about freely and the pens kept clean
Test concentration	(1) At least one concentration must produce an effect (2) The highest test concentration must contain at least 0.1% (1000 ppm) (3) The highest test concentration must be 100 times the highest measured or expected field concentration
Number of test groups	A minimum of 16 pens per test concentration and control group should be used
Number of organisms per chamber	Pairs or groups containing no more than one male
Exposure to test substance	Mix test substance directly into feed
Clinical examinations	Eggs laid; normal eggs; fertile eggs; hatchability; normal young; survival; weight of young; eggshell thickness; residue analysis
Physical and chemical parameters	
Temperature	About 21°C for adults. For hatchlings, the amount and duration of heat is species-specified. A temperature gradient should be established from an appropriate heat source and range down to about 21°C
Humidity	45–70% (higher relative humidities may be appropriate for waterfowl)
Light quality	Should emit a spectrum simulating daylight
Light intensity	65 lux (6 fc)
Photoperiod	For adults: 8 hr light/16 h dark prior to photostimulation; 17 hr light/7 h dark from onset of photostimulation For hatchlings: at least 14 h of light for precocial species
Endpoint	Reproduction

ANIMAL CARE AND USE CONSIDERATIONS

Since the care and well-being of terrestrial vertebrates has been of great public concern, strict guidelines as to husbandry and the humane treatment of these organisms have been produced by various government agencies, notably the National

Table 4.10 Comparison of ASTM and U.S. EPA Standards for Conducting Subacute Dietary Toxicity Tests with Avian Species

Test type	ASTM Avian subacute dietary	EPA Avian subacute dietary
Organisms	Northern bobwhite (*Colinus virginianus*), Japanese quail (*Coturnix japonica*) mallard *(Anas platyrhynchos)* ring-necked pheasant (*Phasianus colchicus*)	Northern bobwhite (*Colinus virginianus*), mallard (*Anas platyrhynchos*)
Age of organism	14 days, 14 days, 5 days and 10 days, respectively	10-14 days and 5-10 days respectively
Experimental design		
Test chamber	Construction materials in contact with birds should not be toxic, nor be capable of adsorbing or absorbing test substances; materials that can be dissolved by water or loosened by pecking should not be used; stainless or galvanized steel, or materials coated with plastics are acceptable; any material or pen shape is acceptable provided the birds are able to move about freely and that pens can be kept clean	Bobwhite: 35 × 100 x 24 Mallards: 70 × 100 x 24 Floors and external walls of wire mesh; ceilings and common walls of galvanized sheeting
Test substance	One concentration should kill more than 0% but less than 50% and one concentration should kill more than 50% but less than 100%. These results can be obtained with four to six treatment levels	Dose levels should attempt to produce mortality ranging from 10–90%
Number of concentrations		4 concentrations minimum, 5 or 6 strongly recommended plus additional groups for control
Number of organisms per group:	Minimum of 10 birds for each test concentration	10 per level
Number of organisms per replicate	Minimum of 5	About 10
Feeding	Test substance is mixed with feed. Birds shall be fed *ad libitum*	Standard commercial game bird or water fowl diet (mash); test substance should be added directly to the diet without a vehicle, if possible
Test duration	Treated diets are available for 5 days then replaced with untreated feed. Birds are held for a minimum of 3 days following treatment	8 days, two phases Phase 1: 5 days treated diet for experimental, "clean" diet for control Phase 2: 3 days observation, clean diet for both groups
Clinical examinations	Body weight (record at beginning and end) and feed consumption	Body weight and feed consumption

Table 4.10   Comparison of ASTM and U.S. EPA Standards for Conducting Subacute Dietary Toxicity Tests with Avian Species

Test type	ASTM Avian subacute dietary	EPA Avian subacute dietary
Physical and chemical parameters		
Temperature	A temperature gradient from approx. 38°C to approx. 22°C should be established in brooders	22-27°C outside, about 35°C inside brooder
Photoperiod	Minimum of 14 h of light	Diurnal recommended, 24 h lighting acceptable
Humidity	45–70% (higher relative humidities may be appropriate for waterfowl)	30–80%
Ventilation	Sufficient to supply 10–15 air changes per hour	Adequate supply should be maintained
Endpoint	Mortality	Mortality

Institutes of Health. These guidelines were not welcomed by many toxicologists during their implementation. The net effect, however, has been in the improvement of research. Now almost all research is reviewed by animal use committees and strict protocols exist that help to ensure the health of the test organisms. In addition these rules help to ensure a more efficient utilization of laboratory animals.

Facilities and animal husbandry are a major consideration with the avian or any other test using a terrestrial vertebrate. Guidelines exist and are promulgated by the United States Department of Agriculture and the National Institutes of Health to ensure that test animals are maintained at an acceptable standard.

An additional consideration when using mammals and birds is the desire to balance the acquisition of data with the pain and suffering of the test organisms. It is crucial to use the fewest numbers of organisms possible and to acquire the maximum amount of data from each toxicity test. Animal use committees are set up to counsel investigators as to the wise use of animals in research.

The first consideration should be a careful examination of the requirement that a certain toxicity test or other research program be undertaken. In environmental toxicology it is often necessary to use the organism in the laboratory as a test organism in order to protect the wild populations. If the research or test methodology is required then there are three other considerations.

Often it is possible to **replace** a toxicity test with an alternative methodology, especially when cellular or mechanistic studies are undertaken. Tissue in laboratory culture, microorganisms, or lower invertebrates can also be used in place of whole-animal studies. In the case of screening tests, there now exists a broad variety of quantitative structure activity models that can predict and actually overestimate acute and chronic toxicity. Compounds that are likely to demonstrate high toxicity can be eliminated from consideration as a product or focused upon for toxicity reduction.

It is also often possible to **reduce** the number of animals used in the evaluation of a chemical or toxic waste site by carefully designing the experiment to maximize the data acquired or by accepting a compromise in the statistical significance and power. Often a slight decrease in the statistical power can result in a large reduction in the number of animals required in a toxicity test.

Finally, it is often possible to **refine** the methodology as to require fewer animals. Biochemical and physiological indicators or toxicant stress or indications of mechanisms can help to reduce the number or even the need for such testing.

Although useful, and forming the backbone of most toxicological research, the single-species toxicity test is not without drawbacks. In the role of providing toxicity data for environmental scenarios, these relatively simple toxicity tests have provided a great deal of information and controversy. The ability to examine the relationships between chemical structure and function is based on a large database produced by comparable toxicity determinations. In addition, the large number of chemicals tested with these methods and organisms provide a relative ranking as to acute toxicity. As will be discussed in detail in following chapters, the usefulness of these tests in predicting environmental effects is questioned. The situations the organisms are in are decidedly not natural and typically are chosen for the cost-effective production of reliable and repeatable toxicity data. Effects at low doses over long periods of time as well as the species-to-species interactions are not generally considered.

FROG EMBRYO TERATOGENESIS ASSAY: FETAX

This toxicity test is one of the few amphibian-based toxicity tests and is summarized in Table 4.11. *Xenopus laevis,* the South African Clawed Frog, is the amphibian species used in this toxicity test. J. Bantle and colleagues (Bantle 1992) have developed and perfected this methodology over the last 10 years. The methodology has been performed in a number of laboratories with repeatable results. This toxicity test has a number of uses. FETAX has been touted as an alternative to performing mammalian teratogenicity test and its correlation with known mammalian teratogens is very good. Teratogenicity of runoff, water collected from lakes and streams, and even elutriates from soil samples have been evaluated using the same basic methodology.

One of the major advantages is the database that has been obtained on the test organism Xenopus. Xenopus is a research organism widely used in developmental research and in the genetics of development. The animals are also easy to mate and large numbers of eggs are produced, ensuring large sample sizes. Compared to mammals, birds, and reptiles, it is easy to observe malformations or other teratogenic effects since the developing embryos are in the open.

FETAX is a rapid test for identifying developmental toxicants. Data may be extrapolated to other species including mammals. FETAX might be used to prioritize hazardous waste samples for further tests which use mammals. Validation studies using compounds with known mammalian and/or human developmental toxicity suggest the predictive accuracy rate compares favorably with other currently available *"in vitro* teratogenesis screening assays" (Bantle 1992). It is important to measure developmental toxicity because embryo mortality, malformation, and growth inhibition can often occur at concentrations far less than those required to affect adult organisms. Because of the sensitivity of embryonic and early life stages, FETAX provides information that might be useful in estimating the chronic toxicity of a test material to aquatic organisms.

Table 4.11   The Frog Embryo Teratogenesis Assay: Xenopus (FETAX)

Test type	96 h static renewal
Organism	*Xenopus laevis*
Age of parent organism	Adult male: at least 2 years of age
	Adult female: at least 3 years of age
Size of parent organism	Adult male: 7.5–10 cm in crown-rump length
	Adult female: 10–12.5 cm in length
Feeding	Adult: three feedings per week of ground beef liver; liquid multiple vitamins should be added to the liver in concentrations from 0.05–0.075 cc/5 g liver
Experimental design	
Test vessel type and size	Adults: large aquarium or fiberglass or stainless steel raceways; side of tank should be opaque and at least 30 cm high.
	Breeding adults: 5- or 10-gallon aquarium fitted with a 1-cm mesh suspended approximately 3 cm from the bottom of the tank; nylon or plastic mesh is recommended; aquarium should be fitted with a bubbler to oxygenate the water; the top of aquarium should be covered with an opaque porous material such as a fiberglass furnace filter
	Embryos: 60-mm glass or 55-mm disposable polystyrene Petri dishes
Test solution volume	Adults: water depth should be 7-14 cm
	Embryos: 10 ml per dish
Exposure to test substance	Continuous throughout test
Replacement of test material	Every 24 h
Number of concentrations	5
Number of replicates per sample	2
Number of organisms per chamber	Adults: 4–6 per 1800 cm² of water surface area
	Breeding adults: 2
	Embryos: 25
Test duration	96 h
Physical and chemical parameters	
Temperature	Adult: 23 ± 3°C
	Embryos: 24 ± 2°C
Photoperiod	12 h light / 12 h dark
pH range	6.5 to 9
TOC	10 mg/l
Alkalinity and hardness	Between 16 and 400 mg/l as CaCO₃
Endpoint	Acute (mortality) and subacute (teratogenesis)

The criticism often presented about the FETAX is that it is a poor representation of native species of amphibians or of other vertebrates. Xenopus is of course not native to the Americas, but it is a typical amphibian and its comparability in teratogenic response to mammalian species has already been documented. Xenopus is also widely available, and the basic methodology can also be transferable to other frogs and toads.

In addition to the American Society for Testing and Materials method, several useful documents are produced by Oklahoma State University in support of the test method. Particularly useful is an atlas of malformations making it easier to score the results of the toxicity test. Given the relative ease of performing the toxicity test and the supporting documentation, FETAX has found a rapid acceptance as a teratogenicity screen in environmental toxicology.

MULTISPECIES TOXICITY TESTS

Toxicity tests using artificially contained communities have long been a resource in environmental toxicology. Many different methodologies have been developed (Table 4.12). Each has particular advantages and disadvantages and none have been demonstrated to faithfully reproduce an entire ecosystem. However, as a research tool to look at secondary effects, bioaccumulation and fate, the various multispecies toxicity tests have been demonstrated to be useful.

The overriding characteristic of a multispecies toxicity test is that it consists of at least two or more interacting species. Which two or more species and their derivation along with the volume and complexity of substrate and heterogeneity of the environment are all matters of debate. Much current theory on the coexistence of species and their interactions emphasizes the role of environmental heterogeneity upon the formation and continuance of a community. Yet in the conduct of a multispecies toxicity test the goal is often to minimize the heterogeneity to allow the performance of traditional hypothesis-testing statistics. On the other hand, including the heterogeneity of nature would require a system so large and complex that it would in essence be a field study with all of the problems assigning cause and effect inherent to those types of studies. It is perhaps more important to use good scientific methodology and emphasize the question being asked as opposed to which multispecies toxicity test is the best mimic for the natural ecosystem. Emphasis upon the specific question will likely lead to the selection of one of the current methods with slight modification as best for that particular situation.

Multispecies toxicity tests range widely in size and complexity. This is the case for both aquatic and terrestrial systems.

In the aquatic arena some of the biodegradation tests are done with volumes of less than a liter. Tests to evaluate community interaction conducted in a laboratory have test vessels ranging in size from 1 l to 55 gal. Larger test systems can also be used outside the laboratory. A recently proposed outdoor aquatic microcosm proposal uses large tanks of approximately 800 l capacities. Larger still are the pond mesocosms used for pesticide evaluations. These systems are designed to mimic farm ponds in size and morphology.

Terrestrial microcosms also see a comparable range in size and complexity. A microbial community living within the soil in a test tube can be used to examine biodegradation. A soil core is comparable in size and utility to the laboratory microcosms described above. In some cases terrestrial microcosms can be established with a variety of plant cover and include small mammals and insects. Field plots are the terrestrial equivalent of the larger outdoor aquatic microcosms. These

Table 4.12   Listing of Current Multispecies Toxicity Tests

Aquatic microcosms
  Benthic-pelagic microcosm
  Compartmentalized lake
  Mixed flask culture microcosm
  Pond microcosm
  Sediment core microcosm
  Ecocore microcosm
  Ecocore II microcosm
  Standard aquatic microcosm
  Stream microcosm
  Waste treatment microcosm
Terrestrial microcosms
  Root microcosm system
  Soil core microcosm
  Soil in a jar
  Terrestrial microcosm chamber
  Terrestrial microcosm system
  Versacore

field plots can vary in size but usually contain a cover crop or simulated ecosystem, and are fenced to prevent escape of the test vertebrates or the migration of other organisms into the test plot. Ecosystems ranging from agroecosystems to wetlands have been examined in this manner. Compared to the aquatic multispecies toxicity tests the terrestrial systems have not undergone the same level of standardization. This is due to the length of time most of these tests require and the specialized nature of most of the test systems rather than any lack of completeness of the method. The development of outdoor multispecies tests for the evaluation of terrestrial effects of pesticides and hazardous waste is a current topic of intense research.

One of the ongoing debates in environmental toxicology has been the suitability of the extrapolation and realism of the various multispecies toxicity tests that have been developed over the last 15 years. One of the major criticisms of small-scale systems is that the low diversity of the system is not representative of natural systems in dynamic complexity (Sugiura 1992). Given the above discussion and the conclusions derived from it, much of this debate may have been misdirected. The small-scale systems have been demonstrated to express complex dynamics. Kersting and van Wungaarden (1992) found that even the three-compartment microecosystem, as developed by Kersting (1984, 1985, 1988), expresses indirect effects as measured by pH changes after dosing with chloropyrifos. Since even full-scale systems cannot serve as reliable predictors of the dynamics of other full-scale systems, it is impossible to suggest that any artificially created system can provide a generic representation of any full-scale system. Debate should probably revert to more productive areas such as improvements in culture, sampling, and measurement techniques or other characteristics of these systems. A more worthwhile goal is probably the understanding of the scaling factors, in a full n-dimensional representation, that should enable the accurate representation of specific ecosystem characteristics. Certain aspects of a community may be included in one system to answer specific

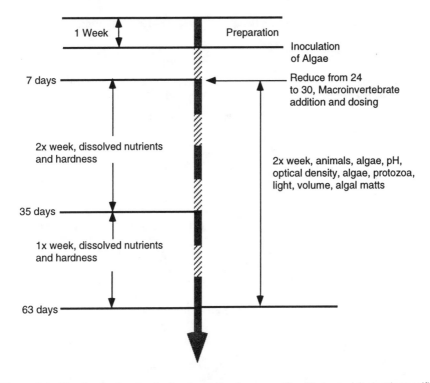

Figure 4.1   Timeline for the standardized aquatic microcosm. The 63-day toxicity test is specific in its sampling requirements, acclimation times, and dosing.

questions that in another system would be entirely inappropriate. If questions as to detritus quality are important then the system should include that particular component. In other words, the system should attempt to answer the particular scientific question.

STANDARDIZED AQUATIC MICROCOSM

The standardized aquatic microcosm (SAM) was developed by Frieda Taub and colleagues to examine the effects of toxicants on multispecies systems in the laboratory. Figure 4.1 illustrates the course of events over the 64 days of the experiment and Table 4.13 provides a tabular overview. The microcosms are prepared by the introduction of ten algal, four invertebrate, and one bacterial species into 3 l of sterile defined medium. Test containers are 4-l glass jars. An autoclaved sediment consisting of 200 g of silica sand and 0.5 g of ground chitin are autoclaved separately and then added to the already autoclaved jar and media.

Numbers of organisms, dissolved oxygen (DO), and pH are determined twice weekly. Nutrients (nitrate, nitrite, ammonia, and phosphate) are sampled and measured twice weekly for the first 4 weeks, then only once weekly thereafter. Room temperature is set at $20 \pm 2°C$. Illumination is set at 79.2 μE m^{-2} s^{-1} PhAR with a range of 78.6 to 80.4 and a 16/8 day/night cycle.

**Table 4.13   Summary of Test Conditions for Standardized Aquatic Microcosms:
Freshwater**

Test type	Multispecies
Organisms	
Type and number of test organisms per chamber	Algae (added on day 0 at initial concentration of 10^3 cells for each algae species): *Anabaena cylindrica*, *Ankistrodesmus* sp., *Chlamydomonas reinhardi* 90, *Chlorella vulgaris*, *Lyngbya* sp., *Nitzschia kutzigiana* (Diatom 216), *Scenedesmus obliquus*, *Selenastrum capricornutum*, *Stigeoclonium* sp., and *Ulothrix* sp. Animals (added on day 4 at the initial numbers indicated in parentheses): *Daphnia magna* (16/microcosm), *Hyalella azteca* (12/microcosm), *Cypridopsis* sp. or *Cyprinotus* sp. (ostracod) (6/microcosm), Hypotrichs [protozoa] (0.1/ml) (optional), and *Philodina* sp. (rotifer) (0.03/ml)
Experimental design	
Test vessel type and size	1-gal (3.8-l) glass jars are recommended; soft glass is satisfactory if new containers are used; measurements should be 16.0 cm wide at the shoulder, 25 cm tall with 10.6-cm openings
Medium volume	500 ml added to each container
Number of replicates	6
Number of concentrations	4
Reinoculation	Once per week add one drop (ca 0.05 ml) to each microcosm from a mix of the ten species; 5×10^2 cells of each alga added per microcosm
Addition of test materials	Add material on day 7; test material may be added biweekly or weekly after sampling
Sampling frequency	2 times each week until end of test
Test duration	63 days
Physical and chemical parameters	
Temperature	Incubator or temperature controlled room is required providing an environment 20 to 25°C with minimal dimensions of $2.6 \times 0.85 \times 0.8$ m high.
Work surface	Table at least 2.6×0.85 m and having a white or light colored top or covering
Light quality	Warm white light
Light intensity	80 µE m^{-2} photosynthetically active radiation s^{-1} (850–1000 fc)
Photoperiod	12 h light/12 h dark
Microcosm medium	Medium T82MV
Sediment	Composed of silica sand (200 g), ground, crude chitin (0.5g), and cellulose powder (0.5 g) added to each container.
pH level	Adjust to pH 7
Endpoint	Many

The test is conducted in a temperature controlled facility on a worktable of approximately 0.85×2.6 m dimensions with light hung 0.56 m from the top of the table. Originally 30 jars are placed under the lights but at day 4 the microcosms are culled to the 24 test systems. Three treatment groups and a control are used.

All data are recorded onto standard computer entry forms, checked for accuracy, and input to the Macintosh compatible data analysis system (SAMS) developed by the University of Washington under contract with the Chemical Research, Development and Engineering Center. Parameters calculated included the DO, DO gain and loss, nutrient concentrations, net photosynthesis/respiration ratio (P/R), pH, algal species diversity, daphnid fecundity, algal biovolume, and biovolume of available algae. The statistical significance of each of these parameters compared to the controls is also computed for each sampling day.

MIXED FLASK CULTURE

The MFC microcosms are smaller systems of approximately 1 l and are inoculated with 50 ml of a stock culture originally derived from a natural system. Over a 6-month period repeated inocula are made into a stock tank so that a number of interactions can be established. At the end of the 6-month period the material from this stock tank is ready for inoculation into the test vessels. Six weeks are allowed for the establishment of the freshwater community followed by an experimental duration of 12 to 14 weeks. In contrast to the SAM, the MFC method relies upon the initial inoculum to provide the prerequisite components of the microcosm community. The protocol requires two species of single-celled green algae or diatoms, one species of filamentous green alga, one species of nitrogen-fixing blue-green alga, one grazing macroinvertebrate, one benthic, detrital feeding macroinvertebrate, and bacteria and protozoa species. Four treatment groups are recommended with five replicates for each group. The MFC has been used for the evaluation of procaryotic organisms introduced into the environment. A summary of this method is found in Table 4.14.

An implicit assumption of the MFC is that the acclimation time is sufficient for coevolution to occur and that coevolution is important to assess the impacts of xenobiotics upon communities. The use of a "natural" inocula should increase species diversity and complexity over a protocol such as the SAM, but the smaller size of the test vessel would tend to decrease species number. Debate also exists as to the applicability of coevolution in the evaluation of test chemicals. If algal populations and others are primarily regulated by density-independent factors then population-specific interspecies interactions may not be particularly important. If ecosystems are loosely connected in an ecological sense then coevolved assemblages may be rare. On the other hand, in enclosed systems that are islands, these relationships may have had an opportunity to occur and coevolved interactions may be important in the assessment of toxicological impacts.

Table 4.14 Summary of Test Conditions for Adaptation of Mixed Flask Culture Microcosms for Testing the Survival and Effects of Introduced Microorganisms

Test type	Multispecies
Organisms	

Number and type
    of organism
	(a) Two species of single-celled green algae or diatoms
	(b) One species of filamentous green alga
	(c) One species of nitrogen fixing blue-green alga
	    (bacteria)
	(d) One grazing macroinvertebrate
	(e) One benthic, detrital feeding macroinvertebrate
	(f) Bacteria and protozoa species

Experimental design	
Test vessel type and size	1-l beakers covered with a large petri dish
Volume/mass	50 ml of acid washed sand sediment and 900 ml of Taub # 82 medium [20], into which 50 ml of inoculum was introduced
Number of groups	4
Number of replicate chambers per group	5
Reinoculation	10 ml of stock community each week
Test duration	12–18 weeks Allow to mature 6 weeks prior to treatment; follow 6–12 weeks after exposure
Physical and chemical parameters	
Temperature	20°C
Photoperiod	12 h light / 12 h dark
Endpoint	Oxygen content, algal densities, microbial activity, respiratory activity, biomass, protozoan population

FIFRA MICROCOSM

Aquatic microcosms too large to be contained in the average laboratory have been routinely manufactured and used to attempt to obtain enough volume to contain fish as grazers or as invertebrate predators. Proposed in late 1991 was a microcosm/mesocosm blend that is substantially larger than the MFC or the SAM experimental units. The experimental protocol is termed the Outdoor Aquatic Microcosm Tests to Support Pesticide Registrations (Table 4.15), but it is also called the FIFRA microcosm to reflect its origin as a pesticide testing methodology. The FIFRA microcosm is a system of approximately 6 m³ in volume for each experimental unit with an inherent flexibility in design. Macrophytes can be included or not, along with a variety of fish species, invertebrates, and a variety of emergent invertebrates. A diagrammatic representation of one system for the examination of the effects of a model herbicide is presented in Figure 4.2.

The flexibility in design is a recognition that this protocol originated to replace larger pond mesocosms mandated by the Office of Pesticide Programs to examine the potential impacts of pesticides to nontarget aquatic organisms. The larger systems were designed to simulate farm ponds and tended to be unwieldy and difficult to sample with a concurrent problem with the data analysis. The FIFRA microcosm was

Table 4.15 Summary of Test Conditions for Conducting Outdoor Aquatic Microcosm Tests to Support Pesticide Registrations

Test type	Multispecies toxicity test
Organisms	Add: bluegill sunfish (*Lepomis macrochirus*), fathead minnow (*Promephales promelas*), channel catfish (*Ictalurus punctatus*), or others may be present (Phytoplankton, periphyton, zooplankton, emergent insects, and benthic macroinvertebrates)
Size of organism	Biomass of fish added to the microcosms should not exceed 2 g/m³ of water.
Experimental design[a]	
Test vessel size and type	Tanks with a surface area of at least 5 m², a depth of at least 1.25 m, and a volume of at least 6 m³ made of fiberglass or some other inert material; smaller tanks could be used for special purposes in studies without fish.
Addition of test material	Allow microcosms to age for approximately 6–8 weeks before adding test material. Apply by spraying across water surface, apply the test material in a soil/water slurry, or apply test material in a water based stock solution
Sampling	Begins approximately 2 weeks after the microcosms are constructed and continues for 2 or 3 months after the last treatment with test material; frequency depends upon the characteristics of test substance and on treatment regime
Physical and chemical parameters	
Temperature	Maintained by partially burying tanks in the ground or immersing in a flat-bottomed pond
Sediment	Obtained from existing pond containing a natural benthic community; added to each microcosm directly on the bottom, in trays, or other containers; sediment should be 5 cm thick
Water	Obtained from healthy, ecologically active pond; water level should be set in the beginning and not allowed to vary more than ±10% throughout study; if water level falls more than 10%, add pond water, fresh well water, or rain water; if water level rises more than 10%, surplus should be released and retained.
Weather	Should be recorded at the study site or records obtained from a nearby weather station; data should include air temperature, solar radiation, precipitation, wind speed and direction, and relative humidity or evaporation

[a] Dosage levels, frequency of test material addition, and number of replicates per dosage level are determined based on the objectives of the study.

an attempt to design a flexible system able to answer specific questions concerning the fate and effects of a material in a more tightly controlled outdoor system.

One of the interesting aspects of the FIFRA microcosm system is the variety of methods used to ensure a uniform temperature among the experimental replicates during the course of the experiment. Basically, two methods have been used. The first method is to bury the test system into the ground and use the ground as an insulator

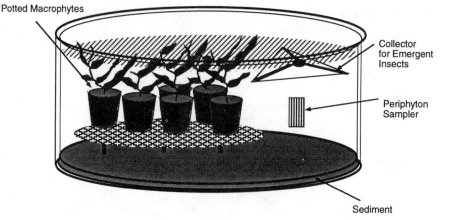

Figure 4.2   FIFRA microcosm experimental unit. An example of a microcosm experimental unit designed to test the effects of a herbicide on an aquatic environment. This particular setup does not include fish since the predatory effects would tend to hide lower trophic level effects upon the invertebrate populations. Typically, a FIFRA microcosm experiment includes fish species, particularly when acetylcholinesterase inhibitors or other toxicants particularly effective against animal species are tested.

and temperature regulator. This has been used extensively. In certain instances water can be used as the insulator. The experimental units are placed in the pond when the water is removed and then replaced as the plumbing and experimental setups have been established. In some locations it may also be important to provide shade and to prevent a deluge from adding sufficient volume to cause an overflow of the test vessels.

Although the FIFRA microcosm has a number of advantages, there are also compromises. The few experiments that have been conducted and the variance in methodologies have not given an accurate representation of the repeatability or replicability of the experiments. In addition, the method is somewhat local-specific since the temperature, diurnal cycle, and to some extent the experimental organisms are controlled by the local environmental conditions. On the other hand, the sensitivity to local conditions can also act as a more accurate model of local fate and effects of the test material.

As of this writing no ASTM or comparable consensus method exists for this larger microcosm system; this is due to the relative newness of the methodology. However, a recent publication (Graney et al. 1994) reviews and discusses the system typically used for the purposes of pesticide registration.

SOIL CORE MICROCOSM

The soil core microcosm (SCM) is one of the first test vehicles developed for the evaluation of xenobiotics on an agroecosystem with it accompanying plants, soil invertebrates, and microbial processes. Table 4.16 summarizes the basic protocol.

Table 4.16 Summary of Test Conditions for Conducting A Terrestrial Soil-Core Microcosm Test

Test type	Multispecies toxicity test
Organisms	Varies; dependent on site being tested
Experimental design	
Microcosm size and type	60-cm-deep by 17-cm-diameter plastic pipe made of ultra-high molecular weight, high-density, and nonplasticized polyethylene and contains an intact soil core covered by homogenized topsoil; tube sits on a Buchner funnel covered by a thin layer of glass wool
Soil volume	40 cm intact soil core; 20 cm homogenized topsoil
Number of replicates	Each cart holds 6–8 microcosms; place microcosms paired for analyses in different carts to ensure that all microcosms are housed under similar conditions.
Number of concentrations	3
Leaching	At least once before dosing and once every 2 or 3 weeks after dosing
Test duration	12 or more weeks
Physical and chemical parameters	
Temperature	Based on season of region being tested; insulated cart is used to prevent drastic temperature changes
Lighting	Based on season of region being tested
Watering	Determined on the basis of site history; use either purified laboratory water or rainwater that has been collected, filtered, and stored in a cooler at 4°C
Endpoint	Many

The SCM is a hybrid methodology with cores derived from an outdoor environment brought into a laboratory setting to more accurately control the environmental variables. In this manner, the intrinsic heterogeneity of the terrestrial ecosystem is preserved although successional changes can occur due to the small size of the experimental unit. Because of the design of the experimental container extensive nutrient and chemical fate analyses can be performed. A typical greenhouse area is required with proper ventilation for the reduction of occupational exposure.

Although a useful methodology and an ASTM standard, few examples of SCM experiments exist in the open literature. This may be due to the somewhat specialized facilities required or the performance of proprietary research that is often unreported.

SUMMARY

This chapter reviewed a wide variety of toxicity tests yet only a small fraction of the toxicity tests that are currently performed or exist. These tests cover the entire range of biological organization that can be expected to fit into a laboratory or outdoor contained setting. There are a few caveats that must be dealt with when dealing with the topic of toxicity testing.

First, there is a tendency to over extrapolate from the results of a few tests that were convenient to perform or mandated by regulation or convention. The danger is extrapolating to situations or to ask questions that the toxicity test was not designed to answer. Examples are numerous. Many single-species tests are extrapolated to establish a safety level to protect a particular habitat or indigenous population. If direct, relatively short-term effects are the points of concern then these tests are probably sufficient; however, if long-term effects are also a concern then other multispecies tests or field studies should be conducted.

Second, there is an element of fashion or style attributed to a method either because of overzealous salesmanship, undue conservatism, or lack of knowledge of alternatives that often comes to play in the selection and review of a test method. The test should be able to stand alone as a means of answering specific questions about the effect of a xenobiotic. Tests that lack an adequate statistical or theoretical foundation should be avoided. Acquisition of data should not be an end onto itself. A well-designed toxicity evaluation should be comprised of toxicity tests that address particular questions that are the basis of the environmental concerns.

Third, many times the toxicity tests are selected on the basis of cost, and this is a valid parameter. A FIFRA mesocosm may cost as much as $750,000 compared to as little as $500 for a *D. magna* acute toxicity test. The danger is from both ends of the spectrum. The more expensive multispecies test is not necessarily better unless it answers specific questions left unanswered by the simpler tests. In fact, the large multispecies tests are performed only after a thorough review and evaluation of simpler testing procedures. Likewise, the simpler and less costly toxicity tests may not adequately address the fate and effects of a xenobiotic, leaving a great deal of uncertainty in the prediction of environmental effects.

APPENDIX: THE NATURAL HISTORY AND UTILIZATION OF SELECTED TEST SPECIES

AQUATIC VERTEBRATES

Coho Salmon (*Oncorhynchus kisutch*)

> **Description:** Body fusiform, streamlined, laterally compressed, usually 18 to 24 in. (457 to 610 mm) in length and 8 to 12 lb in weight as marine adults and 10.8 to 25.8 in. (279 to 656 mm) fork length in Great Lakes freshwater populations; body depth moderate, greater in breeding males.
>
> **Color:** Adults in ocean or Great Lakes are steel-blue to slightly green on dorsal surface, sides brilliant silver, ventral surface white, small black spots on back, sides above lateral line, base of dorsal fin, and upper lobe of caudal fin.
>
> **Distribution:** This species occurs naturally only in the Pacific Ocean and its tributary drainage. It is known in fresh water in North America from Monterey Bay, CA

(in the sea infrequently to Baja California) to Point Hope, AK. In Asia, it occurs from the Anadyr River, USSR, south to Hokkaido, Japan.

Biology: Adults migrate from the sea or lake late in the season and over a prolonged period. Spawning is from early September to early October; segregation into summer and autumn, or autumn and winter runs is more apparent in Asia than in North America; spawning takes place in swifter water of shallow, gravelly areas of river tributaries from October to March, but usually October to November or November to January in North America.

Toxicity Testing: Species can be used as a model salmonid.

Rainbow Trout (*Oncorhynchus gairdneri*)

Description: Body troutlike, elongated, averaged length is 12 to 18 in. (305 to 457 mm); no nuptial tubercles but minor changes to head, mouth, and color especially in spawning males.

Color: Variable with habitat, size, and sexual condition. Stream residents and spawners darker, colors more intense; lake residents lighter, brighter, more silvery.

Systematic notes: Populations in different watersheds were long called by different scientific names and still by different regional common names in the south.

Distribution: Native range was eastern Pacific Ocean and freshwater, mainly west of the Rocky Mountains, from northwestern Mexico (including extreme northern Baja, CA), to the Kuskokwim River, AK; probably native in the drainages of the Peace and Athabasca Rivers east of the Rocky Mountains. Has been widely introduced throughout North America in suitable localities. Also introduced into New Zealand, Australia and Tasmania, South America, Africa, Japan, southern Asia, Europe, and Hawaii.

Biology: Spring spawners, temp. being 50 to 60°F (10.0 to 15.5°C) (FF of C, 184 to 191).

Brook Trout (*Salvelinus fontinalis*)

Description: Average length is 10 to 12 in. (254 to 305 mm); breeding males may develop a hook (or kype) at the front of the lower jaw.

Color: Back is olive-green to dark brown, at times almost black, sides lighter, becoming silvery white below; light green or cream colored wavy lines or vermiculations on top of head and on back, broken up into spots on sides.

Distribution: North American endemic species and under natural conditions occurs only in northeastern North America.

Biology: Brook trout spawn in late summer or autumn, varying with latitude and temperature; a stable and well-defined species (FF of C, 208+).

Goldfish (*Carassius auratus*)

Description: Body stout, thickset, average total length about 5 to 10 in. (127 to 254 mm).

Color: Overall coloration variable, from olive-green through gold (often with black blotches) to creamy white.

Systematic notes: Goldfish hybridize readily with carp.

Distribution: Native to eastern Asia, goldfish originated in China, spread to Japan, parts of Europe, and throughout parts of North America.

Biology: A spring-spawning species and seeks warm, weedy shallows in May or June to deposit its eggs (FF of C, 389 to 390).

Fathead Minnow (*Pimephales promelas*)

Description: Body short, average length about 2 in. (51 mm), thickset, compressed laterally and deep bodied, often with a pronounced belly.
Color: Overall coloration usually dark.
Systematic notes: The fathead minnow varies greatly in many characters throughout its wide geographic range and some populations have been designated as subspecifically distinct.
Distribution: The fathead minnow ranges through most of central North America, from Louisiana and Chihuahua, Mexico, north to the Great Slave Lake drainage, and from New Brunswick on the east to Alberta on the west (FF of C, 480 to 482).

Channel Catfish (*Ictalurus punctatus*)

Description: Average length is 14 to 21 in. (356 to 533 mm), weight is 2 to 4 lb.
Color: Individuals less than 12 to 14 inches (305 to 356 mm) are pale blue to pale olive with silvery overcast; adults with dorsal surface of head and back, and upper side steel-blue to gray, lower sides lighter; ventral surface of head, and body to pelvic fins, dirty white to silver-white; barbels are darkly colored.
Systematic notes: There was, for many years, considerable taxonomic and nomenclatural confusion associated with what we now recognize as this species. Differences in shape and color, now known to be associated with sex, size, season, and locality were once construed to be indicative of several different species or subspecies.
Distribution: Restricted to the fresh waters, and to a limited extent brackish waters, of eastern and central North America.
Biology: Locally abundant in certain parts of Canada but poorly known; very little published information.

Bluegill (*Lepomis macrochirus*)

Description: Has a very deep, compressed body and individuals are usually 7 to 8 in. (178 to 203 mm) in length.
Color: Dorsal surface green-olive to almost brown, with several vague vertical bands extending down sides; upper sides brown to green, shading into brown, orange, or pink; lower sides and abdomen silver to white.
Distribution: Native range of bluegill is restricted to the fresh waters of eastern and central North America; has been introduced throughout the U.S., into Africa, and possibly other areas off the North American continent.
Biology: No detailed account of the life history of a Canadian population; spawning takes place in late spring to early and mid-summer (in Canada) with peak activity in early July (FF of C, 719 to 723).

Green Sunfish (*Lepomis cyanellus*)

Distribution: A deep-bodied, laterally compressed fish, usually not over 5 in. (127 mm) in length in Canada.

Color: Body generally brown to olive with an emerald sheen, darker on dorsal surfaces and upper sides, sides light yellow-green, upper sides with 7 to 12 dark but vague vertical bars; ventral surface yellow to white.

Distribution: Restricted to the fresh waters of east-central North America.

Biology: Spawning occurs in late spring and summer; multiple spawning occurs.

INVERTEBRATES-FRESHWATER

Daphnids (*Daphnia magna, D. pulex, D. pulicaria, Ceriodaphnia dubia*)

Description: Water flea (Cladocera). These are small, laterally flattened forms that usually measure 0.2 to 3 mm. Body is covered by a carapace, but head and antennae are usually apparent. Body does not appear segmented and possesses five or six pairs of legs. Carapace often ends in a spine.

Distribution: Some 135 species of freshwater water fleas are known from North America, where the group is widespread and can be found in most freshwater environments. Most species occur in open waters, where they swim intermittently. The second pair of antennae is used primarily to propel them. Movement is generally vertical, with the head directed upwards. Many of these open-water forms are also known for their vertical migration, which generally consists of upward movement in the dark and downward migration during daylight hours. Some water fleas are primarily benthic.

Daphnia is commonly maintained in laboratories for assaying toxic substances in water. Water fleas are often of great importance in the diets of fishes, especially young fishes, and predaceous insects, such as many of the Diptera larvae.

Amphipods (*Gammarus lacustris, G. fasciatus, G. pseudolimnaeus, Hyalella azteca*)

Description: Scuds (amphipoda) are laterally flattened, often colorful forms that usually measure 5 to 20 mm when mature. Head and first thoracic segment form a cephalothorax. The remainder of the thorax possesses seven pairs of legs, the first two pairs being modified for grasping.

Distribution: Three families and approximately 90 species of scuds occur in North America. The family Talitridae contains one widely distributed North American species, *Hyalella azteca*, which is common in springs, streams, lakes, and ponds. The family Haustoriidae also contains only one species in North America, *Pontooporeeia hoyi*. Somewhat atypical of scuds, this species is confined to the bottom and open waters of deep, cold lakes. The family Gammaridae is the most important group and is divided into about eight genera.

Scuds occur primarily in shallow waters of all kinds. They are benthic and often rest among vegetation and debris or occasionally burrow slightly within soft substrate. They also swim, however, and are sometimes known as "side swimmers". They are generally omnivore-detritivores but rarely predaceous. Several species are restricted to particular spring or cave habitats, whereas others are more widespread in larger surface-water habitats and sometimes occur in very large numbers (aquatic entomology, 389).

> **Gammarus** reach densities of thousands of individuals per square meter where detrital food and cover are abundant
>
> *Hyalella azteca* produce multiple broods during an extended breeding season; warm water species
>
> *G. lacustris* is a cold-water species; a period of short days and long nights (typical of winter) is needed to induce reproduction

Crayfish (*Orconectes* sp., *Combarus* sp., *Procambarus* sp., *Pacifastacus leniusculus*)

Description: Decapoda are somewhat flattened either dorsoventrally or laterally and range in size from 10 to 150 mm. Head and entire thorax form a large cephalothorax covered by a carapace. Cephalothorax possesses five pairs of legs; first two or three pairs are pincer-like at their ends; and first pair is often very robust.

Distribution: The freshwater Decapoda in North America comprise four species of the family Atyidae, which are restricted to certain caves of the southeastern states and coastal streams of California. The family of Astacidae (crayfish) are widely distributed, except that they are not generally found in the Rocky Mountain region. They occur in a wide variety of shallow freshwater habitats, and some live in swamps and wetlands. They are benthic and, at least in daylight hours, usually remain hidden in or burrow under stones and debris. They retreat rapidly backwards when disturbed. Depending on the species, crayfishes may be herbivores, carnivores, detritivores, or omnivores; their very robust first pair of legs (chelae) are used to cut or crush food. These chelae are also used as defensive weapons. Prawns and river shrimps are generally swimmers (aquatic entomology, 390 to 391).

Stoneflies (*Pteronarcys* sp.)

Description: They are all freshwater inhabitants as larvae. As a group they are close relatives of the cockroaches and have retained the primitive condition of possessing tails but demonstrate the advanced ability to fold their wings over the back of the body. Their common name undoubtedly is derived from the fact that individuals of many common species are found crawling or hiding among stones in streams or along stream banks.

Distribution: Close to 500 species are represented in North America. Many stoneflies are known as clean-water insects, since they are often restricted to highly oxygenated water. As such, some are excellent biotic indicators of water quality. Adults of stoneflies can be found throughout the year, some being adapted for winter emergence (AE, 148).

Mayflies (*Baetis* sp., *Ephemerella* sp., *Hexagenia limbata, H. bilineata*)

Over 700 species occurring in North America is possible. As a group, mayflies are one of the most common and important members of the bottom-dwelling freshwater community. Because most species are detritivores and/or herbivores and are themselves a preferred food of many freshwater carnivores, including other insects and fishes, they form a fundamental link in the freshwater food chain. Many species are highly susceptible to water pollution or occur in very predictable kinds of environments. It is for this reason that mayflies have proved very useful in the analysis or biomonitoring of water quality. Several species emerge in mass numbers, and these mass emergences are among the most spectacular in the insect world. In North America, mayflies may also be known locally by such names as willowflies, shadflies, drakes, duns, spinners, fishflies, and Canadian soldiers.

Midges (*Chironomus* sp.)

Larvae are slender, commonly cylindrical and slightly curved forms that usually measure 2 to 20 mm but are occasionally larger. Body has a pair of prothoracic prolegs and a pair of terminal prolegs. Terminal segment usually has a short dorsal pair of tubercles or projections, each with a variable tuft of hairs (dorsal pranal brushes).

Larvae of this very large, common, and geographically widespread family are distinctive.

Pupae of most species live within cylindrical or conical cocoons. Others are free-swimming, and some resemble mosquito larvae. This group is probably the most adapted of all aquatic insects. The larvae of this group are often used as an indicator of environmental quality. Habitats of immatures range from littoral marine waters to mountain torrents, from mangrove swamps to Arctic bogs, and from clear deep lakes to heavily polluted waters. They can be expected in almost all inland waters. Most species are bottom-dwelling, and many live within tubes or loosely constructed silk-lined cases in the substrate. A few build distinctive cases. These benthic forms can occur in extremely high densities; their tube cases sometimes cover large areas of the bottom, virtually becoming substrate themselves for other organisms, such as en-crusting diatoms (AE, 310).

Snails (*Physa integra, P. heterostropha, Amnicola limosa*): (Mollusca, Gastropoda)

Description: These possess a single (univalve), usually drab-colored shell that is either spiraled or coiled or low and conelike. They generally range in size from 2 to 70 mm. Part of the body protrudes from the aperture of the shell and bears a head with a pair of tentacles.

Distribution: The gastropods are well represented in marine, freshwater, and terrestrial environments. Several hundred species of freshwater snails occur in North America. They are benthic organisms that slowly move about on the substrate of almost all shallow freshwater habitats. Some are known to burrow into soft substrates or detritus during periods of drying in vernal habitats or when shallow habitats become frozen solid.

Calcium carbonate is used in the production of the shell, and it is for this reason that many freshwater snails are more common in hard-water habitats, although some do well in soft water. Many feed on the encrusted growths of algae over which they

creep. Others are detritivores or omnivores. Certain freshwater fishes feed extensively on snails, and most marsh fly larvae are predators and parasites of snails.

Planaria (*Dugesia tigrina*): (Platyhelminthes, Turbellaria)

> **Description:** These are soft-bodied, elongated, worm-like forms, usually dorsoven-
> trally flattened or at least flattened ventrally. They are generally less than 1 mm in
> length, but some range to 30 mm. Most are dark colored, and many are mottled.
> Head area is commonly arrowhead-shaped. A pair of dorsal eyespots is usually
> present. Mouth and anus are combined into a single ventral opening usually at about
> midlength along the body.

The phylum Platyhelminthes includes the so-called flat-worms many of which
are parasitic or marine. Most of the free-living, freshwater forms are planarians, and
a few or these are large enough to be considered macroorganisms.

Planarians are usually associated with the substrate of shallow waters. They are
often found on the underside of rocks and detritus. Most are carnivores and scavengers that feed on a variety of soft invertebrates.

INVERTEBRATES: SALTWATER

Copepods (*Acartia clausi, Acartia tonsa*)

> **Description:** These are generally less than 3 mm in length. Body is divided into a
> cephalothorax, thorax, and abdomen. Cephalothorax is covered by a carapace. Six
> pairs of legs are usually present, the first of which is modified for feeding and the
> remaining five pairs for swimming. Body lacks lateral abdominal appendages.

About 180 species of copepods occur in North America. Two groups of copepods
(the Caligoida and Lernaeopodoida) are parasitic on fishes and are highly modified
for this type of existence. The vast majority of copepods are free-living. One genus
(*Cyclopoida*) is parasitic on fishes.

Free-living copepods are planktonic or benthic in a wide variety of freshwater
environments. Some species of cyclopoid and calanoid copepods occur in extremely
high densities. Some of the planktonic copepods have a daily vertical migration in
lakes, similar to that of some water fleas. Parasitic copepods can become a serious
economic problem in fish hatcheries. Many free-living copepods are important in the
food chain of many fishes.

ALGAE

Chlamydomonas reinhardi

Unicellular, green alga which possesses one nucleus, one chloroplast, and several
mitochondria. It is facultatively photosynthetic, and it can grow in the dark with

acetate as its carbon and energy source. It has a sexual life-cycle controlled by two mating type alleles of a single gene, called *mt;* the mating types, and their allele determinants, are called *mt+* and *mt-*, respectively.

Ulothrix sp.

Filamentous member of the Chlorophyta, a multicellular algae which is immobile in the mature state. Reproduction frequently involves the formation and the liberation of motile cells, asexual reproductive cells (zoospores) or gametes. The structure of the motile reproductive cells of multicellular algae thus often reveals their relatedness to a particular group of unicellular flagellates.

Microcystis aeruginosa

Phototroph; Blue-green bacteria

Anabaena flos-aquae

Blue-green bacterium contains gas vacuoles, which accounts for the phase-bright appearance of the vegetative cells.

AVIAN SPECIES

Mallard (Anas platyrhynchos)

Male: Grayish with green head, narrow white ring around neck, ruddy breast, and white tail. **Female:** A mottled brown duck with whitish tail and conspicuous white borders on each side of metallic violet-blue wing-patch. Breeding occurs in western North America east to Great Lakes area; winters from Great Lakes and southern New England south to the Gulf of Mexico.

Species commonly used in acute and chronic toxicity testing as a representative waterfowl.

Northern Bobwhite (Colinus virginianus)

A small, ruddy, chicken-like bird, near the size of a Meadowlark. The male shows a conspicuous white throat and stripe over the eye; the female is buffy. The common habitat is in farming country from Gulf of Mexico north to South Dakota, southern Minnesota, southern Ontario, and southwestern Maine.

This species is extensively used as a model galliform for a variety of acute, chronic, and even field studies. It may be regarded as the white rat of bird toxicity testing.

Ring-Necked Pheasant (*Phasianus colchicus*)

A large chicken-like or gamecock-like bird with a long, sweeping pointed tail. The male is highly colored with a white neck ring; the female is mottled brown with a moderately long pointed tail. The species was introduced to the Americas and is currently established in farming country mainly in the northeastern quarter of the United States.

Larger than the bobwhite, this is another representative galliform not as commonly used as the Northern bobwhite for toxicity testing.

REFERENCES AND SUGGESTED READINGS

ASTM D 4229-84. 1993. Conducting static acute toxicity tests on wastewaters with *Daphnia*. *Annual Book of ASTM Standards*. American Society of Testing and Materials, Philadelphia, PA.

ASTM E 724-89. 1993. Standard guide for conducting static acute toxicity tests starting with embryos of four species of bivalve molluscs. *Annual Book of ASTM Standards*. American Society of Testing and Materials, Philadelphia, PA.

ASTM E 729-88a. 1993. Standard guide for conducting acute toxicity tests with fishes, macroinvertebrates and amphibians. *Annual Book of ASTM Standards*. American Society for Testing and Materials, Philadelphia, PA.

ASTM E 1191-90. 1993. Standard guide for conducting the renewal life-cycle toxicity tests with saltwater mysids. *Annual Book of ASTM Standards*. American Society for Testing and Materials, Philadelphia, PA.

ASTM E 1193-87. 1993. Standard guide for conducting renewal life-cycle toxicity tests with *Daphnia magna*. *Annual Book of ASTM Standards*. American Society of Testing and Materials, Philadelphia, PA.

ASTM E 1197-87. 1993. Standard guide for conducting a terrestrial soil-core microcosm test. *Annual Book of ASTM Standards*. American Society for Testing and Materials, Philadelphia, PA.

ASTM E 1218-90. 1993. Standard guide for conducting static 96-h toxicity tests with microalgae. *Annual Book of ASTM Standards*. American Society for Testing and Materials, Philadelphia, PA.

ASTM E 1241-88. 1993. Standard guide for conducting early life stage toxicity tests with fishes. *Annual Book of ASTM Standards*. American Society for Testing and Materials, Philadelphia, PA.

ASTM E 1295-89. 1993. Standard guide for conducting three-brood, renewal toxicity tests with *Ceriodaphnia dubia*. *Annual Book of ASTM Standards*. American Society for Testing and Materials, Philadelphia, PA.

ASTM E 1366-91. 1993. Standard practice for standardized aquatic microcosms: Freshwater. *Annual Book of ASTM Standards*. American Society for Testing and Materials, Philadelphia, PA.

ASTM E 1367-90. 1993. Standard guide for conducting 10-day static sediment toxicity tests with marine and estuarine amphipods. *Annual Book of ASTM Standards*. American Society for Testing and Materials, Philadelphia, PA.

ASTM E 1383-90. 1993. Standard guide for conducting sediment toxicity tests with freshwater invertebrates. *Annual Book of ASTM Standards*. American Society for Testing and Materials, Philadelphia, PA.

ASTM E 1391-90. 1993. Standard guide for collection, storage, characterization, and manipulation of sediments for toxicological testing. *Annual Book of ASTM Standards*. American Society for Testing and Materials, Philadelphia, PA.

ASTM E 1415-91. 1993. Standard guide for conducting the static toxicity tests with *Lemna gibba* G3. *Annual Book of ASTM Standards*. American Society for Testing and Materials, Philadelphia, PA.

ASTM E 1439-91. 1993. Standard guide for conducting the frog embryo teratogenesis assay — *Xenopus* (FETAX). *Annual Book of ASTM Standards*. American Society for Testing and Materials, Philadelphia, PA.

ASTM E 1463-92. 1993. Standard guide for conducting static and flow-through acute toxicity tests with mysids from the west coast of the Untied States. *Annual Book of ASTM Standards: Water and Environmental Technology*, Vol. 11.04, Philadelphia, PA, pp. 1278-1299.

ASTM D 4229-84. 1984. Conducting static acute toxicity tests on wastewaters with Daphnia. *Annual Book of ASTM Standards*. American Society of Testing and Materials, Philadelphia, PA.

ASTM E 729-88a. 1991. Standard guide for conducting acute toxicity tests with fishes, macroinvertebrates and amphibians. *Annual Book of ASTM Standards*. American Society for Testing and Materials, Philadelphia, PA.

ASTM E 1193-87. 1987. Standard guide for conducting renewal life-cycle toxicity tests with *Daphnia magna. Annual Book of ASTM Standards*. American Society of Testing Materials, Philadelphia, PA.

ASTM E 1197-87. 1987. Standard guide for conducting a terrestrial soil-core microcosm test. *Annual Book of ASTM Standards*. American Society for Testing and Materials, Philadelphia, PA.

ASTM E 1218-90. 1990. Standard guide for conducting static 96-h toxicity tests with microalgae. *Annual Book of ASTM Standards*. American Society for Testing and Material, Philadelphia, PA.

ASTM E 1241-88. 1991. Standard guide for conducting early life stage toxicity tests with fishes. *Annual Book of ASTM Standards*. American Society for Testing and Materials, Philadelphia, PA.

ASTM E 1367-90. 1990. Standard guide for conducting 10-day static sediment toxicity tests with marine and estuarine amphipods. *Annual Book of ASTM Standards*. American Society for Testing and Materials, Philadelphia, PA.

ASTM E 1383-90. 1990. Standard guide for conducting sediment toxicity tests with freshwater invertebrates. *Annual Book of ASTM Standards*. American Society for Testing and Materials, Philadelphia, PA.

Bantle, J.A., J.N. Dumont, R. Finch, and G. Linder. Atlas of abnormalities: A guide for the performance of FETAX. Oklahoma State Publications Department, Peer Reviewed in U.S. Army BRDL.

Callahan, C.A., C.A. Menzie, D.E. Burmaster, D.C. Wilborn, and T. Ernst. 1991. On-site methods for assessing chemical impact on the soil environment using earthworms: A case study at the Baird and McGuire Superfund Site, Holbrook, MA. *Environ. Toxicol. Chem.* 10:817-826.

Callahan, C.A., L.K. Russell, and S.A. Peterson. 1985. A comparison of three earthworm toxicity test procedures for the assessment of environmental samples containing hazardous wastes. *Biol. Fertil. Soils* 1:195-200.

Doherty, F.G. 1983. Interspecies correlations of acute aquatic median lethal concentration for four standard testing species. *Environ. Sci. Technol.* 17:661-665.

Graney, R.L., J.H. Kennedy, and J.H. Rodgers. 1994. *Aquatic Mesocosm Studies in Ecological Risk Assessment.* Lewis Publishers, Boca Raton, FL.

Greene, J.C., C.L. Bartels, W.J. Warren-Hicks, B.R. Parkhurst, G.L. Linder, S.A. Peterson, and W.E. Miller. 1988. Protocols for Short-Term Toxicity Screening of Hazardous Waste Sites. EPA/600/3-88/029. Environmental Research Laboratory, U.S. Environmental Protection Agency, Corvallis, OR.

Kersting, K. 1984. Development and use of an aquatic micro-ecosystem as a test system for toxic substances. Properties of an aquatic micro-ecosystem IV. *Int. Rev. Hydrobiol.* 69: 567-607.

Kersting, K. 1985. Properties of an aquatic micro-ecosystem V. Ten years of observations of the prototype. *Verh. Int. Verein. Limnol.* 22: 3040-3045.

Kersting, K. 1988. Normalized ecosystem strain in micro-ecosystems using different sets of state variables. *Verh. Int. Verein. Limnol.* 23: 1641-1646.

Kersting, K. and R. van Wungaarden. 1992. Effects of Chlorpyifos on a microecosystem. *Environ. Toxicol. Chem.* 11: 365-372.

Linder, G., J.C. Greene, H. Ratsch, J. Nwosu, S. Smith, and D. Wilborn. 1990. Seed germination and root elongation toxicity tests in hazardous waste site evaluation: Methods development and applications. In *Plants for Toxicity Assessment. ASTM STP 1091.* J.G. Pearson, R.B. Foster, and W.E. Bishop, Eds., American Society for Testing and Materials, Philadelphia, PA, 177-187.

McCafferty, W.P. 1981. *Aquatic Entomology: The Fisherman's and Ecologists' Illustrated Guide to Insects and Their Relatives.* Science Books International, Boston, MA.

Petrocelli, S.R. 1985. Chronic Toxicity Tests. In *Fundamentals of Aquatic Toxicology.* G.M. Rand and S.R. Petrocelli, Eds., Hemisphere Publishing Corporation, Washington, D.C., pp. 96-109.

Scott, W.B. and E.J. Crossman. 1973. *Freshwater Fish of Canada.* Ministry of Supply and Services, Canadian Government Publishing Centre, Ottawa, Canada.

Sugiura, K. 1992. A multispecies laboratory microcosm for screening ecotoxicological impacts of chemicals. *Environ. Toxicol. Chem.* 11: 1217-1226.

U.S. Environmental Protection Agency. 1983. Protocols for Bioassessment of Hazardous Waste Sites. EPA/600/2-83/054.

U.S. Environmental Protection Agency. 1987. Methods for Measuring the Acute Toxicity of Effluents to Aquatic Organisms. EPA/600/4-78/012.

U.S. Environmental Protection Agency. 1988. Protocols for Short Term Toxicity Screening of Hazardous Waste Sites. EPA/600/2.

Wang, W. 1987. Root elongation method for toxicity testing of organic and inorganic pollutants. *Environ. Toxicol. Chem.* 6:409-414.

Weber, C.I. 1991. Methods for Measuring the Acute Toxicity of Effluents and Receiving Waters to Freshwater and Marine Organisms. EPA-600/4-90-027. Environmental Monitoring Systems Laboratory, Office of Research and Development, U.S. Environmental Protection Agency, Cincinnati, OH.

STUDY QUESTIONS

1. Discuss the major factor in the performance of a laboratory aquatic toxicity test.
2. Why is the use of a reference toxicant important in the daphnia toxicity test?
3. What are the advantages of the daphnid toxicity test?
4. What is the chronic or partial life cycle toxicity test?
5. Why is the three-brood renewal toxicity test with *Ceriodaphnia dubia* used?
6. How could low concentrations of toxicant in a algal 96-h growth toxicity test lead to a false analysis of toxicity if data is not properly analyzed?
7. Discuss two major problems in conducting acute toxicity tests with aquatic vertebrates and macroinvertebrates.
8. How can terrestrial vertebrate toxicity tests be modified to better assess interspecific variability of a toxic response?
9. Discuss the "replace", "reduce", and "refine" considerations in a required research or test methodology.
10. What are the advantages of the FETAX test?
11. Why have terrestrial systems not undergone the same level of standardization as the aquatic multispecies systems?
12. Discuss coevolution as a component of the mixed flask culture microcosm.
13. Discuss the two methods used to ensure a uniform temperature among experimental replicates during a FIFRA microcosm experiment.
14. Discuss the three caveats to be dealt with in the topic of toxicity testing.

CHAPTER 5

Routes of Exposure and Modes of Action

THE DAMAGE PROCESS

Given sufficiently high concentrations, environmental toxicants can critically influence the physiological processes of a living organism. In order for a pollutant to exert its toxicity on an organism exposed to it, the chemical must first enter the host and reach its target site. Although it is difficult, if not impossible, to generalize the precise mechanism by which each specific pollutant affects living organisms, a few commonalties that are shared by different pollutants are summarized here to provide a general background. The damage processes in plants and in animals are considered separately below.

ATMOSPHERIC POLLUTANTS AND PLANTS

An atmospheric pollutant-induced plant injury may follow a pathway that includes exposure, uptake, transport, storage, metabolism, and excretion (Figure 5.1). To cause injury to any vegetation, an air pollutant must first be taken up by the plant in question. Although the atmospheric concentration of a pollutant is important, the actual amount that gets into the plant is of more concern. The conductance through the stoma, which regulates the passage of ambient air into the cells, is especially critical. Uptake is dependent upon the physical and chemical properties along the gas-to-liquid diffusion pathway. Pollutant flow may be restricted by the physical structures of the leaf or by the chemical reactions that occur following the entry of the pollutant. The leaf orientation and morphology, including epidermal characteristics, and air movement across the leaf are important determinants affecting the initial flux of gases to the leaf surface. More pollutant would enter a leaf when there is some air movement.

Stomatal resistance is a critical factor affecting pollutant uptake. The resistance is determined by stomatal number, size, anatomical characteristics, and the size of the stomatal aperture. Little or no uptake occurs when the stoma is closed. Stomatal opening is regulated by internal CO_2 content, light, temperature, humidity, water

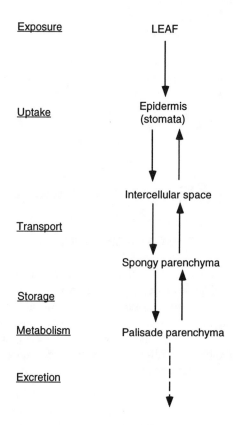

Figure 5.1 Schematic pathway of plant injury induced by atmospheric pollutants.

availability, and nutrient status, particularly, potassium ions. It should be mentioned that, although stomatal resistance is an important factor influencing pollutant uptake, genetic sensitivity of individual species and cultivars are the overriding factors determining plant injury. It should be noted also that the level of a pollutant within the leaf is most critical to plant health.

PLANT INJURY

The epidermis is the first target of atmospheric pollution as the pollutant first passes through the stomata of the epidermal tissue. In passing into the intercellular spaces, a pollutant may dissolve in the surface water of the leaf cells, affecting cellular pH. A pollutant may not remain in its original form as it passes into solution. In fact, it may be converted into a different form, which may be more reactive and toxic than the parent compound. The formation of free radicals following the initial reaction in the cell is an example. The pollutant, either in its original form or in an altered form, may then react with different cellular components such as various membranes and enzymes or their cofactors or coenzymes, thus affecting cell metabolism and causing plant injury. Changes in ultrastructure of various organelles such as chloroplasts and mitochondria can impair photosynthesis and energy metabolism of the plant cell.

As a pollutant moves in the liquid phase from the substomatal regions to the cellular target sites, it may encounter many obstacles along the pathway. Scavenger reactions between endogenous components and the pollutant may occur, influencing the toxicity of the pollutant. For example, ascorbate, which occurs widely in plant cells, may absorb or neutralize a pollutant. On the other hand, an oxidant such as ozone may react with membrane material to form other toxic substances such as aldehydes, ketones, and free radicals, which in turn can adversely affect the cell.

Certain enzymes in the cell may be inhibited when exposed to an air pollutant. For instance, Pb and Cd may inhibit the activity of an enzyme by disrupting its active site containing a sulfhydryl (SH) group. Likewise, sulfur dioxide may oxidize and break apart the sulfur bonds in critical enzymes of the membrane, impairing cellular function.

The net result of all this is an unhealthy plant. Even before visible symptoms are discernible, an exposed plant may be weakened and its growth inhibited. Ultimately, visible symptoms characterizing the effect of specific pollutants may appear, and death of the plant may ensue.

VERTEBRATES

A pollutant may enter an animal through a series of pathways similar to those shown previously (Figure 5.1). The routes may include exposure, uptake, transport, storage, metabolism, and excretion. Figure 5.2 shows the pathways through which a pollutant may pass during its presence in a terrestrial animal or fish.

Exposure

As mentioned earlier, exposure to a pollutant by a host organism constitutes the initial stage in the manifestation of toxicity. In a mammalian organism, exposure of the body occurs through dermal or eye contact, through inhalation, or through ingestion.

Uptake

The immediate and long-term effects of a pollutant are directly related to the mode of entry. The portals of entry for an atmospheric pollutant are the skin, gastrointestinal tract, and lungs.

To be taken up into the body and finally carried to the cell, a pollutant must pass through a number of biological membranes. These include not only the peripheral tissue membranes, but also the capillary and cell membranes. Thus, the nature of these membranes and the chemical and physical properties of the toxicant in question are important factors affecting uptake. The mechanisms by which chemical agents pass through the membranes include: (a) filtration through spaces or pores in membranes; (b) passive diffusion through the spaces or pores, or by dissolving in the lipid material of the membrane; and (c) facilitated transport, whereby specialized transport

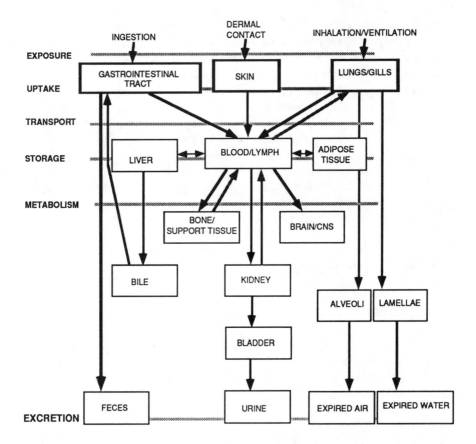

Figure 5.2 Routes of absorption, translocation, and excretion of toxicants in an animal species.

systems carry water-soluble substances across the membrane by a lipid soluble "carrier" molecule, which complexes with the chemical. It can be seen then that, as far as the chemical properties are concerned, **lipophilicity** is the most important factor affecting absorption.

Transport

Once absorbed, a rapid transport of the toxicant throughout the body takes place. A pollutant or a toxicant may be transported via the lymphatic system or blood stream and distributed to various body tissues, including those of storage depots and sites of metabolism or biotransformation.

Storage

The storage depots include the liver, lungs, kidneys, bone, adipose tissue, and others. They may or may not be the sites of the toxic action of the agent. It is possible that a toxicant that is transported to a storage depot may be stored there only temporarily; under certain physiological conditions, the agent may be removed from the depot and translocated again. Similarly, following biotransformation, a toxicant may be transported to a storage depot, or to sites where it is finally excreted. Translocation of a toxicant among tissues may be carried out through binding to a blood protein — a lipoprotein, for example.

Metabolism

The metabolism of toxicants may be carried out at portals of entry or in such organs as the liver, lungs, gastrointestinal tract, and kidney. The liver plays a central role in metabolizing xenobiotics (chemicals foreign to the body). A rich supply of nonspecific enzymes enables the liver to metabolize a broad spectrum of organic substances. The reactions involved in the metabolism of these substances include two phases: Phases I and II. Phase I reactions involve the introduction of a reactive polar group into the xenobiotic through oxidation, reduction, or hydrolysis, forming a primary metabolite. Phase II, on the other hand, involves conjugation reactions in which an endogenous substance combines with the primary metabolite, forming a complex secondary metabolite. An important feature of these reactions is the conversion of lipophilic compounds to more water-soluble and thus more excretable metabolites. While many toxicants are detoxified through these reactions, others may be activated as well (see Chapter 8 for further information).

EXCRETION

The final step involved in the action of a toxicant is excretion from the body. Excretion may take place through the kidneys, lungs, or intestinal tract. A toxicant may be excreted in its original form or as its metabolite(s), depending upon its chemical properties. Excretion is the most permanent means by which toxic substances are removed from the body.

MECHANISMS OF ACTION

The toxic action of chemical agents involves compounds with intrinsic toxicity or activated metabolites. These interact with cellular components at their site of action to initiate toxic effects that may be manifested anywhere in the body. The consequence of such action may be reflected in the inhibition of oxidative metabolism and the central nervous system (CNS), or interaction with nucleic acids resulting in carcinogenesis or injury to the reproductive system. The biological action of a toxicant or a pollutant may be terminated by metabolic transformation, storage, or excretion.

Although the precise mechanism by which each of the many environmental pollutants exerts its toxicity remains to be elucidated, four principal mechanisms are described here. In general, a pollutant may cause an adverse effect on a living organism through: (a) disruption or destruction of cellular structure; (b) direct chemical combination with a cell constituent; (c) its influence on enzymes; and (d) initiation of a secondary action. These are examined below.

DISRUPTION OR DESTRUCTION OF CELLULAR STRUCTURE

A toxicant or a pollutant may exert its injurious effect on a specific organ by causing structural damage to its tissues. For example, airborne pollutants such as SO_2, O_3, NO_2, and fluoride are known to be phytotoxic. Sensitive plants exposed to any of these pollutants at certain concentrations can be structurally damaged, leading to cellular destruction. Evidence suggests that low concentrations of SO_2 can injure epidermal and guard cells, leading to enhanced stomatal conductance and greater entry of the pollutant into the plant (Black and Unsworth 1980). Similarly, after entry into the substomatal cavity of plant leaves, O_3, or the free radicals produced from it, may react with protein or lipid membrane components, disrupting the cellular structure of the leaf (Heath 1980; Grimes et al. 1983).

When inhaled by animals or humans, sufficient quantities of O_3 and sulfuric acid mists can cause damage to surface layers of the respiratory system. Exposure to high levels of ozone may lead to pulmonary edema (Mueller and Hitchcock 1969), i.e., a leakage of fluid into the gas-exchange parts of the lung. This implies that exposure to O_3 can lead to disruption of the lung tissue (see Chapter 7 for further information).

DIRECT CHEMICAL COMBINATION WITH A CELLULAR CONSTITUENT

A pollutant may combine with a cell constituent and form a complex. This often leads to impaired function. For example, carbon monoxide (CO) in the blood readily binds to hemoglobin (Hb) forming carboxyhemoglobin (COHb) as shown below:

$$CO + Hb \rightarrow COHb \tag{5.1}$$

Since hemoglobin in the body is essential in the carbon dioxide-oxygen exchange system between the lungs and the tissues, interference with the functioning of hemoglobin as a result of COHb formation can be detrimental.

Another example is cadmium, a highly toxic heavy metal. Once absorbed, cadmium in the body is mainly bound to the protein metallothionein. This protein is involved in the transport and selective storage of cadmium. A rather selective accumulation of cadmium occurs in the kidneys, leading to eventual tubular dysfunction with proteinuria (Friberg et al. 1974).

EFFECT ON ENZYMES

The most distinguished feature of reactions that occur in a living cell is the participation of protein catalysts called enzymes. As with any catalyst, the basic function of an enzyme is *to increase the rate of a reaction*. All protein enzymes are globular, with each enzyme having a specific function because of its specific globular structure. However, the optimum activity of many enzymes depends on the presence of nonprotein substances called cofactors. The molecular partnership of protein-cofactor is termed a holoenzyme and exhibits maximal catalytic activity. The protein component without its cofactor is termed an apoenzyme and exhibits very low activity, or none at all.

Protein   +   Cofactor   ⇔   Protein – Cofactor (a complex)   (5.2)
 (apoenzyme;   (inorganic ion or   (holoenzyme; optimally active catalyst)
 inactive or   organic substance;
 less active) inactive as a catalyst)

   There are two categories of cofactors: the organic and inorganic. The organic cofactors include several substances of diverse structure, and are usually called coenzymes. Coenzymes are especially important in animal and human nutrition because most of them are vitamins or are substances produced from vitamins. For example, vitamin K after ingestion is unchanged and used directly as vitamin K. The vitamin niacin after being ingested, however, is converted to either of two cofactors — nicotinamide adenine dinucleotide (NADH) or nicotinamide adenine dinucleotide phosphate (NADPH). On the other hand, the inorganic cofactors include several simple inorganic ions such as Mg^{2+}, Mn^{2+}, Zn^{2+}, Ca^{2+}, Fe^{2+}, Cu^{2+}, K^+, and Na^+ ions.
   Several ways in which environmental pollutants may inactivate an enzyme system are described below:

1. A pollutant may combine with the active site or sites of an enzyme thus inactivating it. For example, a heavy metal such as mercury, lead, or cadmium can attach itself to the thiol or SH group on an enzyme molecule, forming a covalent bond with the sulfur atom. This will lead to inactivation of the enzyme if the SH group happens to be the active site of the enzyme. Transaminases and δ-aminolevulinate dehydratase are susceptible to inhibition by lead because they contain the SH group at their active sites.

$$2 \text{ Enz–SH} + Pb^{2+} \rightarrow \text{Enz–S–Pb–S–Enz} + 2 \text{ H}^+ \qquad (5.3)$$

   Additionally, the conformation of an enzyme protein may be altered through disruption of the hydrogen bonds.
2. Many enzymes require cofactors, often cations, for their activity. These ions provide electrophilic centers in the active site. A chemical agent or a pollutant may inhibit an enzyme by inactivating the cofactor involved. For instance, fluoride is known to be a potent inhibitor of enolase, a glycolytic enzyme that requires Mg^{2+} ions for its activity. In the presence of phosphate, fluoride inactivates the Mg^{2+} cofactor, presumably by causing the formation of a magnesium fluorophosphate complex.

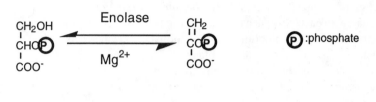

2-Phosphoglycerate   Phosphoenol
                                      pyruvate   (5.4)

3. A pollutant may exert its toxicity through competing with the cofactor for the active site, thus inactivating the enzyme. For example, Be (beryllium) competes with Mg and Mn, and Cd replaces Zn in some enzymes.
4. The activity of an enzyme may be inhibited by the presence of a toxic metabolite. Sodium fluoroacetate, known as rat poison 1080, is extremely toxic to animals. The toxic action, however, is not due to sodium fluoroacetate itself but to a metabolic conversion product, fluorocitrate, formed through a reaction commonly known as "lethal synthesis", as shown in Figure 5.3. The resulting fluorocitrate is toxic because it inhibits aconitase, the enzyme responsible for the conversion of citrate into *cis*-aconitate and then into isocitrate in the Krebs cycle. Inhibition of aconitase results in citrate accumulation. The cycle stops for lack of metabolites, impairing energy metabolism.

SECONDARY ACTION AS A RESULT OF THE PRESENCE OF A POLLUTANT

The presence of a pollutant in a living system may cause the release of certain substances which are injurious to cells. Several examples are given to illustrate this phenomenon.

Subsequent to inhalation of pollen, allergic response occurs in many individuals, leading to a common symptom of hay fever. This is due to the release of histamine, a substance formed from the amino acid histidine through decarboxylation (Figure 5.4). Histamine is made and stored in the mast cell and in many other cells of the body. Release of histamine occurs in anaphylaxis, or as a consequence of allergies; it is also triggered by certain drugs and chemicals. Histamine is a powerful vasodilator and causes dilation and increases permeability of blood vessels. It stimulates secretion of pepsin; it can reduce the blood pressure and can induce shock, if severe enough. In excessive concentrations histamine can cause vascular collapse.

Antihistamines such as diphenylhydramine and antergan are compounds similar to histamine structurally, and can prevent physiologic changes produced by histamine by inhibiting its function.

Another example is seen with the effect of carbon tetrachloride on humans. Once taken up into the body, carbon tetrachloride causes a massive discharge of epinephrine from sympathetic nerve, leading to liver damage. Epinephrine is a potent hormone and is involved in many critical biological reactions in animals and humans, including such diverse functions as stimulation of glycogenolysis, lipolysis, and

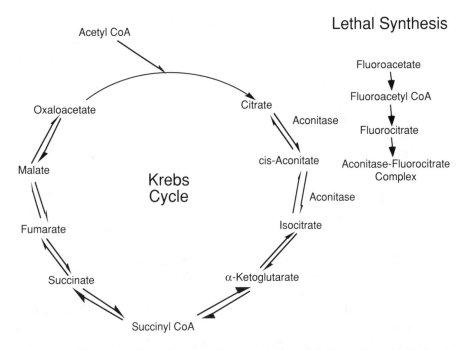

Figure 5.3 Synthesis of fluorocitrate from fluoroacetate through "lethal synthesis". Inhibition of aconitase shuts down Krebs cycle.

glucagon secretion, and inhibition of glucose uptake by muscle, and insulin secretion. It also causes the blood pressure to increase. Like other hormones, epinephrine is rapidly broken down as soon as it performs its function. Metabolism of the hormone takes place mainly in the liver.

A third example involves **chelation**. This is a process wherein atoms of a metal in solution are "sequestered" by ring-shaped molecules, as illustrated in Figure 5.5. The rings of atoms, usually with O, N, or S as electron donor, have the metal as electron acceptor. Within this ring the metal is more firmly gripped than if it were attached to separate molecules. In forming strain-free stable chelate rings, there must be at least two atoms that can attach to a metal ion. The iron in a hemoglobin molecule and the magnesium in a chlorophyll molecule are two examples of this kind. Through chela-

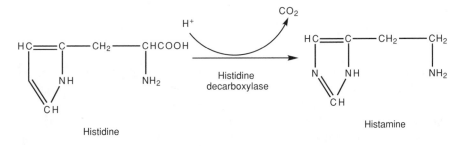

Figure 5.4 Formation of histamine from histidine.

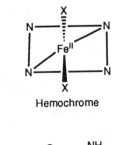

Hemochrome

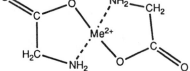

Figure 5.5 Examples of chelation.

tion, some biologically active compounds are absorbed and retained in the body, whereas others may be removed from living systems more readily.

The toxicity of certain chemicals may be the result of chelation. For example, experiments have shown that when rabbits were exposed to CS_2 at 250 ppm, there was a rapid out-pouring of tissue Zn in urine. The loss of body Zn is primarily due to a chemical reaction of CS_2 with free amino groups of tissue protein, to form thiocarbamate and thiazolidone, which could form soluble chelate with Zn (Stokinger et al. 1966) (Figure 5.6). The thiazolidone shown in Figure 5.6 may make copper less available for essential enzyme functions. For example, copper is an essential metal component of several tissue oxidases such as cytochrome oxidase and δ–aminolevulinic acid dehydratase. Removal of copper from the enzyme systems leads to inactivation of the enzymes.

It has been suggested that metal chelation may be one of the mechanisms involved in carcinogenesis. Many carcinogens possess structures or can be metabolized to structures capable of metal-binding. This in turn will aid the entrance of metals into cells. Once inside the cells, interaction between normal metals and abnormal metals can occur thus altering cell metabolism. Certain anticancer agents may function through metal binding, i.e., they may inactivate abnormal metals more than the normal metals within the cells. There are, moreover, numerous toxic environmental chemicals that have either chelate structures or can become so through the usual metabolic processes.

METAL SHIFT

Metal shift refers to the phenomenon in which certain metals shift from one organ to another as a result of the presence of a toxicant. This is among the earliest biological indicators of toxic response. For example, rats fed vanadium (V) at concentrations up to 150 ppm were shown to have iron move into the liver and spleen. When vanadium concentrations were at 250 ppm or above, however, iron moved out of the liver and spleen. As a result, the iron level in the spleen was

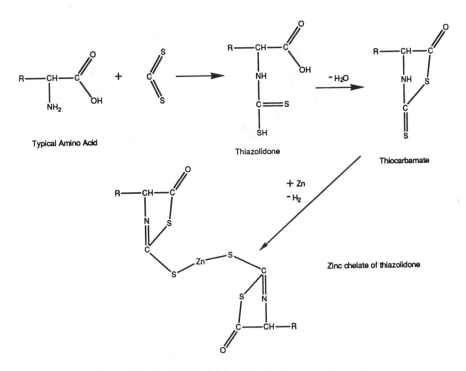

Figure 5.6 Reactions of CS_2 with proteins and amino acids.

decreased to one half to one third of the normal content, while that in the liver, to one-third of the normal content (Furst 1960). These results indicate that treatment with vanadium will lead to depletion of iron in these tissues.

The phenomenon of metal shift in rats exposed to fluoride has recently been reported (Yoshida et al. 1991). Administration of fluoride to rats increased serum Zn levels whereas the levels of Se and Al in the whiskers were decreased.

A similar phenomenon has been observed with rats exposed to ozone. When rats were exposed to the pulmonary irritant for 4 h, the animals showed increased levels of copper, molybdenum, and zinc in the lungs, while these metals were decreased in the liver. This would indicate an altered hemodynamics and changes in cellular permeability in a secondary affected organ.

COMMON MODES OF ACTION IN DETAIL

NARCOSIS

Narcosis is perhaps the most common mode of action of industrial pollutants. An excellent review has been published by Bradbury et al. (1989). A variety of compounds, especially those used as solvents, exhibit this mode of action during the typical toxicity test. Although a common mode of action from the point of view of symptomology, several different molecular mechanisms may be at play.

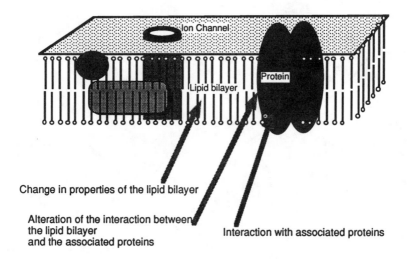

Figure 5.7   Schematic of cell membrane with associated proteins.

Figure 5.7 is a diagram of a typical cellular membrane with the lipid bilayer and its associated proteins. Three sites of actions within the membrane may actually be the place where a molecule exhibits its effect. First, the actual mode of action may be an alteration of the physicochemical properties of the lipid bilayer. Changes to the fluidity or other aspects may sharply alter the passage of molecules through the membrane. Second, the molecule may interact directly with the protein associated with the membrane. Many of the proteins are ion pumps, receptors for regulatory molecules or have some other regulatory function. Finally, the toxicant may alter the interaction of the lipid bilayer with the inserted protein. This change in the bilayer-protein interaction then changes the ability of the protein to perform its function. Each of these modes can be relatively nonspecific and the impact of lipid solubility is obvious. Lipid soluble materials can readily enter the membrane and then alter its function. In fact, most of the models that portray the relationship between structure of the toxicant and the narcotic effect rely extensively if not exclusively on the ratio of the compound's solubility in octanol compared to water.

The fact that not all compounds with narcosis as the mode of action work similarly is depicted in Figure 5.8. Apparently at higher values of log P, the nonpolar compounds demonstrate a lesser slope. Perhaps two different mechanisms are at play.

ORGANOPHOSPHATES

The organophosphates are compounds widely used as insecticides and chemical warfare agents. Although extremely toxic in some cases, these materials are generally short-lived in the environment compared to halogenated organics and related compounds. The toxicity of an organophosphate is related to its leaving group, the

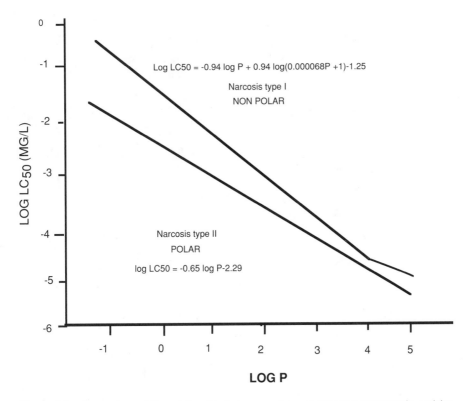

Figure 5.8   Comparison of the relationship between polar and nonpolar compounds, toxicity, and octanol/water partition coefficient. This graph is after Bradbury et al. (1989) and depicts that relationship between log P and fathead minnow 96-h LC_{50} values. Note that the slope of the Type I compounds is continuous until the higher log P values are reached.

double bonded atom, usually O or S, and the phosphorus ligands, the groups surrounding the phosphate in the compound. Several examples of typical organophosphates are shown in Figure 5.9. The more toxic compounds generally have short phosphonate side groups with fluoride or a cyano leaving group. The metabolic replacement of sulfur by oxygen in the liver or other detoxification organ activates the sulfur-containing organophosphate into a much more potent form. The extreme toxicity of these compounds is due to their ability to bind to the amino acid serine, rendering it incapable of participating in a catalytic reaction within an enzyme and the further blocking of the active site by the organophosphate residue. Although many proteins have serine in their active sites and are affected by organophosphates, the acute toxicity of these compounds is usually attributed to their ability to bind to the critical nervous system enzyme acetylcholinesterase.

In normal transmission of a nervous impulse from nerve to nerve, acetylcholine is released into the synapse in order to excite the receiving neuron (Figure 5.10). Unless acetylcholine is rapidly broken down, the receiving nerve is constantly fired resulting in uncoordinated muscle movement, nausea, dizziness, and eventually

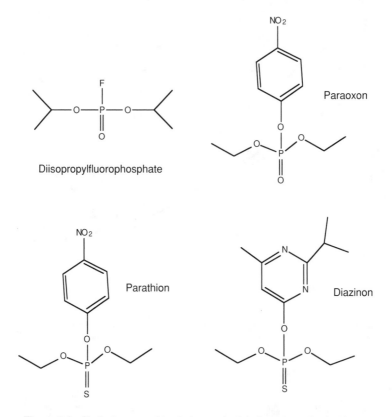

Figure 5.9 Typical organophosphates and related compounds structures.

seizures and unconsciousness. The serine enzyme acetylcholinesterase is responsible for the expedient breakdown of the neurotransmitter acetylcholine.

Typically, acetylcholine is catylitically degraded by the initial binding of the acetylcholine to the serine with a proton donated by the amino acid. This process is graphically demonstrated in Figure 5.11. This results in the release of the choline group with the remainder binding to serine. With the addition of a molecule of water, the serine is reactivated with the release of the acetyl group from the active site.

Organophosphates are able to participate in part of the reaction depicted above. However, as shown in the accompanying figure (Figure 5.12), all does not work as if the organophosphate was acetylcholine. The typical organophosphate is able to enter at the active site and the initial proton donation occurs resulting in the linkage of the serine to the phosphate. This is a two-step process. First a Michaelis complex is formed between the –OH group and the phosphate and then the covalent bond between the serine and phosphate is formed resulting in the loss of a nitrophenol, fluoride, or other leaving group. These reactions are reversible. The next step is an irreversible binding at a glutamyl residue that "ages" the protein. This next step is relatively slower than the initial binding to the organophosphate, but is variable from

Nerve Synapse

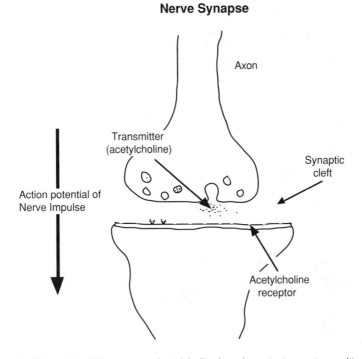

Figure 5.10 Schematic of the synapse. Acetylcholine is an important neurotransmitter and the intervention of acetylcholinesterase prevents subsequent firing of the adjacent neuron.

organophosphate to organophosphate. Compounds typically used as chemical warfare agents have relatively fast aging reactions.

The fact that an organophosphate binds to an enzyme such as acetylcholinesterase can be used to an advantage. Inhibition of acetylcholinesterase and its relative butylcholinesterase is routinely used as an indication of exposure to an organophosphate or other inhibitory compound.

Lastly, organophosphates bind to other proteins and likely affect many other metabolic pathways. It has recently been shown that organophosphates bind to a variety of liver proteins and these protein act, accidentally perhaps, as sinks protecting enzymes of the Central Nervous System (CNS) from exposure. Of course a second dose of an organophosphate soon after would likely be more toxic, not because of the increased toxicity of the molecule, but because of the prior filling of this sink.

MONOHALOACETIC ACIDS

Monhaloacetic acids are compounds derived from acetic acid with the substitution of a halogen to replace one of the hydrogens. Chloroacetate, fluoroacetate, iodoacetate, and bromoacetate are compounds that vary in toxicity and mode of

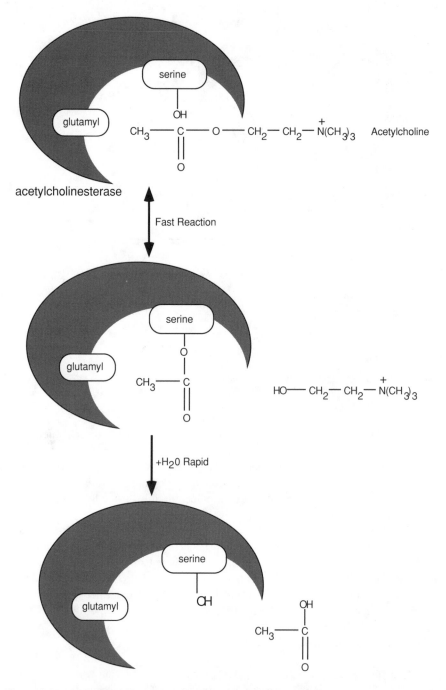

Figure 5.11 Normal hydrolysis of acetylcholinesterase. The amino acid serine is important in the donation of a proton used in the catalytic process. The proton is regenerated during the reaction.

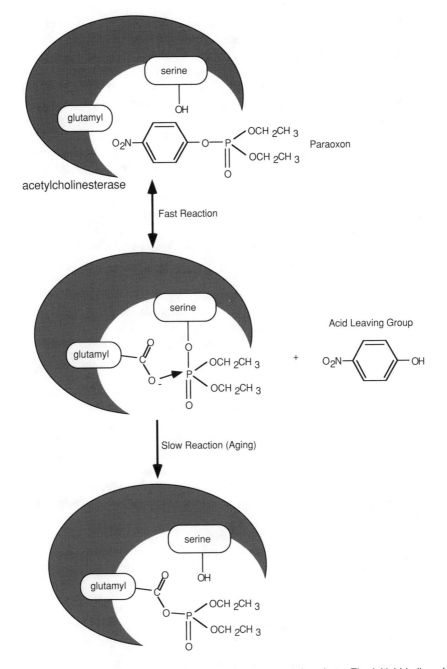

Figure 5.12 Inhibition of acetylcholinesterase by an organophosphate. The initial binding of the organophosphate to the active site prevents the normal substrate from entering the active site. The aging process subsequently binds the organophosphate to the active site, permanently inactivating the enzyme.

action although they are closely related. Sodium fluoroacetate was a widely used mammalian pesticide known as compound 1080. Chloroacetic acid is used as a feedstock and is manufactured in large amounts.

Hayes compared the toxicity of the chloroacetate, fluroacetate, and iodoacetate in rats. The 24-h LD_{50} values were 108, 5, and 60 mg/kg, respectively. LD_{90} doses were delivered to rats and the time until death (LT) was determined. The LT_{50} for chloroacetate, fluroacetate, and iodoacetate was 130, 310, and 480 min, respectively. Based upon this comparative study, fluoracetate was the most toxic, iodoacetate the intermediate, and chloroacetic acid was the least toxic of the three compounds. Bromoacetic acid is not as well studied although it is a potent enzyme inhibitor.

Although the monohaloacetic acids have similar chemical properties and structure, the unique properties of the halogen cause very different physiological effects. Figure 5.13 is a spatial representation of the four monohaloacetic acids compared to acetic acid. As shown in the figure, fluoroacetic acid and acetic acid are very similar in configuration. The small size of the fluorine atom allows fluroacetate to mimic acetate in the TCA cycle as described previously in this chapter. Briefly, the fluoro-acetate is metabolized in the TCA cycle to the point where fluorocitric acid is synthesized in the place of citric acid. Aconitase accepts the molecule into the active site but the strong electronegativity of the fluorine prevents the enzyme from cata-lyzing the reaction or dislodging the molecule. Since there is competition for the active site of the enzyme, fluorocitrate is a competitive inhibitor of aconitase and the inhibition is reversible.

In contrast, iodoacetic and bromoacetic acids inhibit enzymes by alkylating sulfhydryl (–SH) and amino (–NH_2) groups. This involves the replacement of the hydrogen atom by the acetic acid group –CH_2COOH. This reaction prevents these proton donor groups from participating in the biochemical reactions requiring the addition of the proton. Enzymes containing these proton donor groups are inhibited. Examples of such enzymes are guinea pig monoamine oxidase, GAPD, and various enzymes involved in glycolysis. Iodoacetic and bromoacetic acids do not enter the TCA cycle due to the relatively large size of the halogen. However, since competition for the active site of the affected enzyme does not occur, they are irreversible inhibitors of enzyme function.

Chloroacetate is an intermediate case. Apparently –SH groups and acetate oxida-tion are affected. The relatively small chlorine atom may allow chloroacetic acid to slowly enter the TCA cycle and inhibit aconitase while at the same time alkylating –SH groups.

INTRODUCTION TO QSAR

Quantitative structure activity relationships (QSAR) are a method of estimating the toxic properties of a compound using the physical and structural construction of a compound. These properties and the knowledge that similar compounds typically have similar modes of action make QSAR a possibility. In many instances no toxicity

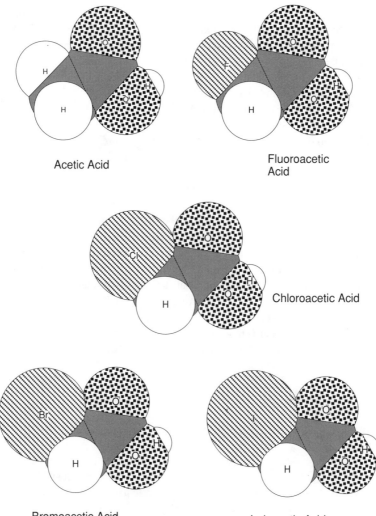

Figure 5.13 Relative configurations of the monohaloacetic acids and acetic acid.

data are available for a compound for a variety of reasons. Perhaps the most interesting one is in the evaluation of proposed compounds of which only small amounts or none at all are available. QSAR can be instrumental in selecting compounds with the desired properties but with low toxicity to the environment.

Each substructure of a molecule contributes to its toxicity in a specific way and the QSAR equation describes this contribution. Models of this type have proven to be successful in the estimation of carcinogenicity, mutagenicity, and rat, mouse, daphnia, and fathead minnow acute toxicity, and in establishing toxicological relationships across species boundaries.

Toxicity data are generally of two types. First, most toxicity data are continuous, that is they may have virtually any numerical value. LD_{50}, NOEC, EC_{50}, and EC_{10} data are all examples of data that are continuous. Second, discriminate data exist. These data place the result into categories such as mutagenic/not mutagenic, carcinogenic/not carcinogenic, and so forth. These two basic types of toxicological determinations require models different in structure.

Continuous toxicity data can be generally described using a regression-type model as depicted in Figure 5.14. This is a simple linear regression model using only one parameter to describe the toxicity. The resulting expression used to describe the relationship between toxicity and the parameter is a typical linear equation:

$$y = mx + b \tag{5.5}$$

where y is the estimate of toxicity, m is the slope of the line, x is the numeric expression of the predictive parameter, and b is the constant value that represents the y intercept of the line. This equation can be generalized to use as many dimensions as there are parameters that contribute to the estimate of toxicity. Table 5.1 portrays such an equation in tabular form. Note that the form is the basic linear equation.

Discriminate data are either/or situations and can be depicted similar to the continuous type variables (Figure 5.15). However, the black and white squares depict dichotomous data. The goal is to derive a line that separates the two groups and this is known as a discriminant analysis. The resulting equation is similar in basic form to the linear regression depicted above.

CONSTRUCTION OF QSAR MODELS

Three sets of traditional models for toxicity using regression and discriminate analysis are generally produced. General models are often produced relying upon chemical parameters such as log P. Models are also often produced that attempt to describe a particular subset of compounds unique in their composition or mode of action. Models can also be produced that incorporate toxicity data from other species or other types of biological measurements.

The first groups of models are generally constructed using molecular connectivity indices, kappa environmental descriptors, electronic charges and substructural keys. In many instances the octanol/water coefficient (log P) has been used; however, our experience has been that models based upon log P do not model well a biological endpoint for a heterogeneous series of compounds. The attempt is made in these models to broadly map the relationship between toxicity and general chemical parameters. These models have proven successful in predicting toxicity in a number of toxicity tests including rat oral LD_{50}, Daphnia EC_{50}, and fathead minnow to name a few. In addition to modeling continuous endpoints, this approach has also been found to be useful in predicting categorical endpoints such as mutagenicity, carcinogenicity, and skin irritation.

Regression Analysis

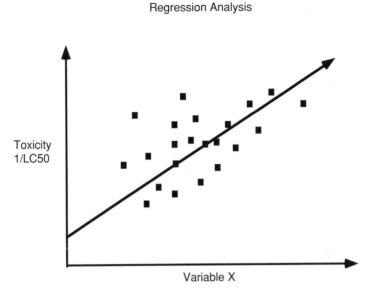

Figure 5.14 Linear regression model for continuous data in QSAR analysis. The model is a simple linear regression with toxicity plotted against the physical or structural variable being used for the estimate.

Table 5.1 Daphnia EC$_{50}$ Equation for Model Incorporating Molecular Connectivity Indices and Substructural Keys

Key	Coefficient
Primary amine bound to aromatic ring atom	1.0167
Primary amine bound to aliphatic or alicycle carbon	1.0343
Aliphatic alcohol	−0.5294
Oxygen-substituted aryl ester	−0.7801
Benzene	1.0320
Secondary or tertiary diphatic alcohol	−1.0058
1,1-Dichloro(non-beta phynyl)	0.8091
1,1-Divinyl chloride (non-beta phynyl)	1.0021
Secondary or tertiary amine bound to electron-releasing groups only	1.3375
One or more electron-releasing groups and four or more electron-withdrawing groups on a single benzene ring	0.7820
Three carbon fragments between two functional groups (electron withdrawing, electron releasing, or combination)	0.9442
NH substituted with one electron-releasing and one electron-withdrawing group	1.3467
Ethane or ethylene between two electron-releasing groups	−0.1819
Valence path MCI, order 2	0.3515
Valence path MCI, order 4	0.1198
Sum simple and valence chain MCI, order 6	0.3621
Intercept	2.2578

After Enslein et al. 1989.

Discriminate Analysis

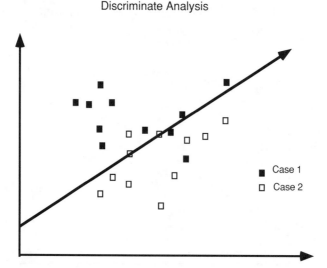

Figure 5.15   Discriminant analysis. In this case the goal is to differentiate data that are in two categories, Case 1 and Case 2. Case 1 could be mutagenic, and Case 2 not mutagenic. Many toxicological measurements are categorical in nature.

Occasionally, compounds with distinctive modes of action are better modeled apart from the general case. An example of such compounds are the acetylcholinesterase inhibitors. These compounds are very specific in their inhibition of serine enzymes. In the instance of predicting Daphnia EC_{50} it was found that the organophosphates were outliers that biased the regression and were better removed from the general model and treated separately. Another class of specialized models are those grouped by chemical class. These have proven popular because of their relative simplicity, but the data sets upon which they are built are usually small.

The third set of models would be interspecies models similar to those used for the extrapolation of rat oral LD_{50} to *Daphnia magna* EC_{50}. These interspecies models have been shown to be very accurate when the size of the database is taken into account and may prove useful when mammalian data are the only available toxicity data for a compound. Sets of these models may have a great deal of utility in interspecies estimations made necessary by the lack of data with wild species.

TYPICAL QSAR MODEL DEVELOPMENT

All three types of models are produced using similar methodologies. The basic methodology for the construction of a multiparameter QSAR is presented in Figure 5.16. Among the most difficult aspect is the acquisition of a reliable and consistent database. The reliability of the database cannot be overemphasized since all subsequent processes are totally dependent upon the size and quality of this data. Published open literature, government reports, contractor data, and premanufacturing notices

Procedure for Constructing Qsar Models

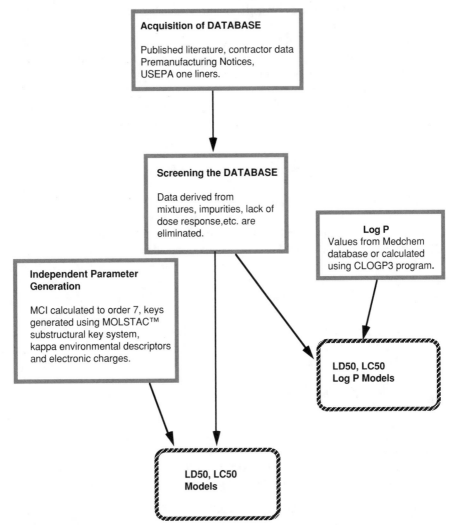

Figure 5.16 The developmental process for the construction of a structure activity model.

submitted as part of TSCA all have been useful in supplying the raw data for the modeling process. Next, the data are evaluated according to preset guidelines to ensure the consistency of the data. Often guidelines such as those set by American Society for Testing and Materials, the U.S. Environmental Protection Agency, and programs such as GENETOX are used to establish criteria for the inclusion of data. Data derived from mixtures, compounds with known impurities, and experiments that do not show a dose response are eliminated from the data set. An attempt is made to include as wide a variety of classes of compounds as possible in order to describe as much of molecular space as possible. In interspecies models only the intersection of the appropriate species are used. The size of the intersection determines the accuracy of interspecies model construction. In studies conducted to date the number of compounds in this intersection have been small; however, the power of including a toxicity endpoint increases the predictive power of the model when compared to models with chemical endpoints alone.

Because a molecule is the unit of toxicity, not mass in mg/kg, it is generally necessary and desirable to transform the LD_{50} and LC_{50} values into molar form as follows:

$$\log 1/C = \log (\text{mol w} \times 1000/LD_{50} \text{ or } LC_{50})$$

where C is the molar concentration.

A variety of parameters are included into the QSAR equation. Log P is a commonly used parameter and is obtained from Medchem or estimated using the CLOGP3 computer program. Molecular weight is calculated. In interspecies models the LD_{50} or LC_{50} value is incorporated as a typical parameter. Molecular connectivity indexes, electronic charge distributions, and kappa environmental descriptors have been proven recently as powerful predictors of toxicity. The efficacy of these values lies in the fact that each of these parameters describes a molecule in a fashion similar to that actually seen by the molecular receptors that initiate a toxic response. Substructural keys are identified with the help of the MOLSTAC™ substructural key system. MOLSTAC™ consists of five classes of descriptors:

1. Identification of the longest continuous chain of atoms (excluding hydrogen) in the molecule
2. Identification of carbon chain fragments
3. Identification of ring systems, including combinations such as the rings forming the bay region of certain carcinogens
4. Identification of chemically or biologically or both functional substructural fragments
5. Identification of electron-donating and electron-withdrawing substructural keys

Multiple regression is used to generate the final equation. Figure 5.17 outlines the derivation of the QSAR equation. After database assembly potential parameters are examined using simple statistics for the detection of problematical distributions that may have to be transformed. Next a stepwise regression analysis is performed. F-scores of at least 1.7 are necessary for the parameter to be included into the final equation. Care is taken to avoid spurious correlations or collinearity difficulties.

The initial regression is examined for robustness from the standpoint of both influential chemicals and poorly behaved parameters. Ridge regression, Cook's distance, partial correlations, and principal components are used to evaluate the regression. After the poorly behaved parameters are removed, another analysis of the regression is performed. Usually several parameters are removed during this process.

Model Development Process

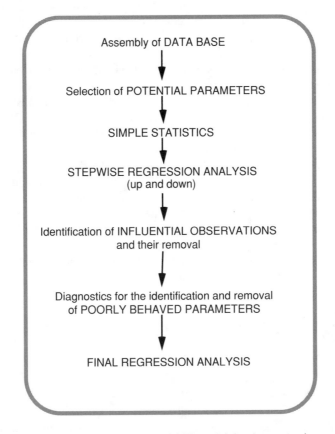

Figure 5.17   The statistical processes of QSAR model development using regression.

Validation of the predictions of the model is one of the most difficult aspects of environmental QSAR development due to the comparatively small size of the database. Cross-validation has been useful in validating the effectiveness of the model. In this method, one compound is removed from the database, the equation is recalculated, and the toxicity of the omitted compound is estimated. The process is repeated for all compounds in the data set and the results are tabulated. In this manner a calculation of the accuracy of prediction of continuous data and the rate of misclassification for categorical data can be compiled. A more useful estimate of the validity of the QSAR model is its ability to predict the toxicity of new compounds. Generally, this is difficult to accomplish in a statistically significant way due to the slow accumulation of new data that meet the criteria used in the modeling process and the associated expense.

ESTIMATION OF TOXICITY USING QSAR

The example of the toxicity estimation using QSAR is based on the TOPKAT system developed by Health Designs, Inc., and is the computer program most familiar to the authors (other software is available). The process of estimation is straightforward when the equations are incorporated into the TOPKAT program. The structure to be evaluated is input using a linear notation, SMILES, for the two-dimensional structure of the compound. The model to be used is specified and loaded along with the accompanying database for validation process. The TOPKAT program searches for parameters and calculates the regression score and the resultant LD_{50} estimate. Using the TOPKAT program, an evaluation of the reliability of the estimate is made looking for similar compounds in the database. The results are reported with a comment on the terms that contributed to the estimate and a comparison of the estimate to literature values for similar compounds.

An example of the process is the estimation of the toxicity to *D. magna* of the simple organic isopropylamine. The compound was given a unique identification and that is usually the Chemical Abstracts Service (CAS) number for easy identification. The chemical structure is then represented in SMILES and the model selected. In the case of the *D. magna* model the estimate was

Key	Cross Product
Primary amine (noncyclic) r-NH2 (R = alkyl)	0.961
Valence Adjusted Path MCI order 1	0.437
Constant term	2.287
Total	3.685

That is, the estimate of EC_{50} was log(1000/Molar)=3.685 or 12.2 mg/l.

The compound was examined using the structural key and other indices to test how well the keys used in the modeling process described isopropylamine. The computer search of these keys confirmed that isopropylamine was well described by the model.

The next step is the validation process. Validation is simply an examination of the model with compounds for which toxicity data are available and that were estimated by the QSAR equation. This process provides an indication of how well the model predicts the toxicity of compounds similar to the unknown. In this estimate six compounds were used as comparisons.

Compound	Actual EC_{50}	Predicted EC_{50}
2-Ethylhexylamine	2.2	4.44
Allyamine	110.0	14.1
Cyclohexylamine	80.0	6.9
n-Butylamine	75.0	30.8
Ethanolamine	140.0	49.6
Ethylamine	110.0	12.0

In general the model overestimated the toxicity of these compounds. Toxicity tests performed with isopropylamine confirmed that the estimated toxicity was an overestimate. The 48-h *D. magna* EC_{50} was found to be 89.4 mg/l with the pH uncontrolled and 195.3 mg/l with the pH adjusted to a normal range. The importance of the confirmation step is crucial. The performance of the model can be measured and the overestimation of the isopropylamine toxicity was consistent with past performance.

Another crucial aspect of the confirmatory process is the test of how well described and represented the molecule is in the map of the chemical-toxicity space that the regression equation represents. If the substructural key does not exist in the database used to build the model then it is unlikely that the compound can be accurately estimated. In addition, if compounds similar to the test compound do not exist then a comparison as was done above cannot be conducted and a measure of the performance of the model with compounds similar to the test material cannot be made. This type of validation requires a large database and a substructural search algorithm, and should be included in a QSAR estimate.

Other types of QSAR models are under development. Perhaps most intriguing is the ability to actually use molecular models of proteins and the organic compound in question to examine at the molecular level the interactions giving rise to toxicity. Widespread use of such models is unlikely to occur due to the enormous amount of data necessary on protein structure, charge distribution, and the properties of the test compound, and the expense of the software and hardware necessary to perform the analysis.

REFERENCES AND SUGGESTED READINGS

Ashford, J.R. and J.M. Cobby. 1974. A system of models for the action of drugs applied singly or jointly to biological organisms. *Biometrics* 30:11-31.

Belsley, D.A., E. Kuth, and R.W. Welsch. 1980. *Regression Diagnostics.* Wiley, New York.

Black, V.J. and M.H. Unsworth. 1980. Stomatal responses to sulfur dioxide and vapor pressure deficit. *J. Exp. Bot.* 31:667-677.

Bradbury, S.P., R.W. Carlson, and T.R. Henry. 1989. Polar narcosis in aquatic organisms. In *Aquatic Toxicology and Hazard Assessment:* 12th Volume. ASTM STP 1027. U.M. Cowgill and L.R. Williams, Eds., American Society for Testing and Materials, Philadelphia, PA, pp. 59-73,

Brown, V.M. 1968. The calculation of the acute toxicity of mixtures of poisons to rainbow trout. *Wat. Res.* 2:723-733.

Calamari, D. and R. Marchetti. 1973. The toxicity of mixtures of metals and surfactants to rainbow trout (*Salmo gairdneri* Rich.). *Wat. Res.* 7:1453-1464.

Calamari, D. and J.S. Alabaster. 1980. An approach to theoretical models in evaluating the effects of mixtures of toxicants in the aquatic environment. *Chemosphere* 9:533-538.

Christensen, E.R. and C.Y. Chen. 1991. Modeling of combined toxic effects of chemicals. *Toxic Subst. J.* 11:1-63.

Costanza, M.C. and A.A. Afifi. 1979. Comparison of stopping rules in forward stepwise discrimination analysis. *J. Am. Stat. Assoc.*, 74:777-785

Enslein, K., A. Ralston, and H.S. Wilf. 1977. *Statistical Methods for Digital Computers.* Wiley, New York.

Enslein, K., T.M. Tuzzeo, B.W. Blake, J.B. Hart, and W.G. Landis. 1989. Prediction of *Daphnia magna* EC$_{50}$ values from rat oral LD$_{50}$ and structural parameters. In *Aquatic Toxicology and Environmental Fate:* 11th Volume. ASTM STP 1007. G.W. Suter and M.A. Lewis, Eds., American Society for Testing and Materials, Philadelphia, PA, pp 397-409.

Friberg, L., M. Piscator, G.F. Nordberg, and T. Kjellstrom. 1974. *Cadmium in the Environment*, 2nd ed. CRC Press, Cleveland.

Furst, A. 1960. Chelation and cancer: A speculative review. In *Metal Binding in Medicine.* M.J. Seven, Ed., J.B. Lippincott Co., Philadelphia. pp. 344

Gray, H.L. and W.R. Schucany. 1972. *The Generalized Jackknife Statistic.* Marcel Dekker, New York.

Grimes, H.D., K.K. Perkins, and W.F. Boss. 1983. Ozone degrades into hydroxyl radical under physiological conditions. *Plant Physiol.* 72:1016-1020.

Heath, R.L. 1980. Initial events in injury to plants by air pollutants. *Annu. Rev. Plant Physiol.* 31:395-432.

Herbert, D.W.M. and J.M. VanDyke. 1964. The toxicity to fish of mixtures of poisons. *Ann. Appl. Biol.* 53:415.421.

Hewlett, P.S. and R.L. Plackett. 1959. A unified theory for quantal responses to mixtures of drugs: Noninteractive action. *Biometrics* 591-610.

Kier, L.B. and L.H. Hall. 1986. *Molecular Connectivity in Structure-Activity Analysis.* Wiley, New York.

Konemann, H. 1981. Fish toxicity test with mixtures of more than two chemicals: A proposal for a quantitative approach and experimental results. *Toxicology* 19:229-238.

Marking, L.L. and V.K. Dawson. 1975. Method for assessment of toxicity or efficacy of mixtures of chemicals. *U.S. Fish. Wildl. Serv. Invest. Fish Control* 647:1-8.

Marking, L.L. and W.L. Mauck. 1975. Toxicity of paired mixtures of candidate forest insecticides to rainbow trout. *Bull. Environ. Contam. Toxicol.* 13:518-523.

Marking, L.L. 1977. Method for assessing additive toxicity of chemical mixtures. In *Aquatic Toxicology and Hazard Evaluation,* F.L. Mayer and J.L. Hamelink, Eds., American Society for Testing and Materials, Philadelphia, PA, pp. 99-108.

Marking, L.L. 1985. Toxicity of chemical mixtures. In *Fundamentals of Aquatic Toxicology,* G.M. Rand and S.R. Petrocelli, Eds., Hemisphere Publishers, New York, pp. 164-176.

Marquardt, D.W. and R.D. Snee. 1975. Ridge regression in practice. *Am. Stat.* 29:3-20.

Mueller, P.K. and M.J. Hitchcock. 1969. Air quality criteria — toxicological appraisal for oxidants, nitrogen oxides, and hydrocarbons. *Air Pollut. Contr. Ass.* 19:670-676.

Stokinger, H.E., J.T. Mountain, and J.R. Dixon. 1966. Newer Toxicologic Methodology. Effect on industrial hygiene activity. *Arch. Environ. Health* 13:296-305.

Yoshida, Y., K. Kono, M. Watanabe, and H. Watanabe. 1991. Metal shift in rats exposed to fluoride. *Environ. Sci.* 1:1- 9.

STUDY QUESTIONS

1. Which is most critical to plant health when an atmospheric pollutant is introduced: ambient concentration or pollutant concentration within the leaf?
2. Describe the route by which photosynthesis and energy metabolism of a plant cell are impaired beginning with the pollutant passing through the stomata of the epidermal tissue.
3. List six routes by which a pollutant may enter an animal. What is the most common means of entry into the body system for a toxicant?
4. What is the most important chemical property factor affecting absorption of a pollutant?
5. What role does the liver play in affecting a pollutant which has entered an animal?
6. What is the most permanent method of removing toxic substances from the body?
7. Describe the four principal mechanisms by which environmental pollutants exert toxicity.
8. How can pollutants inactivate an enzyme system?
9. Name three examples of secondary action resulting from pollutant presence.
10. What is metal shift?
11. What are the three sites of action within the membrane in narcosis?
12. The toxicity of an organophosphate is related to what chemistry?
13. Organophosphate acute toxicity is usually attributed to the ability to bind to what enzyme?
14. What is "aging" of a protein by an organophosphate?
15. Give an example of another binding site of organophosphates in an organism.
16. What are monohaloacetic acids? Describe the mode of action of fluoroacetic acid, iodoacetic and bromoacetic acids, and chloroacetic acid.
17. What is QSAR?
18. What are the two general types of toxicity data? How are they modeled?
19. Describe the three sets of traditional models for toxicity using regression and discriminate analysis.
20. Describe the developmental process for the construction of a structure-activity model. What is the importance of the reliability of the database?
21. What is MOLSTAC™ and how is it used?
22. Describe the statistical processes of the QSAR model.
23. Explain cross-validation of the QSAR model.
24. Describe the TOPKAT system.
25. What are two examples of problems that may be encountered when a compound or molecule is tested for description and representation in the map of the chemical-toxicity space represented by the regression equation.

CHAPTER 6

Factors Modifying the Activity of Toxicants

Just as there are a large number of toxicants in our environment, so are there many factors that affect the toxicity of these substances. The major factors affecting the toxicity of environmental chemicals include physicochemical properties of the chemicals, exposure time, environmental factors, interaction, biological factors, and nutritional factors. The parameters that modify the toxic action of toxicants are examined in this chapter.

PHYSICOCHEMICAL PROPERTIES OF TOXICANTS

Characteristics such as whether a pollutant is solid, liquid, or gas, or whether the pollutant is soluble in water or in lipid, whether organic or inorganic material, whether ionized or nonionized, etc. can affect the ultimate toxicity of the pollutant. For example, since membranes are more permeable to a nonionized than an ionized substance, a nonionized substance will generally have a higher toxicity than an ionized substance.

One of the most important factors affecting pollutant toxicity is the concentration of the pollutant in question. Even a generally highly toxic substance may not be very injurious to a living organism if its concentrations remain very low. On the other hand, a common pollutant such as carbon monoxide can become extremely danger-ous if its concentrations in the environment are high. As mentioned earlier, exposure to high levels of pollutants often results in acute effects, while exposure to low concentrations may cause chronic effects. Once a pollutant gains entry into a living organism and reaches a target site, it may exhibit an action. The effect of the pollutant, then, is a function of its concentration at the locus of its action. For this reason, any factors capable of modifying internal concentration of the chemical agent can alter the toxicity.

TIME AND MODE OF EXPOSURE

Exposure time is another important determinant of toxic effects. Normally, one can expect that for the same pollutant the longer the exposure time, the more detrimental the effects. Also, continuous exposure is more injurious than intermittent exposure, with other factors being the same. For example, continuous exposure of rats to ozone for a sufficient period of time may induce pulmonary edema. But when the animals were exposed to ozone at the same concentration intermittently, no pulmonary edema may be observed. The mode of exposure, i.e., continuous or intermittent, is important in influencing pollutant toxicity because living organisms often can recover homeostatic balance during an 'off' phase of intermittent exposure than if they are exposed to the same level of toxicant continuously. In addition, organisms may be able to develop tolerance at low continuous doses, as well as after an intermittent dose.

ENVIRONMENTAL FACTORS

Environmental factors such as temperature, light, and humidity also influence the toxicity of pollutants.

TEMPERATURE

Numerous effects of temperature changes on living organisms have been reported in the literature (Krenkel and Parker 1969). Thermal pollution has been a concern in many industries, particularly with power plants. Thermal pollution is the release of effluent that is at a higher temperature than the body of water it is released in. Vast amounts of water are used for cooling purposes by steam-electric power plants. Cooling water is often discharged at an elevated temperature, causing river water temperatures to be raised to such an extent that the water may be incompatible for fish life.

Temperature changes in a volume of water affect the amount of dissolved oxygen (DO). The amount of DO present at saturation in water decreases with increasing temperature. On the other hand, the rate at which most chemical reactions occur increases with increased temperatures. Many enzymes have a peak temperature range. Above and below that range they are much more inefficient at catalyzing reactions. An elevated temperature leads to faster assimilation of waste and therefore faster depletion of oxygen. This oxygen depletion also adversely affects the ability of fish and other animals to survive in these heated waters. Additionally, subtle behavioral changes in fish may result from temperature changes too small to cause injury or death.

Temperature also affects the response of vegetation to air pollution. Generally, plant sensitivity to oxidants increases with increasing temperature up to 30°C. Soybeans are more sensitive to ozone when grown at 28 than at 20°C, regardless of exposure temperature, or ozone doses (Dunning et al. 1974). The response of pinto bean to a 20 and 28 °C growth temperature was found to be dependent on both exposure temperature and ozone dose.

HUMIDITY

Generally, the sensitivity of plants to air pollutants increases as relative humidity increases. However, the relative humidity differential may have to be greater than 20% before differences are shown.

LIGHT INTENSITY

The effect of light intensity on plant response to air pollutants is difficult to generalize because of several variables involved. For example, light intensity during growth affects the sensitivity of pinto bean and tobacco to a subsequent ozone exposure. Sensitivity increased with decreasing light intensities within the range of 900 to 4000 foot-candles (fc) (Dunning and Heck 1973). In contrast, the sensitivity of pinto bean to PAN (peroxyacyl nitrate), a gaseous pollutant, increased with increasing light intensity. PAN will be discussed in more detail in Chapter 7. Plants exposed to pollutants in the dark are generally not sensitive. At low light intensities, plant response is closely correlated with stomatal opening. However, since full stomatal opening occurs at about 1000 fc, light intensity must have an effect on plant response beyond its effect on stomatal opening.

INTERACTION OF POLLUTANTS

Seldom are living organisms exposed to a single pollutant. Instead, they are exposed to combinations of pollutants simultaneously. In addition, the effect of pollutants is dependent on many factors including portals of entry, action mode, metabolism, and others previously described. Exposure to combinations of pollutants may lead to manifestation of effects different from those that would be expected from exposure to each pollutant separately. The combined effects may be synergistic, potentiative, or antagonistic, depending on the chemicals and the physiological condition of the organism involved.

SYNERGISM AND POTENTIATION

These terms have been variously used and defined but, nevertheless, refer to toxicity greater than would be expected from the toxicities of the compounds administered separately. It is generally considered that, in the case of potentiation, one compound has little or no intrinsic toxicity when administered alone, while in the case of synergism both compounds have appreciable toxicity when administered alone. For example, smoking and exposure to air pollution may have synergistic

Table 6.1   Synergistlc Effect of Ozone and Sulfur
Dioxide on Tobacco Bel W3 Plants

| Duration, h | Toxicants, ppm | | Leaf damage, % |
	O_3	SO_2	
2	0.03	—	0
2	—	0.24	0
2	0.031	0.24	38

effects, resulting in increased lung cancer incidence. The presence of particulate matter such as sodium chloride (NaCl) and sulfur dioxide (SO_2), or SO_2 and sulfuric acid mist simultaneously, would have potentiative or synergistic effects on animals.

Similarly, exposure of plants to both O_3 and SO_2 simultaneously is more injurious than exposure to either of these gases alone. For example, laboratory work indicated that a single 2- or 4-h exposure to O_3 at 0.03 ppm and to SO_2 at 0.24 ppm did not injure tobacco plants. Exposure for 2 h to a mixture of 0.031 ppm of O_3 and 0.24 ppm of SO_2, however, produced moderate (38%) injury to the older leaves of Tobacco Bel W3 (Menser and Heggestad 1966) (Table 6.1).

ANTAGONISM

Antagonism may be defined as that situation in which the toxicity of two or more compounds present or administered together, or sequentially, is less than would be expected when administered separately. Antagonism may be due to chemical or physical characteristics of the xenobiotics, or it may be due to the biological actions of the chemicals involved. For example, the highly toxic metal cadmium (Cd) is known to induce anemia and nephrogenic hypertension as well as teratogenesis in animals. Zinc (Zn) and selenium (Se) act to antagonize the action of Cd (Hill 1988). Antagonism may include cases where the lowered toxicity is caused by inhibition or induction of detoxifying enzymes.

Physical means of antagonism can also exist. For example, oil mists have been shown to decrease the toxic effects of O_3 and NO_2 or certain hydrocarbons in experimental mice. This may be due to the oil dissolving the gas and holding it in solution, or the oil containing neutralizing antioxidants.

TOXICITY OF MIXTURES

Evaluating the toxicity of chemical mixtures is an arduous task and direct measurement through toxicity testing is the best method for making these determinations. However, the ability to predict toxicity by investigating the individual components and predicting the type of interaction and response to be encountered is tantamount. These mathematical models are used in combination with toxicity testing to predict the toxicity of mixtures (Brown 1968; Calamari and Marchetti 1973; Calamari and Alabaster 1980; Herbert and VanDyke 1964; Marking and Dawson 1975; Marking and Mauck 1975).

Elaborate mathematical models have been used extensively in pharmacology to determine quantal responses of joint actions of drugs (Ashford and Cobby 1974; Hewlett and Plackett 1959). Calculations are based on knowing the "site of dosage", "site of action", and the "physiological system", which are well documented in the pharmacological literature. Additionally, numerous models exist for predicting mixture toxicity but require prior knowledge of pairwise interactions for the mixture (Christensen and Chen 1991). Such an extensive database does not exist for most organisms used in environmental toxicity testing, precluding the use of these models.

Simpler models exist for evaluating environmental toxicity resulting from chemical mixtures. Using these models, toxic effects of chemical mixtures are determined by evaluating the toxicity of individual components. These include the Toxic Units, Additive (Marking 1977), and Multiple Toxicity Indices (Konemann 1981). These models, working in combination, will be most useful for the amount of data that is available for determining toxicity of hazardous waste site soil to standard test organisms.

The most basic model is the Toxic Unit model which involves determining the toxic strength of an individual compound, expressed as a "toxic unit". The toxicity of the mixture is determined by summing the strengths of the individual compounds (Herbert and VanDyke 1964) using the following model:

$$= \frac{P_S}{P_{T_{50}}} + \frac{Q_S}{Q_{T_{50}}} \tag{6.1}$$

where S represents the actual concentration of the chemical in solution and T_{50} represents the lethal threshold concentration. If the number is greater than 1.0, less than 50% of the exposed population will survive; if it is less than 1.0, greater than 50% will survive. A toxic unit of 1.0 = incipient LC_{50} (Marking 1985).

Building on this simple model, Marking and Dawson (1975) devised a more refined system to determine toxicity based on the formula:

$$\frac{A_m}{A_i} + \frac{B_m}{B_i} = S \tag{6.2}$$

where A and B are chemicals, i and m are the toxicities (LC_{50}s) of A and B individually and in a mixture, and S is the sum of activity. If the sum of toxicity is additive, S = 1; sums that are less than 1.0 indicate greater than additive toxicity, and sums greater that 1.0 indicate less than additive toxicity. However, values greater than 1.0 are not linear with values less than 1.0.

To improve this system and establish linearity Marking and Dawson (1975) developed a system in which the index represents additive, greater than additive, and less than additive effects by zero, positive, and negative values, respectively. Linearity was established by using the reciprocal of the values of S that were less than 1.0, and a zero reference point was achieved by subtracting 1.0 (the expected sum for simple additive toxicity) from the reciprocal $[(1/S) - 1]$. In this way greater than

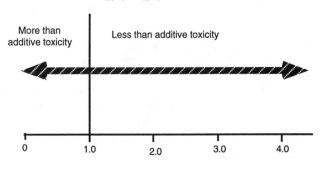

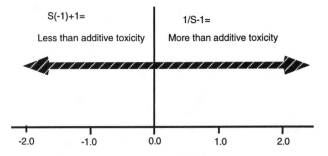

Figure 6.1 Graphical representation of the sum of toxic contributions. In the top part of the figure the sum of toxic contributions is counterintuitive, the more than additive toxicity has a ratio of less than one and the proportions are nonlinear. With the corrections in the corrected sum of toxic contributions the less than additive toxicity is less than one with the more than additive toxicity greater than one.

additive toxicity is represented by index values greater than 1.0. Index values representing less than additive toxicity were obtained by multiplying the value of S that were greater than 1.0 by −1 to make them negative, and a zero reference point was determined by adding 1.0 to this negative value [S(−1)+1]. Therefore, less than additive toxicity is represented by negative index values (Figure 6.1). A summary of this procedure is as follows:

$$\frac{A_m}{A_i} + \frac{B_m}{B_i} = S \text{, the sum of biological effects} \tag{6.3}$$

$$\text{Additive Index} = 1/S - 1.0 \text{ for } S \leq 1.0 \text{ and} \tag{6.4}$$

$$\text{Additive index} = S\,(-1) + 1.0 \text{ for } S \geq 1.0 \tag{6.5}$$

Although the toxic units and additive index are useful in determining toxicity, in some cases they have disadvantages. Their values depend on the relative proportion of chemicals in the mixture. Also, because of the logarithmic form of the concentration in log-linear transformations such as Probit and Logit, it is desirable to have a toxicity index that is logarithmic in the toxicant concentration. For these reasons H. Konemann (1981) introduced a Multiple Toxicity Index (MTI):

$$MTI = 1 - \frac{\log M}{\log m_o} \qquad (6.6)$$

where $m_o = M/f_{max}$; f_{max} = largest value of z_i/Z_i in the mixture; z_i = concentration of toxicant i in the mixture; Z_i = concentration of toxicant i, acting singly, giving the desired response (endpoint); $M = \sum_{i=1}^{n} z_i/Z_i$ = sum of toxic units giving the desired response; and n = number of chemicals in the mixture.

When the concentration z_i of each chemical relative to its effective concentration Z_i, when acting alone, is a constant f for all chemicals, $f = z_i/Z_i$, the above equation reduces to:

$$MTI = 1 - \frac{\log M}{\log n} \qquad (6.7)$$

Even the simplest model requires prior knowledge of the LC_{50} for each compound acting singly. The Additive Toxicity and Multiple Toxicity Indices require an LC_{50} for the specific mixture as well as the singular compounds. Therefore, access to a large database or the ability to estimate toxicity will be extremely important. Of these two methods the corrected sum of toxic contributions derived by Marking and Dawson appears to be the easiest to implement and to interpret.

MIXTURE ESTIMATION SYSTEM

The usefulness of these equations is (1) in the estimation of the toxicity of a mixture and (2) the setting of priorities for cleanup by establishing the major contributor to the toxicity of the mixture. The major disadvantage to the implementation is that these equations are not set up for easy use and the lack of environmental toxicity data. A combination of implementation of the selected methodology into a computer program coupled to a large database and quantitative structure activity relationships estimation system should make these evaluations of mixture toxicity efficient and useful. The components of such a system might be

- The front end for data input, namely the available toxicity data for the components, CAS numbers for the compounds with an unknown toxicity, and the toxicity of the mixture if known. Concentrations of each material are also input.
- A system for searching the appropriate databases for toxicity data or SAR models for estimating the desired parameter. The QSAR system should provide adequate warnings for the appropriateness of the model and its coverage in the database from which the equation was derived.

• A processor that incorporates the data from the literature and the QSAR along with the concentration of the compounds. An estimate of the toxicity of the mixture or identification of the major contributors will be the generated output.

The difficulty in estimating the toxicity of mixtures using any of these models is the difficulty of establishing interaction terms. All of the models require actual toxicity tests to estimate these terms. Even in a simple mixture of four components this requires six toxicity tests of the pairwise combinations and four three-component tests to examine interactive terms. Perhaps the best that could be done in the short term is to establish interaction terms between classes of compounds and use those as models.

Initially, it would be desirable to use a simple model incorporating a linear relationship. Since the data are lacking for the determination of interactive effects, a simple additive toxic units model would make the fewest assumptions and require the minimal amount of data. Such a model would simply consist of

$$A_c/A_i + B_i/B_t + C_i/C_t + \ldots = MT \tag{6.8}$$

where A_c = environmental concentration of compound A, A_i = concentration resulting in the endpoint selected, for example a EC_{50} or LC_{10}, and MT is the mixture toxicity as a fraction with 1 equal to the mixture having the effect as the endpoint selected.

It is certainly possible to make these estimations routine given the uncertainties in the interaction terms and the lack of toxicity data. Properly designed, such a system should allow the rapid and routine estimation of mixtures within the limitations presented above.

BIOLOGICAL FACTORS AFFECTING TOXICITY

PLANTS

In plants, the most widely studied and probably the most important factor affecting response to air pollutants is genetic variation. Plant response varies between species of a given genus and between varieties within a given species. Such variation is a function of genetic variability as it influences morphological, physiological, and biochemical characteristics of plants. Gladiolus has long been recognized to be extremely sensitive to fluoride. Varietal differences in fluoride response in gladiolus have also been observed. Plants show differences in their susceptibility to different pollutants. For instance, some plants may be sensitive to O_3 but relatively resistant to SO_2, while in others the opposite may be true. The sensitivity of plants to atmospheric pollutants such as O_3 is known to be related, in part, to stomatal opening and closure.

Leaf maturity also affects the sensitivity of plants to air pollutants. Generally, young tissues are more sensitive to PAN and hydrogen sulfide, and maturing leaves are most sensitive to the other airborne pollutants.

ANIMALS

Genetic, developmental, health status, sex variation, and behavior are among the important factors affecting the response of animals and humans to pollutant toxicity (Hodgson 1980).

Genetic Factors

Similar to the plants previously discussed, different species of animals respond differently to a given dose of a chemical or an environmental pollutant. In experimental animals, species variation as well as variation in strains within the same species occurs. In humans, such factors as serum, red blood cell, and immunological disorders, and genetically induced malabsorption can contribute to differences in their response to environmental stresses. Individuals with malabsorption syndrome, for example, may suffer nutritional deficiencies, which in turn may lead to an increased susceptibility to environmental chemicals.

Developmental Factors

Immature immune system, aging, pregnancy, and immature detoxication systems are included in this category. For example, lack of γ-globulin to cope with invading bacteria and viruses, decline in renal function as a result of aging, lack of receptors needed in hormonal action, greater stresses encountered by pregnant women to metabolize and detoxify foreign chemicals not only for themselves but for the fetus, and immature hepatic MFO system in the young, are all contributing factors to varying responses exhibited by the individuals to xenobiotics.

Diseases

Diseases in the heart, lungs, kidney, and liver predispose a person to more severe consequences following the exposure to pollutants. As previously shown, organs such as these are responsible for storage, metabolism, and excretion of environmental pollutants. Diseases in any of these organs would lead to impaired functioning and decreased ability to cope with xenobiotics. For instance, cardiovascular and respiratory diseases of other origins decrease the individual's ability to withstand superimposed stresses. An impaired renal function will certainly affect the kidney's ability to excrete toxic substances or their metabolites. As mentioned earlier, the liver plays a vital role in detoxication of foreign chemicals, in addition to its role in the metabolism of different nutrients. Liver disorders, therefore, will seriously impair detoxication processes.

Lifestyle

Smoking, drinking, and drug habits are some examples of lifestyle that can affect human response to environmental chemicals. Research has shown that smoking acts synergistically with many environmental pollutants. A smoker may thus be at a higher risk than a nonsmoker when exposed to an additional environmental stress. For example, asbestos workers or uranium miners who smoke have been shown to exhibit higher lung cancer death rates than workers who do not smoke.

Sex Variation

The rate of metabolism of foreign compounds varies with the difference in sex of both humans and animals. For example, response to chloroform ($CHCl_3$) exposure by experimental mice shows a distinct sex variation. Male mice are highly sensitive to $CHCl_3$; death often results following their exposure to this chemical. The higher sensitivity of male mice to certain toxic chemicals may be due to their inability to metabolize the chemicals as efficiently as the female mice. Interestingly, death rates of male mice resulting from exposure to $CHCl_3$ is affected by different strains as well (Table 6.2).

NUTRITIONAL FACTORS

The importance of nutrition as a major factor affecting the toxicity of chemicals has been recognized in recent years. Results obtained from human epidemiological and animal experimental studies strongly support the relationship between nutrition and pollutant toxicity. For example, laboratory animals fed low-protein diets have been reported to be more susceptible to the toxicity of chemicals under test. The interaction between nutrition and environmental pollutants is complex, and understanding its nature is a great challenge in the study of both toxicology and nutritional biochemistry. It may be mentioned that a new area of study called **nutritional toxicology** has emerged in the recent years.

The relationship between nutrition and toxicology falls into three major categories: (1) the effect of nutritional status on the toxicity of drugs and environmental chemicals; (2) the additional nutritional demands that result from exposure to drugs and environmental chemicals; and (3) the presence of toxic substances in foods (Parke and Loannides 1981).

Generally, nutritional modulation can alter rates of absorption of environmental chemicals, thus affecting circulating level of those chemicals. It can cause changes in body composition, leading to altered tissue distribution of chemicals. Dietary factors can also influence renal function and pH of body fluids, resulting in altered toxicity of chemicals. In addition, responsiveness of the target organ may be modified as a result of changing nutrition.

Fasting/Starvation

This is the most severe form of nutritional modulation. The effect of fasting or starvation, generally, is decreased metabolism and clearance of chemicals, resulting in increased toxic effects. Studies showed that the effect of fasting on microsomal

Table 6.2   Effect of CHCl$_3$ Exposure on Death
             Rate of Various Strains of Male Mice

Strains	Death rate (%)
DBA-2	75
DBA-1	51
CsH	32
BLAC	10

oxidase activity is species-, substrate-, and sex-dependent, i.e., some reactions are decreased in male rats and increased in females, while others may not be affected at all. Animal studies also showed that glucuronide conjugation was decreased under starvation.

Proteins

Many different chemical compounds induce the MFO in the liver and other tissues. Induction of the MFO is associated with increased biosynthesis of new protein. The most potent inducers are substrates whose rates of metabolism are low, so that they remain associated with the enzyme for long periods of time. In humans, severely limited protein intake is usually accompanied by inadequate intake of all other nutrients; thus, it is difficult to designate specific pathological conditions to protein deficiency per se. Protein deficiency causes impaired hepatic function and hypoproteinemia, resulting in decreased hepatic proteins, DNA, and microsomal cytochrome P-450, as well as lowered plasma binding of xenobiotics. Conjugation is also influenced, but the effect is less consistent. Removal of toxicants from the body may be impaired, leading to an increased toxicity, although exceptions do exist.

The effect of proteins on pollutant toxicity includes both quantitative and qualitative aspects. Experiments show that animals fed proteins of low biological value exhibited a lowered microsomal oxidase activity; when dietary proteins were supplemented with tryptophan, the enzyme activity was enhanced. Alteration of xenobiotic metabolism by protein deprivation may lead to enhanced or decreased toxicity, depending on whether metabolites are more or less toxic than the parent compound. For example, rats fed a protein-deficient diet show decreased metabolism but increased mortality with respect to pentobarbital, parathion, malathion, DDT, and toxaphene (Table 6.3). On the other hand, rats treated under the same conditions may show a decreased mortality with respect to heptachlor, CCl_4, and aflatoxin. It is known that in the liver heptachlor is metabolized to the epoxide, which is more toxic than heptachlor itself, while CCl_4 is metabolized to CCl_3^-, a highly reactive free radical. As for aflatoxin, the decreased mortality is due to reduced binding of its metabolites to DNA.

Carbohydrates

A high-carbohydrate diet usually leads to a slower rate of detoxication. The microsomal oxidation is generally depressed when the carbohydrate/protein ratio is

Table 6.3 Effect of Protein on Pesticide Toxicity

Compounds	Casein content of diet	
	3.5%	26%
	LD_{50}, mg	
Parathion	4.86	37.1
Diazinon	215	466
Malathion	759	1401
Carbaryl	89	575
DDT	45	481
Chlordane	137	217
Toxaphene	80	293
Endrin	6.69	16.6
Diuron	437	2390
Captan	480	12,600

Note: Male rats fed for 28 days from weaning on diets of varying casein contents, then given an oral dose of pesticides.

increased. In addition, the nature of carbohydrates also affects oxidase activity. Since dietary carbohydrates influence body lipid composition, the relationship between carbohydrate nutrition and toxicity is often difficult to assess. However, environmental chemicals can affect, and be affected by, body glucose homeostasis in several different ways. For example, poisoning by chemicals may deactivate hepatic glucose 6-phosphatase by damaging the membrane environment of the enzyme. Compounds that are metabolized by the liver to glucuronyl conjugates are more hepatotoxic to fasted animals than fed animals. Low hepatic glycogen contents may also lead to a greater vulnerability of fasted animals to xenobiotics such as acetaminophen, whose metabolism is associated with depletion of the glutathione (GSH) component of the hepatic antioxidant defense system.

Lipids

Dietary lipids may affect the toxicity of environmental chemicals by delaying or enhancing their absorption. The absorption of lipophobic substances would be delayed and that of lipophilic substances accelerated.

The endoplasmic reticulum contains high amounts of lipids, especially phospholipids, rich in polyunsaturated fatty acids. Lipids may influence the detoxication process by affecting the cytochrome P-450 system because phosphatidylcholine is an essential component of the hepatic microsomal MFO system (Parke and Loannides 1981). A high-fat diet may favor more oxidation to occur, as it may contribute to more incorporation of membrane material.

Types of lipids can also affect toxicant metabolism, as a high proportion of phospholipids is unsaturated due to the presence of linoleic acid (18:2) in the β-position of triacylglycerol. Dietary 18:2 is important in determining the normal levels of hepatic cytochrome P-450 concentration and the rate of oxidative demethylation in rat liver.

Significant as it is, higher doses of linoleic acid decrease hepatic cytochrome P-450 and MFO activity (Hietanen et al. 1978), and unsaturated fatty acids added to rat and rabbit liver microsomes *in vitro* inhibit MFO activity with Type I substrates (e.g., *p*-nitroanisole), probably because the fatty acids act as competitive substrates (Di Augustinem and Foutsm 1969).

Dietary lipids play a unique role in the toxicity of chlorinated hydrocarbon pesticides. Dietary lipids may favor more absorption of these pesticides, but once these chemicals are absorbed into the body, they may be stored in the adipose tissue without manifestation of toxicity. For this reason, obesity in humans is considered protective against chronic toxicity of these chemicals. Similarly, the body fat in a well-fed animal is known to store organochlorine pesticides. Fat mammals, fish, and birds are thus more resistant to DDT poisoning than their thinner counterparts. In times of food deprivation, however, stored organic materials such as DDT and PCB can be mobilized from fat deposits, and reach concentrations potentially toxic to the animal.

The role of dietary lipids in affecting pollutant toxicity has been fairly well defined for a few specific chemicals including lead, fluoride, and hydrocarbon carcinogens. For example, high-fat diets are known to increase Pb absorption and retention. Competitive absorption of Pb and Ca exists and this is probably due to competition for the Ca binding protein (CaBP) whose synthesis is mediated by vitamin D, a fat-soluble vitamin. In earlier studies, a high-fat diet was shown to result in an increased body burden of fluoride, leading to enhanced toxicity. This is attributed to delaying of gastric emptying caused by high dietary fat. As a consequence, enhanced fluoride absorption may result, and thus increased body burden of fluoride. Dietary fat does not increase metabolic toxicity of fluoride itself. Aflatoxin, a toxin produced by the mold *Aspergillus flavus*, is a potent liver cancer-causing agent. A high-fat diet offers protection from lethal effects of the toxin, presumably through dissolution of the carcinogen.

Vitamin A

Interest in vitamin A and its synthetic analogs as potential factors in the prevention and treatment of certain types of cancer has been growing. In addition, there is evidence that vitamin A may be related to pollutant toxicity. Recent epidemiological studies in humans with a sample of 8000 men in Chicago showed a low lung cancer incidence in those with a high vitamin A level in the diet, while the incidence was higher in those people with a low dietary vitamin A. Experimental studies show that rats exposed to PCB, DDT, and dieldrin caused a marked reduction in liver vitamin A store, suggesting that the metabolism of vitamin A may be affected by exposure to these organochlorines. In another study, rats deficient in retinol were shown to have a lowered liver cytochrome P-450 activity. The effect of vitamin A deficiency on detoxication, however, depends on several factors such as substrate, tissue, and animal species.

It should be pointed out that intakes of high amounts of vitamin A can lead to serious toxicity, whereas β-carotene, a precursor of vitamin A, is relatively nontoxic. Epidemiological studies have suggested that β-carotene helps in cancer prevention, and β-carotene may exert this effect independent of its role as a precursor of vitamin A. There is a growing interest in the antioxidant functions of carotenoids. Like vitamins E and C discussed later, carotenoids react with free radicals and singlet molecular oxygen ($^1O_2$). It is important that antioxidant functions are associated with lowering DNA damage, malignant transformation, and other cell damages (Sies et al. 1992).

Vitamin D

The role that vitamin D plays in the prevention of rickets and osteomalacia is widely known. Studies have shown the mechanism involved in the conversion of vitamin D into its metabolically active form is responsible for the maintenance of calcium homeostasis. Cholecalciferol (vitamin D) is first hydroxylated in the liver to 25-hydroxy-D_3. This is then converted in the kidney to 1,25-dihydroxy-D_3, the "hormone-like" substance that is the active form of the vitamin. The 25-hydroxylation of cholecalciferol requires NADPH, O_2, and an enzyme whose properties are similar to those of microsomal MFO (I. Bjorkhelm et al. 1979). In addition, 25-hydroxy-D_3 has been shown to competitively inhibit some cytochrome P-450 reactions *in vitro*.

Vitamin E

Vitamin E (α-tocopherol), as a potent membrane-protecting antioxidant, protects against toxicants causing membrane damages through peroxidation. Male rats supplemented with daily doses of 100 mg tocopheryl acetate and exposed to 1.0 ppm O_3 have been shown to survive longer than vitamin E-deficient rats. The action of O_3 is attributed in part at least to free radical formation. In addition, there is sufficient evidence that vitamin E protects phospholipids of microsomal and mitochondrial membranes from peroxidative damage by reacting with free radicals. Because lipid peroxidation is associated with decrease in oxidase activities, it is expected that the enzyme activity is affected by dietary vitamin E. Maximum activity has been observed when diets included both polyunsaturated fatty acids and vitamin E.

Nitrosamine is known to be carcinogenic; it leads to liver cancer. Relationships between vitamin E and nitrosamines are attributed to the inhibitory effect of the vitamin on nitrosamine formation, i.e., vitamin E competes for nitrite, a reactant in the formation of nitrosamine.

Vitamin C

Vitamin C (ascorbic acid) is found in varying amounts in almost all of our body tissues. In particular, high contents are found in adrenal and pituitary glands, eye lens,

Table 6.4   Ascorbic Acid Content of Adult Human Tissues

Tissue	Ascorbic acid (mg/100 g wet tissue)
Pituitary glands	40–50
Leucocytes	35
Adrenal glands	30–40
Eye lens	25–31
Brain	13–15
Liver	10–16
Spleen	10–15
Pancreas	10–15
Kidneys	5–15
Heart muscle	5–15
Lungs	7
Skeletal muscle	3–4
Testes	3
Thyroid	2
Plasma	0.4–1.0
Saliva	0.07–0.09

and various soft tissues (Table 6.4). Ascorbic acid is a potent antioxidant and participates in a large number of cellular oxidation-reduction reactions. While vitamin E is membrane protective, vitamin C acts preferentially in the cytoplasm. Thus, vitamin C protects against superoxide formation in the cytosol. Its relationship to drug metabolism as well as pollutant toxicity has attracted attention in the recent years. For example, vitamin C-deficient guinea pigs have been shown to have an overall deficiency in drug oxidation, with marked decreases in N- and O-demethylations, and in the contents of cytochrome P-450 and cytochrome P-450 reductase (Parke and Loannides 1981). Administration of ascorbate to the deficient animals for 6 days reversed these losses of MFO activity. The effect of vitamin C appears to be tissue-dependent (Kuenzig et al. 1977).

Recent research suggests that vitamin C may reduce the carcinogenic potential of some chemicals. It has been demonstrated that a variety of experimental tumors of the gastrointestinal tract, liver, lung, and bladder can be produced by nitroso compounds (Narisawa et al. 1976; Mirvish et al. 1975), which are produced by the reaction of nitrite with secondary and tertiary amines, amides, or others:

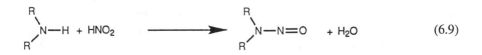

$$\begin{array}{c} R \\ \diagdown \\ N\!\!-\!\!H \ + \ HNO_2 \\ \diagup \\ R \end{array} \longrightarrow \begin{array}{c} R \\ \diagdown \\ N\!\!-\!\!N\!\!=\!\!O \\ \diagup \\ R \end{array} + \ H_2O \qquad (6.9)$$

The nitrosation of several secondary and tertiary amines can be blocked *in vitro* by the addition of vitamin C. The vitamin appears to compete for the nitrite, thus inhibiting nitrosation. It has been demonstrated that vitamin C does not react with amines, nor does it enhance the rate of nitrosamine decomposition. However, it reacts very rapidly with nitrite and nitrous acid. It has been suggested that the vitamin decreases the available nitrite by reducing nitrous acid to nitrogen oxides, leading to inhibition of the nitrosation reaction:

$$2\ HNO_2 + Ascorbate \rightarrow Dehydroascorbate + 2\ NO + 2\ H_2O \qquad (6.10)$$

Although little or no evidence is available that a similar effect occurs in humans, it has been suggested that, in view of our increasing exposure to various drugs and xenobiotics, the current RDA (Recommended Dietary Allowances) for ascorbic acid may be inadequate (Zannoni 1977). For instance, the average American is thought to ingest approximately 70 µg Cd/day, 0.9 mg As/day, 4.1 mg nitrite/day, in addition to exposure to ambient air containing CO, O_3, Pb, cigarette smoke, and others (Calabrese 1980). Recommendations for increasing the RDA for vitamin C to meet such additional needs, however, has not received general support. Moreover, it is known that a dietary excess of vitamin C can produce various adverse effects, based on a nutritional and clinical point of view.

Minerals

About 20 mineral elements are considered to be essential in human nutrition, and seven of these, including calcium, phosphorus, sodium, potassium, magnesium, sulfur, and chlorine are called macrominerals, while the rest are often referred to as trace elements. Mineral nutrition influences toxicology in different ways. Interactions are common concerning the effects of the trace elements on detoxication. It is recognized that trace mineral elements, like the macrominerals, can influence absorption of xenobiotics. Divalent cations can compete for chelation sites in intestinal contents as well as for binding sites on transport proteins. As is well known, competitive absorption of Pb and Ca occurs, and this is probably due to competition for binding sites on intestinal mucosal proteins mediated by vitamin D.

Zinc is known to provide protection against Cd and Pb toxicities (Sandstead 1980). Absorption of Zn is facilitated by complexing with picolinic acid, a metabolite of the amino acid tryptophan. Although both Cd and Pb form complexes with picolinic acid, the resulting complexes are less stable than the Zn complex.

Cytochrome P-450 requires iron for its biosynthesis; thus, deficiency of Fe might lead to decrease in MFO activity. It has been shown that the villous cells of rat duodenal mucosa rapidly lose their cytochrome P-450 content and MFO activity when dietary Fe is deficient (Hoensch et al. 1975). Selenium is antagonistic to both Cd and Hg thus reducing their toxicity. In addition, Se enhances vitamin E function in the prevention of lipid peroxidation. The mechanisms involved in the functioning of these two trace nutrients are different, however. Whereas vitamin E is thought to function as a membrane-bound antioxidant, acting as free radical scavenger, Se participates at the active site of glutathione peroxidase and thus as part of the enzyme. This enzyme protects membrane lipids by catalyzing the destruction of H_2O_2 and organic hydroperoxides before they can cause membrane disruption.

As previously mentioned, increasing evidence suggests that oxygen radicals play a major role in the pathophysiology of many diseases including cancer. Reference was also made that antioxidants play a vital role in counteracting these radicals. Like other antioxidants such as vitamins C and E and β-carotene, selenium is considered by many to exhibit anticarcinogenic action. It should be noted that only relatively small amounts of selenium are needed to meet known physiological functions, and, like other essential nutrients, selenium is toxic when consumed in excess.

REFERENCES AND SUGGESTED READINGS

Ashford, J.R. and J.M. Cobby. 1974. A system of models for the action of drugs applied singly or jointly to biological organisms. *Biometrics* 30:11-31.

Bjorkhelm, I., I. Holmberg, and K. Wikvall. 1979. 25-Hydroxylation of vitamin D_3 by a reconstituted system from rat liver microsomes. *Biochem. Biophys. Res. Commun.* 90:615-622.

Brown, V.M. 1968. The calculation of the acute toxicity of mixtures of poisons to rainbow trout. *Wat. Res.* 2:723-733.

Calabrese, E.J. 1980. *Nutrition and Environmental Health.* Vol. 1. John Wiley & Sons, Inc. New York, NY, pp. 452-455.

Calamari, D. and J.S. Alabaster. 1980. An approach to theoretical models in evaluating the effects of mixtures of toxicants in the aquatic environment. *Chemosphere* 9:533-538.

Calamari, D. and R. Marchetti. 1973. The toxicity of mixtures of metals and surfactants to rainbow trout (*Salmo gairdneri* Rich.). *Wat. Res.* 7:1453-1464.

Christensen, E.R. and C.Y. Chen. 1991. Modeling of combined toxic effects of chemicals. *Toxic Subst. J.* 11:1-63.

Di Augustinem, R.P. and J.R. Foutsm. 1969. The effects of unsaturated fatty acids on hepatic microsomal drug metabolism and cytochrome P-450. *Biochem. J.* 115:547-554.

Dunning, J.A. and W.W. Heck. 1973. Response of pinto bean and tobacco to ozone as conditioned by light intensity and/or humidity. *Environ. Sci. Technol.* 7:824-826.

Dunning, J.A., W.W. Heck, and D.T. Tingey. 1974. Foliar sensitivity of pinto bean and soybean to ozone as affected by temperature, potassium nutrition, and ozone dose. *Water Air Soil Pollut.* 3:305-313.

Herbert, D.W.M. and J.M. VanDyke. 1964. The toxicity to fish of mixtures of poisons. *Ann. Appl. Biol.* 53:415-421.

Hewlett, P.S. and R.L. Plackett. 1959. A unified theory for quantal responses to mixtures of drugs: Non-interactive action. *Biometrics* 591-610.

Hietanen, E., O. Hanninen, M. Laitinen, and M. Lang. 1978. Regulation of hepatic drug metabolism by elaidic and linoleic acids in rats. *Enzyme* 23:127-134.

Hill, C.H. Interactions among trace elements. 1988. In *Essential and Toxic Trace Elements in Human Health and Disease.* A. S. Prasad, Ed., Alan R. Liss, Inc. New York, pp. 491-500.

Hodgson, E. 1980. Chemical and environmental factors affecting metabolism of xenobiotics. In *Introduction to Biochemical Toxicology.* E. Hodgson and F.E. Guthrie, Eds., Elsevier, New York, pp. 143-161.

Hoensch, H., C.H. Woo, and R. Schmid. 1975. Cytochrome P-450 and drug metabolism in intestinal villous and crypt cells of rats: Effect of dietary iron. *Biochem. Biophys. Res. Commun.* 65:399-406.

Konemann, H. 1981. Fish toxicity test with mixtures of more than two chemicals: A proposal for a quantitative approach and experimental results. *Toxicology* 19:229-238.

Krenkel, P.A. and F.L. Parker, Eds. 1969. *Biological Aspects of Thermal Pollution.* Vanderbilt University Press, Nashville, TN.

Kuenzig, W., V. Tkaxzevski, J.J. Kamm, A.H. Conney, and J.J. Burns. 1977. The effect of ascorbic acid deficiency on extrahepatic microsomal metabolism of drugs and carcinogens in the guinea pig. *J. Pharmacol. Exp. Ther.* 201:527-533.

Marking, L.L. and V.K. Dawson. 1975. Method for assessment of toxicity or efficacy of mixtures of chemicals. *U.S. Fish. Wildl. Serv. Invest. Fish Control* 647:1-8.

Marking, L.L. and W.L. Mauck. 1975. Toxicity of paired mixtures of candidate forest insecticides to rainbow trout. *Bull. Environ. Contam. Toxicol.* 13:518-523.

Marking, L.L. 1977. Method for assessing additive toxicity of chemical mixtures. In *Aquatic Toxicology and Hazard Evaluation,* F.L. Mayer and J.L. Hamelink, Eds., American Society for Testing and Materials, Philadelphia, PA, pp. 99-108.

Marking, L.L. 1985. Toxicity of chemical mixtures. In *Fundamentals of Aquatic Toxicology,* G.M. Rand and S.R. Petrocelli, Eds., Hemisphere Publishing Co., New York, pp. 164-176.

Menser, H.A. and H.E. Heggestad. 1966. Ozone and sulfur dioxide synergism. Injury to tobacco plants. *Science* 153:424-425.

Mirvish, S.S., A. Cardesa, L. Wallcave, and P. Shubik. 1975. Induction of mouse lung adenomas by amines or ureas plus nitrite and by N-nitroso compounds: effect of ascorbate, gallis acid, thio cyanate, and caffeine. *J. Natl. Cancer Inst.* 55:633-636.

Narisawa, T., C.Q. Wong, R.R. Moronpot, and J.H. Weisburger. 1976. Large bowel carcinogenesis in mice and rats by several intrarectal doses of methylnitrosourea and negative effect of nitrite plus methylurea. *Cancer Res.* 36:505-510.

Parke, D.V. and C. Loannides. 1981. The role of nutrition in toxicology. *Annu. Rev. Nutr.* 1:207-234.

Sandstead, H.H. 1980. Interactions of toxic elements with essential elements: Introduction. *Ann. N.Y. Acad. Sci.* 355:282-284.

Sies, H., W. Stahl, and A.R. Sundquist. 1992. Antioxidant functions of vitamins. Vitamins E and C, beta-carotene, and other carotenoids. *Ann. N.Y. Acad. Sci.* 669:7-20.

Zannoni, V.G. 1977. Ascorbic acid and liver microsomal metabolism. *Acta Vitaminol. Enzymol.* 31:17-29.

STUDY QUESTIONS

1. Which substance will have a higher toxicity — ionized or nonionized? Why?
2. Exposure to high levels of pollutants results in effects, low concentrations result in effects.
3. Describe why intermittent exposure to a pollutant may not be as detrimental as continuous exposure.
4. Name two effects temperature changes (thermal pollution) have on living organisms.
5. How can humidity levels and light intensity affect pollutants' effects?
6. Describe synergistic, potentiative, and antagonistic effects resulting from the interaction of pollutants.
7. Describe the toxic unit model.
8. How is a value for additive toxicity found?

9. What is the multiple toxicity index? What are the component parts of the equation used to calculate the index?
10. What are the two uses of the toxicity equations?
11. What are plants' most important factor affecting response to air pollutants? What is another factor for plant sensitivity?
12. Name five important factors affecting the response of animals to pollutant toxicity.
13. What effects can nutritional modulation have on response to pollutant toxicity?
14. What effect does a high-carbohydrate diet have on detoxification? What effect do dietary lipids have?
15. What are several possibilities of mechanisms involved in vitamin A action in relation to carcinogenesis?
16. Discuss the relationships of vitamin E and vitamin C with nitrosamine.

CHAPTER 7

Inorganic Gaseous Pollutants

In this section five of the major gaseous air pollutants are considered, i.e., sulfur oxides (SOx), nitrogen oxides (NOx), ozone (O_3), carbon monoxide (CO), and fluoride (F). Although much of the fluoride emitted into the atmosphere from various sources is in the particulate form, fluoride is included in our discussion here with other inorganic gaseous pollutants mainly because gaseous fluoride causes most damages to living organisms, especially plants.

SULFUR OXIDES

Sulfur oxides include both sulfur dioxide (SO_2) and sulfur trioxide (SO_3), of which SO_2 is more important as an air pollutant. Sulfur trioxide may be formed in the furnace by reaction between sulfur and O_2 or SO_2 and O_2. Sulfur dioxide is probably the most dangerous of all gaseous pollutants on the basis of amounts emitted.

SOURCES OF SO₂

Sulfur oxide emission results from the combustion of sulfur-containing fossil fuels, particularly coal and oil. The sulfur content of coal ranges from 0.3 to 7% and the sulfur is in both organic and inorganic forms, while in oil sulfur content ranges from 0.2 to 1.7% and its sulfur is in organic form. The most important sulfur compound in coal is iron disulfide (FeS_2) or pyrite. When heated at high temperatures, pyrite undergoes the following reactions:

$$FeS_2 + 3\ O_2 \rightarrow FeSO_4 + SO_2 \tag{7.1}$$

$$4\ FeS_2 + 11\ O_2 \rightarrow 2\ Fe_2O_3 + 8\ SO_2 \tag{7.2}$$

In the smelting process, sulfide ores of copper, lead, and zinc are oxidized (roasted) to convert a sulfide compound into an oxide. For example, zinc sulfide (ZnS) undergoes the oxidation process in a smelter, forming ZnO and SO_2:

$$2 \text{ ZnS} + 3 \text{ O}_2 \rightarrow 2 \text{ ZnO} + 2 \text{ SO}_2 \qquad (7.3)$$

In the U.S. SO_2 emission from stationary sources and industry accounts for about 95% of all SO_2 emission.

CHARACTERISTICS OF SO_2

SO_2 is highly soluble in water, with a solubility of 11.3 g/100 ml. Once emitted into the atmosphere, SO_2 may undergo oxidation in the gaseous phase, forming H_2SO_4 aerosol. Gaseous SO_2 may also become dissolved in water droplets and, following oxidation, form H_2SO_4 aerosol droplets. Both forms of H_2SO_4 thus produced may be removed by deposition to the earth's surface (Figure 7.1).

Recent studies have shown that the photochemistry of the free hydroxyl radical controls the rate at which many trace gases, including SO_2, are oxidized and removed

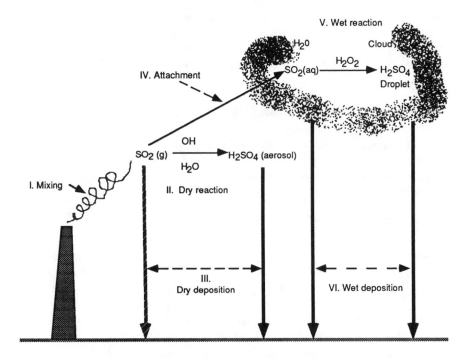

Figure 7.1 SO_2 transport, transformation, and deposition processes. Initially SO_2 is mixed into the atmosphere (I). Gaseous SO_2 may undergo oxidation with subsequent formation of H_2SO_4 aerosol (II). Both gaseous SO_2 and H_2SO_4 aerosol may be deposited at the earth's surface (III). Gaseous SO_2 may become dissolved in a water droplet (IV). The dissolved SO_2 can be oxidized in solution to form H_2SO_4 aerosol droplets (V). The H_2SO_4 aerosol and the H_2SO_4 droplet may be removed to the earth's surface by wet deposition (VI). (From Fox, D.L. 1986. The transformation of pollutants. In: *Air Pollution*, 3rd ed. Vol. VI, A.C. Stern, Ed., Academic Press, New York. pp. 86-87. With permission.)

from the atmosphere. The photochemistry involving the OH radical is illustrated in Figure 7.2.

EFFECTS ON PLANTS

For SO_2, the stomatal pores are the main entry ports to the internal air spaces of plant leaves. Absorption takes place mainly by gaseous diffusion through these pores. The number of stomata and size of aperture are important factors affecting the uptake of SO_2. Other factors such as light, humidity, wind velocity, and temperature are involved as well, as these influence the turgidity of guard cells. Low concentrations of SO_2 can injure epidermal and guard cells, leading to increased stomatal conductance and greater entry of SO_2 into the plant. Following the uptake by plant leaves, SO_2 is rapidly translocated through the plant and affects photosynthesis, transpiration, and respiration, the three major functions of plant leaves. A slight increase in both net photosynthesis and transpiration may occur at low SO_2 concentrations for short time periods, followed by a decrease in both processes. Higher SO_2 concentrations induce immediate decreases in these processes. Plant injuries may be manifested by leaf chlorosis and spotty necrotic lesions. Damage to mesophyll cells is commonly observed in microscopic studies.

Once within the substomatal air spaces of the leaf, SO_2 comes into contact with cell walls of the mesophyll cells. Sulfur dioxide readily dissolves in the intercellular water to form sulfite (SO_3^{2-}), bisulfite (HSO_3^-), and other ionic species. Both SO_3^{2-} and HSO_3^- have been shown to be phytotoxic, as they affect many biochemical and physiological processes (Malhotra and Hocking 1976). Both SO_3^{2-} and HSO_3^- have a lone pair of electrons on the sulfur atom that strongly favor reactions with electron-deficient sites in other molecules. The phytotoxicity of SO_3^{2-} and HSO_3^- can be overcome by conversion of these species to less toxic forms such as SO_4^{2-}. Oxidation of HSO_3^- to the less toxic sulfate can occur by both enzymatic and nonenzymatic reactions. Several factors, including cellular enzymes such as peroxidase and cytochrome oxidase, metals, ultraviolet light, and O_2^-, stimulate the oxidation of SO_2. In the presence of SO_3^{2-} and HSO_3^-, more O_2^- is formed by free-radical chain oxidation. Other free radicals can be formed as well. These oxidizing radicals can have detrimental effects on the cell.

Plant metabolism is affected by SO_2 in a variety of ways, e.g., stimulation of phosphorus metabolism (Plesnicar 1983) and reduction in foliar chlorophyll concentration (Lauenroth and Dodd 1981). Carbohydrate concentrations were increased by low levels of SO_2 and decreased by higher levels (Koziol and Jordon 1978). Effects of SO_2 on enzyme systems have been investigated in many studies. The enzymes studied include alanine and aspartate aminotransferases, glutamate dehydrogenase, malate dehydrogenase, glycolate oxidase, glyceraldehyde-3-phosphate dehydrogenase, glucose-6-phosphate dehydrogenase, fructose-1,6-bisphosphatase, and ribulose-5-phosphate kinase. Enzyme activity may be increased or decreased by exposure to SO_2 at different concentrations. As mentioned previously, there are differences in sensitivity of plant species to SO_2 under similar biophysical conditions. This suggests

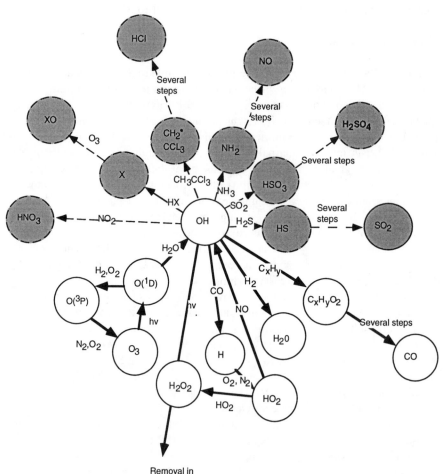

Figure 7.2 Photochemistry of the OH radical controls the trace gas concentration. The photochemistry of the free hydroxyl radical controls the rate at which many trace gases
are oxidized and removed from the atmosphere. Processes that are of primary
importance in controlling the concentration of OH in the troposphere are indicated
by solid lines in the schematic diagram; those that have a negligible effect on OH
levels but are important because they control the concentrations of associated
reactions and products are indicated by broken lines. Circles indicate reservoirs of
species in the atmosphere; arrows indicate reactions that convert one species to
another, with the reactant or photon needed for each reaction indicated along each
arrow. Multistep reactions actually consist of two or more sequential elementary
reactions. HX = HCl, HBr, HI, or HF. C_xH_y denotes hydrocarbons. (From Chameides
and Davis, *Chem. Eng. News, 60 (40):38-52,* 1982. American Chemical Society.
With permission.)

that delicate biochemical and physiological differences operating in different plant species could affect the sensitivity of a particular plant to SO_2.

EFFECT ON ANIMALS

Although SO_2 is an irritating gas for the eyes and upper respiratory tract, no major injury from exposure to any reasonable concentrations of this gas has been demonstrated in experimental animals. Even exposure to pure gaseous SO_2 at concentrations 50 or more times the ambient values produced little distress (Alarie et al. 1970; Alarie et al. 1973). Concentrations of 100 or more times are required to kill small animals. Mortality is associated with lung congestion and hemorrhage, pulmonary edema, thickening of the interalveolar septa, and other relatively nonspecific changes of the lungs. For example, mice exposed to 10 ppm SO_2 for 72 h showed necrosis and sloughing of the nasal epithelium (Giddens and Fairchild 1972). The lesions were more severe in animals with preexisting infection. Other symptoms include decreased weight gains, loss of hair, nephrosis in kidneys, myocardial degeneration, and accelerated aging.

Many studies have demonstrated increase in the response of animals to SO_2 in the presence of particulate matter and elevations of relative humidity. Thus, H_2SO_4 mist and some particulate sulfates enhance the reactions of animals to SO_2, suggesting that alteration of SO_2 to a higher oxidation state may increase its irritability in animals. These interactions have important implications in air pollution control, as the rate of conversion of SO_2 to acid sulfates may have greater health significance than the concentration of SO_2 in the air.

EFFECT ON HUMANS

Sulfur dioxide is rapidly absorbed in the nasopharynx of humans. Humans exposed to 5 ppm of the gas showed increased respiratory frequency and decreased tidal volume. Similar to observations made with animals, human exposure to SO_2 alters the mode of respiration, as manifested by increased frequency, decreased tidal volume, and lowered respiratory and expiration flow rates. Synergism and elevated airway resistance with SO_2 and aerosols of water and saline have been demonstrated.

It was previously thought that SO_2 and black suspended particulate matter interacted and that both had to be elevated in order to exhibit health effects. New findings and analyses have changed such perceptions concerning the health effects of this group of pollutants. Emitted SO_2 is generally thought to be oxidized slowly by atmospheric oxygen to SO_3, which readily combines with water to form H_2SO_4. Ultimately the aerosol reacts with atmospheric particles or surfaces to form sulfates. The World Health Organization recommended that the air quality standards reflect the joint presence of SO_2 and the resulting acid sulfates. Recent experimental and epidemiological data do not provide evidence for a specific effect of sulfate aerosol. However, airway reactivity is variable among subjects. Individuals with airway hyperactivity, e.g., asthmatics, have been shown to exhibit increased pulmonary flow resistance when exposed to SO_2 by mouthpiece, while the increase was less with nasal breathing (Frank et al. 1962). Exercise augments responses to the pollutants.

NITROGEN OXIDES

FORMS AND FORMATION OF NITROGEN OXIDES

Six forms of nitrogen oxides are present, i.e., nitrous oxide (N_2O), nitric oxide (NO), nitrogen dioxide (NO_2), nitrogen trioxide (N_2O_3), nitrogen tetroxide (N_2O_4), and nitrogen pentoxide (N_2O_5). Of these, NO_2 is the major toxicant because of its relatively high toxicity and its ubiquity in ambient air, while N_2O, N_2O_3, and N_2O_4 have low relative toxicities and air pollution significance. Basic chemical reactions involved in the formation of NO_2 are shown below:

$$\overset{1210°C}{N_2 + O_2 \rightarrow 2\ NO} \tag{7.4}$$

$$2\ NO + O_2 \rightarrow 2\ NO_2 \tag{7.5}$$

The NO formed in Equation 7.4 persists when temperature is rapidly lowered, as is the case in ambient air. The reaction shown in Equation 7.5 is one of the few reactions that are slowed down with increase in temperature.

Major Reactive N Species in the Troposphere

Several reactive N species including NO, NO_2, and HNO_3 occur in the troposphere. Among these species, NO_2 is of particular environmental concern because it plays a complex and important role in the production of photochemical oxidants and acidic deposition. NO_2 is a unique air pollutant in that it absorbs UV light energy, whereby it is decomposed and forms NO and atomic oxygen. The energetic oxygen atom reacts with molecular oxygen to form O_3. The ozone then reacts with NO to form molecular oxygen and NO_2, thus terminating the photolytic cycle of NO_2 (Figure 7.3). It is clear that, as far as the cycle is concerned, there is no net loss or gain of chemical substances.

However, for reasons to be described in the next section, in actuality O_3 accumulates. Several other reactions also occur, resulting in the production of photochemical smog. In addition to NO and NO_2, nitric acid is also an important N compound in the troposphere. It is formed mainly from NO_2 and OH radicals. Nitric acid is also formed through a secondary reactive pathway, whereby NO_2 is first oxidized to NO_3 by O_3. The NO_3 then reacts with a molecule of NO_2, forming N_2O_5. The N_2O_5 thus formed combines with a molecule of water, yielding HNO_3. The resultant HNO_3 may be precipitated through rain or dry deposition. These reactions and others are shown in Figure 7.4.

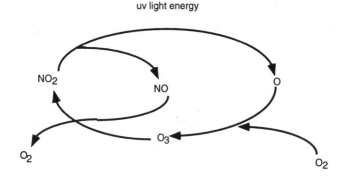

Figure 7.3 The photolytic cycle of NO_2.

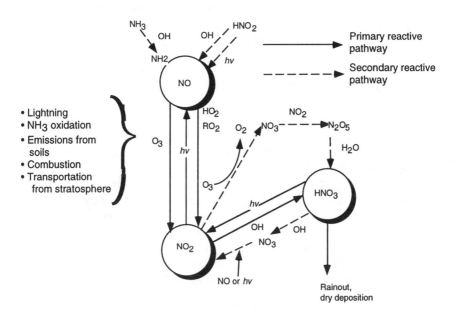

Figure 7.4 Major reactive nitrogen species in the troposphere. From Chameides and Davis, *Chem. Eng. News, 60 (40):38-52*, 1982. American Chemical Society. With permission.)

EFFECTS ON PLANTS

Plants absorb gaseous NOx through stomata. NO_2 is more rapidly absorbed than NO, mainly because NO_2 reacts rapidly with water, while NO is almost insoluble. The absorbed NO_2 is then converted to NO_3^- and NO_2^- before being utilized in plant metabolism. The NO_2 injury to plants may be due to either acidification or a photooxidation process (Zeevart 1976). Symptoms exhibited by plants exposed to NO_2 are similar to those from SO_2, but much higher concentrations are required to cause acute injury. However, decreased photosynthesis has been demonstrated even at concentrations that do not produce visible injury. The combined effect of NO and NO_2 gases appears to be additive.

Photosynthetic inhibition caused by NOx may be due to competition for NADPH between the processes of nitrite reduction and carbon assimilation in chloroplasts. NO_2 has been shown to cause swelling of chloroplast membranes (Wellburn et al. 1972). Biochemical and membrane injury may be caused by ammonia produced from NO_3^-, if it is not utilized soon after its formation. Plants can metabolize the dissolved NOx through their NO_2 assimilation pathway:

$$NOx \rightarrow NO_3^- \rightarrow NO_2^- \rightarrow NH_3 \rightarrow amino\ acids \rightarrow proteins$$

Other biochemical pathways affected by NOx include inhibition of lipid biosynthesis, oxidation of unsaturated fatty acids *in vivo*, and stimulation of peroxidase activity. It should be noted that the general sense of the plant scientists is that nitrogen oxides play little to no role, by themselves, in causing damage to plants at current ambient concentrations.

EFFECTS ON HUMANS AND ANIMALS

Studies of the pathological and physiological effects of NO_2 on animals have been conducted at concentrations much higher than those found in ambient air. The toxic action of NO_2 is mainly on the deep lung and peripheral airway. Exposure of various species of animals to 10 to 25 ppm of NO_2 for 24 h resulted in bits of fibrin in the airway, increased number of macrophages, and altered appearance of the cells in the distal airway and adjacent pulmonary alveoli. Terminal bronchioles showed hyperplasia and hypertrophy, loss of cilia, and disturbed ciliagenesis. Large crystaloid depositions also occurred in the cuboidal cells. Continuous exposure for several months produced thickening of the basement membranes, resulting in narrowing and fibrosis of the bronchioles. Emphysema-like alterations of the lungs developed, followed by death of the animals (Freeman and Haydon 1964).

Physiological Effects

NO_2 is rapidly converted to nitrite (N_2^-) and nitrate (N_3^-) ions in the lungs, and these ions are found in the blood and urine shortly after exposure to 24 ppm of NO_2 (Orehek et al. 1976). Increased respiration was shown in some studies. Other physiological alterations include a slowing of weight gain and decreased swimming ability in rats, alteration in blood cellular constituents such as polycythemia, lowered hemoglobin content, thinner erythrocytes, leukocytosis, and depressed phagocytic activity. Methemoglobin formation occurred only at high concentrations. Methemoglobinemia is a disorder manifested by high concentrations of methemoglobin in the blood. Under this condition, the hemoglobin contains Fe^{3+} ions and is thus unable to reversibly combine with molecular oxygen. As mentioned previously, although almost all the studies done were conducted by using much higher concentrations of NO_2 than are found in ambient air, a few papers did deal with low NO_2 concentrations. Orehek et al. (1976) showed that in asthmatic subjects exposure to 0.1 ppm of NO_2 significantly aggravated the hyperreactivity in the airway. While at the prevailing concentrations of NO_2 the health effects are generally considered insignificant, NO_2 pollution may be an important aspect of indoor pollution. Evidence suggests that gas cooking and heating of homes, when not vented, can increase NO_2 pollution, and that such exposures may lead to increased respiratory problems among young children.

Biochemical Effects

Extracts of lung lipids from rats exposed to NO_2 have been shown to involve oxidation. Lipid peroxidation was more severe in animals fed a diet deficient in vitamin E (Roehm et al. 1971). In contrast to ozone, reaction of NO_2 with fatty acids appears to be incomplete and phenolic antioxidants can retard the oxidation from NO_2. Exposure to NO_2 may cause changes in the molecular structure of lung collagen. In a series of papers, Buckley and Balchum (1965, 1967a, 1967b) have demonstrated that exposure for 10 weeks or longer at 10 ppm or for 2 h at 50 ppm increased both tissue oxygen consumption and LDH and aldolase activity. Stimulation of glycolysis has also been reported.

OZONE

SOURCES

Ozone is a natural constituent of the upper atmosphere; trace amounts naturally exist in the lower atmosphere. Formation of O_3 in the upper atmosphere occurs in steps, i.e., a molecule of oxygen being split into atomic oxygen, and the resultant atomic oxygen reacting with another oxygen molecule to form ozone:

$$O_2 \xrightarrow{h\nu} O + O \tag{7.6}$$

$$O + O_2 \rightarrow O_3 \tag{7.7}$$

Ozone in the lower atmosphere is also produced as a result of modern technology. Equipment that produces sparks, arcs, static discharge, and ultraviolet and other ionizing radiation, including commercial applications such as air purifiers and deodorizers in homes, hospitals, and offices and closed environmental systems such as aerospace cabins and submarine chambers are some examples.

By far the most important source of O_3 contributing to environmental pollution is that found in photochemical smog. As shown in the previous section on NOx, disruption of the photolytic cycle of NO_2 (Figure 7.3 and Equations 7.8 through 7.10) by atmospheric hydrocarbons is the principal cause of photochemical smog.

$$\overset{h\nu}{NO_2 \rightarrow NO + O} \qquad (7.8)$$

$$O + O_2 \rightarrow O_3 \qquad (7.9)$$

$$Net: NO_2 + O_2 \rightleftarrows NO + O_3 \qquad (7.10)$$

In the above equations, theoretically, the back reaction proceeds faster than the initial reaction, so that the resulting O_3 should be removed from the atmosphere. But free radicals formed from hydrocarbons and other species present in the urban atmosphere react with and remove NO, thus stopping the back reaction. This leads to O_3 accumulation. Free radicals are noncharged fragments of stable molecules, for example, hydroxy radical, $OH\cdot$, hydroperoxy radical, $HO_2\cdot$, atomic oxygen, O^1D, and higher homologs, $RO\cdot$ and $RO_2\cdot$, where R is a hydrocarbon group. Free radicals participate in chain reactions in the atmosphere. The OH–HO_2 chain is particularly effective in oxidizing hydrocarbons and NO. Some examples illustrating these reactions are shown below:

$$OH\cdot + RH \rightarrow R\cdot + H_2O \qquad (7.11)$$

$$R\cdot + O_2 \rightarrow RO_2\cdot \qquad (7.12)$$

$$RO_2\cdot + NO \rightarrow RO\cdot + NO_2 \qquad (7.13)$$

$$RO\cdot + O_2 \rightarrow R'CHO + HO_2 \qquad (7.14)$$

$$HO_2 + NO \rightarrow NO_2 + OH\cdot \qquad (7.15)$$

It is noticeable that the process starts with an $OH\cdot$ radical. After one pass through the cycle, two molecules of NO are oxidized to NO_2. The $OH\cdot$ radical formed in the last step (Equation 7.15) can start the cycle again. On the other hand, O_3 can also be formed from O_2 reacting with hydrocarbon free radicals:

$$\overset{O_2}{R\cdot + O_2 \rightarrow R\cdot O_2\cdot \rightarrow O_3 + RO\cdot} \qquad (7.16)$$

Photochemical Smog

The hydrocarbon free radicals such as $RO_2\cdot$ in Equation 7.12 can react further with other hydrocarbons or with such species as NO, NO_2, O_2, and O_3. The hydrocarbon free radical ROO· can react with O_2 and NO_2 leading to the production of peroxyacyl nitrate (PAN), one of the main components of photochemical smog.

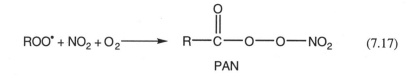

$$ROO^{\bullet} + NO_2 + O_2 \longrightarrow R\text{—}\overset{\overset{\displaystyle O}{\|}}{C}\text{—}O\text{—}O\text{—}NO_2 \qquad (7.17)$$

PAN

Peroxyacyl nitrate can also be formed from a reaction involving $RO_3\cdot$ and NO_2:

$$RO_3\cdot + NO_2 \rightarrow RO_3NO_2 \qquad (7.18)$$

Clearly, a large number of chemical reactions occur in the atmosphere resulting in the formation of many secondary air pollutants. In areas where abundant sunshine is available coupled with unique topographical conditions, as is the case in Los Angeles, accumulation of these pollutants occurs, leading to photochemical smog formation. This is a problem that many large cities in the world are confronted with. Principal components of photochemical smog include O_3 (up to 90%), NOx (mainly NO_2, about 10%), PAN (0.6%), free radical oxygen forms, and other organic compounds such as aldehydes, ketones, and alkyl nitrates.

EFFECTS ON PLANTS

By far, ozone is the most important of all the phytotoxic pollutants. A large volume of literature has been published dealing with both field and experimental studies on the influence of O_3 on higher plants. Highlights of the experimental results include: (1) either an increase or a decrease in plant growth (Blum and Heck 1980); (2) reduction in size, weight, and number of fruits (Henderson and Reinert 1979); (3) reduction in shoot and root growth (Grunwald and Endress 1984; Letchworth and Blum 1977); (4) reduction in seed oil (Grunwald and Endress 1984); (5) reduction in growth ring size (McLaughlin et al. 1982); (6) reduction in net photosynthesis (Blum et al. 1983; Izuta et al. 1991); (7) reduction in unsaturated fatty acids (Perchorozicz and Ting 1974); (8) increase in membrane permeability (Pauls and Thompson 1981); (9) increase in respiration (Dugger and Ting 1970); and (10) altered intermediate metabolism.

The effect of O_3 on plant metabolism is complex, and contradictory results have been reported. However, it is well established that photochemical oxidants such as O_3 and PAN can oxidize SH groups, and such oxidation may be sufficient to cause loss of enzyme activity. For example, several enzymes involved in carbohydrate metabolism, such as phosphoglucomutase and glyceraldehyde-3-phosphate dehydrogenase, have been shown to be inhibited by O_3. The hydrolysis of reserve starch was inhibited by exposure to 0.05 ppm O_3 for 2 to 6 h in cucumber, bean, and monkey flower (Dugger and Ting 1970), suggesting an inhibition of amylase or phosphorylase.

While decrease in glyceraldehyde-3-phosphate dehydrogenase activity suggests inhibition of glycolysis, increase in the activity of glucose-6-phosphate dehydrogenase and 6-phosphogluconate dehydrogenase reported by some workers (Tingey et al. 1975) implies increased activity of the pentose phosphate pathway. Recent studies showed that mung bean seedlings exposed to 250 ppb of O_3 for 2 h caused a marked decrease in invertase activity (Yu 1994). In addition to carbohydrates, lipids are also affected by exposure to O_3. Lipid synthesis, requiring NADPH and ATP, for example, is known to proceed at a lower rate, presumably because O_3 lowers the total energy of the cell.

EFFECTS ON HUMANS AND ANIMALS

Ozone and other oxidants cause respiratory and eye irritation. The TLV (threshold limit value) for O_3 in industry is 0.1 ppm. Exposure to 0.6 to 0.8 ppm O_3 for 60 min resulted in headache, nausea, anorexia, and increased airway resistance. Exposure at 0.7 to 0.9 ppm in experimental animals may predispose or aggravate a response to bacterial infection. Coughing, chest pain, and a sensation of shortness of breath were shown in the exposed subjects who were exercised (Batesm and Hazucha 1973). Morphological and functional changes occur in the lung in experimental animals subjected to prolonged exposure to O_3. Such changes as chronic bronchitis, bronchiolitis, and emphysematous and septal fibrosis in lung tissues have been observed in mice, rabbits, hamsters, and guinea pigs exposed daily to O_3 at concentrations slightly above 1 ppm. Thickening of terminal and respiratory bronchioles was the most noticeable change. In the small pulmonary arteries of rabbits exposed to O_3, for example, the walls were thicker and the lumina were narrower than those of the controls. The mean ratio of wall thickness-to-lumen diameter was 1:4.9 for the control, while that of the exposed animals was 1:1.7 (P'an et al. 1972). Other physiological effects include dryness of upper airway passages, irritation of mucous membranes of nose and throat, bronchial irritation, headache, fatigue, and alterations of visual response.

There is suggestive evidence that O_3 exposure accelerates aging processes. Some investigators suggest that aging is due to irreversible cross-linking between macromolecules, principally proteins and nucleic acids. Animals exposed to 0.1 ppm O_3 may increase their susceptibility to bacterial infections. Exposed mice may have congenital abnormalities and neonatal deaths.

Development of hyperreactivity following O_3 exposure in humans and dogs has been shown. The most characteristic toxic effect of relatively high-level O_3 exposure is pulmonary edema (Mueller and Hitchcock 1969), a leakage of fluid into the gas-exchange parts of the lung. This effect was seen at concentrations only slightly above that observed in community pollution in Los Angeles, CA.

It has long been known that humans as well as animals develop tolerance to O_3. Tolerance refers to increased capacity of an organism that has been preexposed to the oxidant to resist the effects of later exposures to ordinarily lethal (or otherwise injurious) doses of the same agent. Rodents exposed to 0.3 ppm, for example, would become "tolerant" to subsequent exposures of several ppm, which would produce massive pulmonary edema in animals exposed for the first time. Some human subjects exposed to 0.3 ppm at intervals of a day or so showed diminished reactivity with later exposures. This response is designated as **adaptation**.

BIOCHEMICAL EFFECTS

Research on the biochemical effects of O_3 has been extensive. Among the many postulated mechanisms that have been advanced concerning the toxicity of O_3, the following are noted: (1) reactions with proteins and amino acids; (2) reactions with lipids; (3) formation of free radicals; (4) oxidation of sulfhydryl compounds and pyridine nucleotides; (5) influence on various enzymes; and (6) production of more or less nonspecific stress, with the release of histamine.

Ozone interacts with proteins and some amino acids causing alteration. For instance, the lysozyme in tears of individuals exposed to smog has been reported to be 60% less than the normal. Concentrations of protein sulfhydryl and nonprotein sulfhydryl in the lungs of rats exposed to 2 ppm O_3 for 4 to 8 h have been shown to be decreased. Mudd et al. (1969) showed that aqueous solutions of amino acids such as tyrosine, histidine, and tryptophan were oxidized by O_3. Methionine, for example, was oxidized to methionine sulfoxide. A number of investigators have shown that O_3 could cause the oxidation of the SH group, and that addition of SH compounds was protective. The activities of several enzymes have been shown to be either enhanced or depressed in animals exposed to O_3. These include decrease in glucose-6-phosphate dehydrogenase, glutathione reductase, and succinate-cytochrome c reductase in the lungs of rats exposed to 2 ppm O_3 for 4 to 8 h and increase in glucose-6-phosphate dehydrogenase, 6-phosphogluconate dehydrogenase, and isocitrate dehydrogenase.

Balchum et al. (1971) have provided evidence supporting the concept that the peroxidation or ozonization of unsaturated fatty acids in biological membranes is a primary mechanism of the deleterious effects of O_3. The hypothesis is based on the tendency of O_3 to react with the ethylene groups of unsaturated fatty acids, resulting in the formation of free radicals. The free radicals can, in the presence of molecular oxygen, cause peroxidation of unsaturated fatty acids. In support of this hypothesis is the evidence that after O_3 exposure there was a relative decrease in unsaturated fatty acids as compared to saturated fatty acids, and the more unsaturated the fatty acid, the greater the loss. Furthermore, a deficiency of vitamin E increases the toxicity of O_3 for the rat (Goldstein et al. 1970).

Another chemical pathway leading to O_3-dependent unsaturated fatty acid oxidation is through incorporation of O_3 into the fatty acid double bond, resulting in ozonide formation. This process is generally known as ozonolysis:

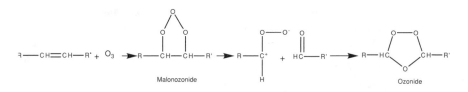

Ozone is also known to oxidize glutathione and pyridine nucleotides NADH and NADPH. The ozonization of NAD(P)H may proceed in the nicotinamide ring as follows:

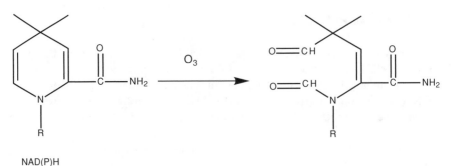

NAD(P)H

Since the intracellular ratios of NADH/NAD$^+$, NADPH/NADP$^+$, and ATP/adenylates are carefully regulated by the cell, loss of the reduced nucleotide can be compensated by faster operation of the Krebs cycle. But, the cell can only make up for a net loss of all nucleotides by an increase in synthesis. The oxidation of NADPH or NADH results in elevated enzyme activity, and this permits the cell to restore the initial ratio of the nucleotides. NADPH oxidation increases the activity of the pentose phosphate pathway. Such an increase also occurs following the oxidation of GSH as shown below. Oxidation of either NADPH or GSH, therefore, may be responsible for the apparent increase in the enzymes found in the pentose phosphate pathway after repeated O_3 exposure.

$$\text{2 GSH + [O]} \xrightarrow{\text{GSH peroxidase}} \text{GSSG + H}_2\text{O} \tag{7.19}$$

$$\text{GSSG + 2 NADPH} \xrightarrow{\text{GSH reductase}} \text{2 GSH + 2 NADP}^+ \tag{7.20}$$

$$\text{Glucose-6-phosphate + NADP}^+ \xrightarrow{\text{G-6-P dehydrogenase}} \text{6-Phosphogluconolactone + NADPH} \tag{7.21}$$

CARBON MONOXIDE

Carbon monoxide is an odorless, colorless, and tasteless gas that is found in high concentrations in urban atmosphere. While it is of no importance to plants, carbon monoxide (CO) is highly important for humans and animals as far as their health is concerned. High concentrations of CO are often found in urban environments.

Historically, early exposures began from fires and then from coal for domestic heating. Combustion associated with developing industry, explosions, fires in mines, and illumination gas prepared from coal have all been sources of exposure. The migration of agricultural populations to cities increased the proportion of the population exposed, as well as the number of persons generating CO.

With the emergence of automobiles propelled by the internal combustion engine, the CO emitted from the exhaust pipe has become the major source for human exposure. Serious problems exist with occupational exposure to increased ambient CO for fire fighters, traffic police, toll-booth attendants, coal miners, coke oven and smelter workers, and transportation mechanics.

FORMATION OF CO

Formation of CO occurs usually through one of the following three processes.

1. Incomplete combustion of carbon or carbon-containing compounds. This occurs when available oxygen is less than the amount required for complete combustion in which carbon dioxide is the product, or when there is poor mixing of fuel and air. In the reactions shown below, the rate of reaction in Equation 7.22 is much greater than that in Equation 7.23.

$$2\ C + O_2 \rightarrow 2\ CO \tag{7.22}$$

$$2\ CO + O_2 \rightarrow 2\ CO_2 \tag{7.23}$$

2. Reactions between CO and carbon-containing materials at high temperature. This occurs at elevated temperature, common in many industrial devices such as blast furnaces.

$$CO_2 + C \rightarrow 2\ CO \tag{7.24}$$

The CO produced in this way is beneficial and used in certain applications, as in the blast furnace where CO acts as a reducing agent in the production of iron from Fe_2O_3 ores as shown below. Some CO may escape into the atmosphere, however.

$$3\ CO + Fe_2O_3 \rightarrow 2\ Fe + 3\ CO_2 \tag{7.25}$$

3. Dissociation of CO_2 at high temperature. Carbon dioxide dissociates into CO and O at high temperature as follows:

$$CO_2 \xrightarrow{\text{high temp.}} CO + O \tag{7.26}$$

High temperatures favor the dissociation of CO_2. For example, at 1745°C the dissociation is 1%, while at 1940°C, it is 5%.

TOXICOLOGICAL EFFECTS

The most widely known physiological effect of CO is interference with O_2 transfer. This interference is brought about by the combination of CO gas with hemoglobin (Hb), leading to the formation of carboxyhemoglobin, HbCO or COHb:

$$Hb + 4\ O_2 \rightarrow Hb(O_2)_4 \tag{7.27}$$

$$Hb + 4\ CO \rightarrow Hb(CO)_4 \tag{7.28}$$

$$Sum:\ Hb(O_2)_4 + 4\ CO \rightarrow Hb(CO)_4 + 4\ O_2 \tag{7.29}$$

The dissociation constant, K, for Equation 7.29 is 210 at 37°C. In other words, CO has a more than 200-times-greater affinity for combination with Hb than O_2 does. A binding site on a Hb molecule cannot be occupied by both CO and O_2. Although increases in oxygen concentrations can shift the equilibrium in Equation 7.29 to the left, recovery of Hb is slow, while the asphyxiating effect of putting Hb out of business is rapid. Normal or background level of blood carboxyhemoglobin is about 0.5%. The CO is derived from both the CO in ambient air and the CO produced by the body during catabolism of heme (a component of Hb). The equilibrium percentage of COHb in the bloodstream of a person continually exposed to an ambient CO concentration of less than 100 ppm can be calculated from the following equation:

Percent COHb in blood = 0.16 x (CO conc. in the air in ppm) + 0.5 (7.30)

Based on COHb levels, various health effects may be expected to occur. Table 7.1 summarizes demonstrated health effects associated with COHb levels.

Carbon monoxide also inhibits function of alveolar macrophages. This inhibition can lead to weakened tissue defenses against airborne bacterial infection. Maternal CO poisoning during pregnancy has been shown to cause fetal death because of lack of O_2 in fetal circulatory system. Carbon monoxide poisoning causing unconsciousness for 30 min to 5 h does not cause permanent damage to the mother but can cause brain damage, mental deficiency, or death to the fetus. Severity of damage is related to the month of pregnancy; the fetus is particularly vulnerable shortly before birth.

The half-life of COHb is 4 h at rest in room air. It is shortened to 60 to 90 min if 100% oxygen is given using a face mask. Since more than 2 h at 100% oxygen can cause pulmonary oxygen toxicity, the oxygen concentration should be reduced to 60% at 2 h.

MECHANISM OF ACTION

As previously mentioned, CO competes with O_2 for binding of hemoglobin, but, in addition, it binds other proteins such as myoglobin, cytochrome c oxidase, and

Table 7.1   COHb Levels and Demonstrated Toxicological Effects

COHb level (%)	Demonstrated effects
< 1.0	No apparent effect
2–4	Impairment of visual function
5–10	Impairment of visual perception, of manual dexterity, of learning, and impairment of performance of certain intellectual tasks
	Increased coronary blood flow
	Impairment in response to certain psychomotor tests
	Decreased night vision and peripheral vision
20–30	Nausea, weakness (particularly in the legs),occasional vomiting
30–35	Clouding of mental alertness occurs with increasing weakness
35–45	Collapse and coma
> 50	Death (in young people)

cytochrome P-450. Carbon monoxide also impairs the facilitated diffusion of O_2 to the mitochondria, shifting the oxyhemoglobin dissociation curve to the left. Alteration of oxyhemoglobin dissociation curve by COHb occurs in such a manner that O_2 is released to tissues with great difficulty and at a lower O_2 tension.

HUMAN EXPOSURE TO CO

Exposure to CO comes mainly from three sources: (1) CO in the surrounding ambient environment particularly from exhaust gases (automobile, industrial machinery), suicidal and accidental intoxication (e.g., house fires, >50,000 ppm), and home environmental problems such as defective furnaces, charcoal burning in poorly vented houses, or garages connected to living quarters, and space heaters in campers; (2) occupational exposure such as experienced by fire fighters (>10,000 ppm CO), traffic police, coal miners, coke oven and smelter workers, toll-booth attendants, and transportation mechanics; and (3) cigarette smoking. Smokers have higher COHb levels than nonsmokers (Table 7.2). Nonsmokers may be subjected to inhalation of CO from cigarette smoke in confined places.

Table 7.2   Blood COHb Levels of Smokers

Category of smokers	Median equilibrium blood COHb (%)
Never smoked	1.3
Ex-smoker	1.4
Pipe and/or cigar smokers only	1.7
Light cigarette smoker	
(<1/2 pack/day; noninhaler)	2.3
(<1/2 pack/day; inhaler)	3.8
Moderate smoker	
(1/2–2 packs/day; inhaler)	5.9
Heavy smoker (> 2 packs/day; inhaler)	6.9

FLUORIDE

ENVIRONMENTAL SOURCES AND FORMS OF FLUORIDE

Fluorine is the lightest element in Group VII of the periodic table, with atomic number 9 and an atomic weight of 18.998. It has a single isotope and its valence in all naturally occurring compounds is 1. Fluoride (F) is ubiquitous: it occurs in minerals and soils, air, natural waters, and foods.

- **Minerals and soils** — The F content of rocks is 0.06 to 0.09% by weight of upper layer of lithosphere. The chief F-containing minerals are: fluorspar (CaF_2), cryolite (Na_3AlF_6), and fluorapatite ($Ca_{10}F_2(PO_4)_6$).
- **Natural waters** — The F content of natural waters in the northeastern part of the U.S. ranges from 0.02 to 0.1 ppm, while it is about 0.2 ppm in the West and Midwest. River waters contain 0.0 to 6.5 ppm, with an average of 0.2 ppm. Groundwaters contain from 0.1 to 8.7 ppm, depending on the rocks from which the waters flow. Seawater contains about 1.2 ppm F.
- **Foods** — Virtually all foods contain trace amounts of fluoride. Fluoride-containing foods and beverages are, therefore, the most important sources of F intake. In the U.S., an adult takes in about 0.2 to 0.3 mg/day from foods, whereas the intake from drinking water ranges from 0.1 to 0.5 mg. In communities where the water supply is fluoridated, F intake may amount to 1 to 2 mg/day. Plants can absorb F from soil, water, or air. Most plants contain 0.1 to 10 ppm F (on dry basis); forage plants generally contain 5 to 10 ppm (dry basis). Fluoride contents in plants vary with plant species. Several species of plants are F accumulators. For instance, tea leaves may contain as high as 760 ppm F, camellia, 620 ppm, and elderberry, 3600 ppm, all on dry basis. Tea beverage, however, may contain less than 0.5 mg F per cup. The F contents of some foods are shown in Table 7.3.
- **Air** — The air in U.S. residential and/or rural communities contains less than 0.04 to 1.2 ppb (0.03 to 0.90 µg F/m^3).
- **Industrial sources of fluoride in the environment** — The major sources of fluoride emissions into the atmosphere from industrial sources include steel industry, phosphate fertilizer industry, ceramics industry (brick, tile, glass, etc.), aluminum industry, combustion of coal (F content of coal: 0.001 to 0.048% in U.S., average 0.008%), nonferrous metal foundries, and welding operations.

FLUORIDE POLLUTION

Fluoride emitted into the atmosphere from volcanism, industries, and other sources include both gaseous and particulate forms. They also contribute F to surface waters. The forms of F emitted from various industrial processes include hydrogen fluoride, cryolite, calcium fluoride, and silicon tetrafluoride (SiF_4). Some heavy discharges of F into the atmosphere and waters have occurred in connection with phosphate and aluminum industries. Combustion of coal in power plants can emit considerable quantities of F into the atmosphere. In addition to deposition into surface waters, airborne F may eventually be deposited onto the ground and taken up by soils, plants, and animals. Figure 7.5 shows environmental transfer of F.

Table 7.3 Fluoride Content of Selected Foods.

Food	Fluoride content (ppm on dry basis)
Meats	0.01–7.7
Fish	0.10–24
Cheese	0.13–1.62
Butter	0.4–1.50
Rice and peas	10
Cereal and cereal products	0.10–0.20
Vegetables and tubers	0.10–2.05
Citrus fruits	0.04–0.36
Sugar	0.10–0.32
Coffee	0.2–1.6
Tea (U.S. brands)	av. 60

Adapted from Committee on Biologic Effects of Atmospheric Pollutants. 1971.

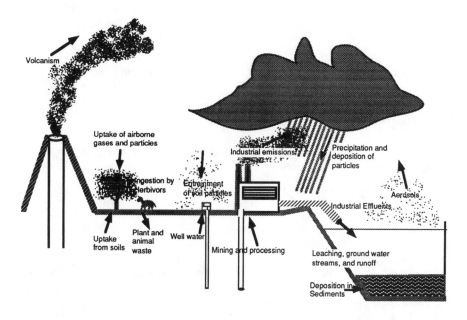

Figure 7.5 Environmental transfer of fluoride and other elemental pollutants.

EFFECTS ON PLANTS

Fluoride is by far the most important of the source-specific pollutants. Fluoride-induced effects in plants may be viewed based on four levels of biologic organization including ecosystem, organism, tissue or organ, and cellular levels. Fluoride ions accumulate in plant leaves mainly as a result of diffusion through the stomata from the atmosphere. Very little fluoride enters the plant system from the soil unless the soil is very acidic. In contrast to other major air pollutants such as SO_2, NOx, O_3, PAN, etc. discussed in the previous sections, F accumulates in the leaf tips and margins of many species as a result of translocation in the transpiration stream. Plants growing near F-emitting sources can accumulate high levels of F in leaves. Gaseous forms of F such as HF and SiF_4 are taken up by leaves much more rapidly than are particulate fluorides. Although plants differ widely in their susceptibility to F injury, accumulation of high levels of F in leaves normally leads to chlorosis or necrosis, which in turn can lead to reduced growth and yield.

Fluoride causes both structural and functional changes adversely in plant cells. These include altered cellular and subcellular membranes with subsequent injuries. Many metabolic processes such as glycolysis, Krebs cycle reactions, photosynthesis, protein synthesis, lipid metabolism, and others have been shown to be affected by fluoride. Much of the action of fluoride on these processes can be attributed to fluoride-dependent inhibition of enzymes. A large number of enzymes have been shown to be affected by fluoride. Some examples include enolase, phosphoglucomutase, phosphatase, hexokinase, PEP carboxylase, pyruvate kinase, succinic dehydrogenase, malic dehydrogenase, pyrophosphatase, phytase, nitrate reductase, mitochondrial ATPase, urease (Miller et al. 1983), lipase (Yu et al. 1987), and amylase (Yu et al. 1988).

Inhibition of enzymes are often manifested by compositional changes in plant tissues, such as altered concentrations of soluble sugars, organic acids, and amino acids (Yang and Miller 1963).

EFFECT ON ANIMALS

Excessive intakes of F can injure domestic animals, leading to acute or chronic effects depending on F concentrations.

- **Acute effects** — Fluoride has caused serious injurious effects on livestock in the U.S. and other industrialized countries. The sources of the pollutant are mostly limited to phosphate-fertilizer manufacturing, and aluminum fluorohydrocarbon, and heavy metal production. Safe levels of soluble F in animal rations range from 30 to 50 mg/kg for cattle and from 70 to 100 mg/kg for sheep and swine. Such physiological effects as gastroenteritis, muscular weakness, pulmonary congestion, nausea, vomiting, diarrhea, chronic convulsions, necrosis of mucosa of the digestive tract, anorexia, cramping, collapse, and respiratory and cardiac failure are observed, leading to eventual death of the animals.
- **Chronic effects** — The two most conspicuous and thoroughly studied manifestations of chronic F-poisoning are dental and skeletal fluorosis. Once absorbed in the animal body, F has a great affinity for developing and mineralizing teeth. Such affinity of fluorides for developing and mineralizing teeth can either enhance tooth development or induce dental lesions, depending on the amounts of fluorides ingested. Dental lesions are manifested by abnormal enamel matrix such as chalkiness, mottling, and hypoplasia (a thin enamel). It is noteworthy that dental lesions will not be seen in animals brought into endemic fluorosis areas after their permanent teeth have erupted (Shupe and Olson 1983). In skeletal fluorosis, the affected bones lose their normal, hard, smooth luster and appear rough, porous, and chalky white.

Table 7.4 Fluoride Tolerances (in ppm) in Livestock Diets

	Breeding or lactating animals	Finishing animals
Dairy and beef heifers	30	100
Dairy cows	30	100
Beef cows	40	100
Sheep	50	160
Horses	60	—
Swine	70	—
Turkeys	100	—
Chicken	150	—

From Committee on Biologic Effects of Atmospheric Pollutants. 1971. *Fluorides*. National Academy of Sciences, Washington, D.C., pp. 295.

Lameness or stiffness is an intermittent sign of fluoride toxicity. The clinical basis for the lameness is not well understood. Appetite is normally impaired and this may result in decreased weight gain, cachexia, and lowered milk yield. The animals may become more susceptible to other environmental stresses and suffer a decrease in longevity.

A number of factors influence the manifestation of dental and skeletal fluorosis. Some examples are the amount and the bioavailability of F ingested, duration of ingestion, species of animals involved (Table 7.4), age at time of excessive F ingestion, nutritional and general health status of animals, mode of F exposure (i.e., continuous or intermittent), presence of synergistic or antagonistic substances, presence of other stress factors, and individual biologic response.

Several nutrients such as proteins, Ca, and vitamin C influence the severity of F toxicity. Adverse effects of F are alleviated by these nutrients. For example, both vitamin C and Ca have been shown to decrease the toxicity in guinea pigs (Hodge and Smith 1965). Laboratory experiments showed that mice fed a low-protein (4%) diet deposited 500% more F in tibia than control animals fed a regular diet containing 27% protein, and that supplemental vitamin C greatly reduced the F deposition in the bone (Yu and Hwang 1986). It should be mentioned that mice produce vitamin C as well.

EFFECTS ON HUMAN HEALTH

- **Daily intake** — Daily intake of F from food is about 0.2 to 0.3 mg, from water 0.1 to 0.5 mg (1 to 2 mg if water is fluoridated), and varying quantities from beverages (F content of wine, 0 to 6.3 ppm; beer, 0.15 to 0.86 ppm; milk, 0.04 to 0.55 ppm). The amount of F inhaled from air is about 0.05 mg/day.
- **Absorption** — Absorption of F from the gastrointestinal tract occurs through a passive process; it does not involve active transport. Absorption is rapid and probably occurs in the lumen. The rate of absorption is dependent on the compounds involved. Even at low levels of F intake, appreciable amounts of F will in time accumulate in calcified tissues. The effectiveness of low levels of F intake in

reducing dental caries in humans, rats, and some other species of animals has been
well recognized. With humans, water supplies containing 1 ppm F have been
widely known to reduce by more than 50% the incidence of dental caries in
individuals who consume F from infancy.

- **Acute intoxication** — The lethal dose of inorganic F has been estimated to be in
  the range of 2.5 to 5 g for a 70 kg man, or approximately 50 mg/kg, a dose similar
  to the LD_{50} for several animal species. The cause of death is probably related to the
  prompt binding of serum Ca and Mg by F. Clinical symptoms include excessive
  salivation, perspiration, nausea, painful spasms of limbs, stiffness, chronic convul-
  sion, necrosis of mucosa of digestive tract, and heart failure.
- **Chronic effects** — Fluoride accumulates in the skeleton during prolonged, high-
  level exposures. Radiological evidence of hypermineralization (osteofluorosis) is
  shown when bone concentrations reach about 5000 ppm F (Hodge and Smith 1965).
  Coupled with other environmental factors such as nutrition and health status, the
  patient may suffer severe skeletal dysfunction. In addition, vomiting and neurologi-
  cal complaints have been reported. Increases of serum and urinary F levels usually
  occur. In some parts of the world, such as India and China where the water supply
  (from wells) in many villages and towns contain high levels of F, osteofluorosis is
  common. In China alone it is estimated that about 20 million people may be
  suffering chronic F poisoning (Yu and Tsunoda 1988).

BIOCHEMICAL EFFECTS OF FLUORIDE

While it is clear that the action of F on living organisms is complex and involves
a variety of enzymes, the mode of action of F ions on these enzymes is not so clear.
Nevertheless, several mechanisms have been suggested for the observed F-induced
inhibition of enzymes. These include: (1) formation of complex with metalloenzymes,
(2) removal of a metal cofactor from an enzyme system, and (3) binding to the free
enzyme or to the enzyme substrate complex (Miller et al. 1983). Using a model
system, Edwards and coworkers (Edwards et al. 1984) indicated that fluoride could
disrupt the hydrogen bonding of protein molecules. Because hydrogen bonding is
important in the maintenance of the tertiary structure of protein molecules, such
disruption would result in enzyme inhibition.

REFERENCES AND SUGGESTED READINGS

Alarie, Y., R.J. Kantz, C.E. Ulrich, A.A. Krumm, and W.M. Busey. 1973. Long-term continu-
    ous exposure to sulfur dioxide and fly ash mixtures. *Environ. Health* 27:251-253.
Alarie, Y., C.E. Ulrich, W.M. Busey, H.E. Swann, Jr., and N.H. MacFarland. 1970. Long-term
    continuous exposure of guinea pigs to sulfur dioxide. *Arch. Environ. Health* 21:769-777.
Balchum, O.J., J.S. O'Brian, and B.D. Goldstein. 1971. Ozone and unsaturated fatty acids.
    Arch. Environ. Health 22:32-34.
Batesm, D.V. and M. Hazucha. 1973. In Proceedings of the Conference on Health Effects of
    Air Pollutants, Ser. 93-15. National Academy of Sciences/National Research Council for
    the U.S. Senate Committee on Public Works., U.S. Govt. Printing Office, Washington,
    D.C.

Blum, U. and W.W. Heck. 1980. Effects of acute ozone exposures on snap bean (*Phaseolus vulgaris* cultivar BBL-290) at various stages of its life cycle. *Environ. Exp. Bot.* 20:73-86.

Blum, U., E. Mrozek, Jr., and E. Johnson. 1983. Investigation of ozone (O_3) effects on ^{14}C distribution in Ladino clover. *Environ. Exp. Bot.* 23:369-378.

Buckley, R.D. and O.J. Balchum. 1965. Acute and chronic exposures to nitrogen dioxide. *Arch. Environ. Health* 10:220-223.

Buckley, R.D. and O.J. Balchum. 1967a. Effects of nitrogen dioxide on lactic dehydrogenase isozymes. *Arch. Environ. Health* 14:424-428.

Buckley, R.D. and O.J. Balchum. 1967b. Enzyme alterations following nitrogen dioxide exposure. *Arch. Environ. Health* 14:687-692.

Committee on Biologic Effects of Atmospheric Pollutants. 1971. *Fluorides*. National Academy of Sciences, Washington, D.C., pp. 295.

Dugger, W.M. and I.P. Ting. 1970. Air pollution oxidants — Their effects on metabolic processes in plants. *Annu. Rev. Plant Physiol.* 21:215-234.

Edwards, S.L., T.L. Poulos, and J. Kraut. 1984. The crystal structure of fluoride-inhibited cytochrome c peroxidase. *J. Biol. Chem.* 259:12984-12988.

Frank, N.R., M.O. Amdur, J. Worcester, and J.L. Whittenberger. 1962. Effects of acute controlled exposure to SO_2 on respiratory mechanics in healthy male adults. *J. Appl. Physiol.* 17:252-258.

Freeman, G. and G.B. Haydon. 1964. Emphysema after low-level exposure to NO_2. *Arch. Environ. Health* 8:125-128.

Giddens, Jr. W.E. and G.A. Fairchild. 1972. Effects of sulfur dioxide on the nasal mucosa of mice. *Arch. Environ. Health* 25:166-173.

Goldstein, B.D., R.D. Buckley, R. Cordinos, and O.J. Balchum. 1970. Ozone and Vitamin E. *Science* 169:605-606.

Grunwald, C. and A.G. Endress. 1984. Fatty acids of soybean seeds harvested from plants exposed to air pollutants. *J. Agric. Chem.* 32:50-53.

Henderson, W.R. and R.A. Reinert. 1979. Yield response of 4 fresh market tomato (*Lycopersicon esculentum*) cultivars after acute ozone exposure in the seedling stage. *J. Am. Soc. Hortic. Sci.* 104:754-759.

Hodge, H.C. and F.A. Smith. 1965. In *Biological Properties of Inorganic Fluorides*. J.H. Simmons, Ed., Vol. IV., Academic Press, New York.

Horvath, S.M., J.A. Gliner, and L.J. Folinsbee. 1981. Adaptation to ozone: duration of effect. *Am. Rev. Respir. Dis.* 123:496-499.

Izuta, T., S. Funada, T. Ohashi, H. Miyake, and T. Totsuka. 1991. Effects of low concentrations of ozone on the growth of radish plants under different light intensities. *Environ. Sci.* 1:21-33.

Koziol, M.J. and C.F. Jordon. 1978. Changes in carbohydrate levels in red kidney bean (*Phasedus vulgaris* L.) exposed to sulfur dioxide. *J. Exp. Bot.* 29:1037-1043.

Lauenroth, W.K. and J.L. Dodd. 1981. Chlorophyll reduction in western wheatgrass (*Agropyron smithii Rydb.*) exposed to sulfur dioxide. *Water Air Soil Pollut.* 15:309-315.

Letchworth, M.B. and U. Blum. 1977. Effects of acute ozone exposure on growth, modulation, and nitrogen content of Ladino clover. *Environ. Pollut.* 14:303.

Malhotra, S.S. and D. Hocking. 1976. Biochemical and cytological effects of sulphur dioxide on plant metabolism. *New Phytol.* 76:227-237.

McLaughlin, S.B., R.K. McCornathy, E. Duvick, and L.K. Mann. 1982. Effects of chronic air pollution stress on photosynthesis, carbon allocation, and growth of white pine trees (*Pinus strobus*). *For. Sci.* 28:60-70.

Miller, G.W., M.H. Yu, and J.C. Pushnik. 1983. Basic metabolic and physiologic effects of fluorides on vegetation. In *Fluorides — Effects on Vegetation, Animals and Humans*. J.L. Shupe, H.B. Peterson, and N.C. Leone, Eds., Paragon Press, Inc., Salt Lake City, UT.

Mudd, J.B., R. Leavitt, A. Ongun, and T.T. McManus. 1969. Reaction of ozone with amino acids and proteins. *Atm. Environ.* 3:669-682.

Mueller, P.K. and M. Hitchcock. 1969. Air quality criteria — toxicological appraisal for oxidants, nitrogen oxides, and hydrocarbons. *J. Air Pollut. Contr. Ass.* 19:670-676.

Orehek, J., J.P. Massari, P. Gaylord, C. Grimaud, and J. Charpin. 1976. Effect of short-term, low-level nitrogen dioxide exposure on bronchial sensitivity of asthmatic patients. *J. Clin. Invest.* 57:301-307.

P'an, A.Y.S., J. Beland, and J. Zygmund. 1972. Ozone-induced arterial lesions. *Arch. Environ. Health* 24:229-232.

Parker, C.M., R.P. Sharma, and J.L. Shupe. 1979. The interaction of dietary vitamin C, protein, and calcium with fluoride: effects in guinea pigs in relation to breaking strength and radio density of bone. *Clin. Toxicol.*, 15:301-311.

Pauls, K.P. and J.E. Thompson. 1981. Effects of *in vitro* treatment with ozone on the physical and chemical properties of membranes. *Physiol. Plant.* 53:255-262.

Perchorozicz, J.T. and I.P. Ting. 1974. Ozone effects on plant cell permeability. *Am. J. Bot.* 61:787-793.

Plesnicar, M. 1983. Study of sulfur dioxide effects on phosphorus metabolism in plants using 32p as indicator. *Int. J. Appl. Radiat. Isot.* 34:833-835.

Roehm, J.N., J.G. Hadley, and D.B. Menzel. 1971. Oxidation of unsaturated fatty acids by ozone and nitrogen dioxide — A common mechanism of action. *Arch. Environ. Health* 23:142-148.

Shupe, J.L. and A.E. Olson. 1983. Clinical and pathological aspects of fluoride toxicosis in animals. In *Fluorides — Effects on Vegetation, Animals and Humans*. J.L. Shupe, H.B. Peterson, and N.C. Leone, Eds., Paragon Press, Inc., Salt Lake City, UT. pp. 319-338.

Tingey, D.T., R.C. Fites, and C. Wickliff. 1975. Activity changes in selected enzymes from soybean leaves following ozone exposure. *Physiol. Plant.* 33:316-320.

Wellburn, A.R., O. Majernik, and F.A.M. Wellburn. 1972. Effects of SO_2 and NO_2 polluted air upon the ultrastructure of chloroplasts. *Environ. Pollut.* 3:37-49.

Yang, S.F. and G.W. Miller. 1963. Biochemical studies on the effects of fluoride on higher plants. *Biochem. J.* 88:505-509.

Yu, M.H. 1994. Personal communication.

Yu, M.H. and S.H.L. Hwang. 1986. Influence of proteins and ascorbic acid on fluoride-induced changes in blood composition and skeletal fluoride deposition in mice. In *Fluoride Research 1985*. H. Tsunoda, and M. H. Yu, Eds., Elsevier, Amsterdam. pp. 203.

Yu, M.H. and H. Tsunoda. 1988. Environmental fluoride problems in China. *Fluoride*, 21:163-166.

Yu, M.H., M.O. Shumway, and A. Brockbank. 1988. Effects of NaF on amylase in mung bean seedlings. *J. Fluorine Chem.* 41:95-100.

Yu, M.H., R. Young, and L. Sepanski. 1987. Inhibition of lipid metabolism in germinating mung bean seeds by fluoride. *Fluoride* 20:113-117.

Zeevaart, A.J. 1976. Some effects of fumigating plants for short periods with NO_2. *Environ. Pollut.* 11:97-108.

STUDY QUESTIONS

1. What is the most dangerous gaseous pollutant and why?
2. How does SO_2 affect a plant's structure and function? What affects SO2 uptake by a plant? How is plant metabolism affected?
3. At what levels does SO_2 affect experimental animals? What does it affect?
4. What condition of SO_2 might have a greater health significance than the air concentration of SO_2?
5. What effect does SO_2 have on humans?
6. Which form of nitrogen oxide is the major toxicant and why?
7. How do gaseous NO_2 affect plants?
8. How does NO_2 affect animals?
9. What is the most important source of O_3 which contributes to environmental pollution? What causes this source?
10. How do photochemical oxidants affect plant enzyme activity and lipid synthesis?
11. Describe the effects oxidants have on humans and animals.
12. What is adaptation to O_3?
13. Discuss the five mechanisms postulated for O_3 toxicity?
14. How does CO formation occur? What is an important physiological effect of CO?
15. Describe fluoride-induced effects in plants based on the four levels of biologic organization.
16. What are the principal mechanisms suggested as the mode of action of fluoride ion on plant metabolic enzymes?
17. What are the acute, chronic, and biochemical effects of fluoride on animals?
18. What are the chronic effects of fluoride accumulation in humans?

CHAPTER 8

Biotransformation, Detoxification, and Biodegradation

INTRODUCTION

As mentioned in Chapter 5, following the entry into a living organism and translocation, a foreign chemical may be stored, metabolized, or excreted (Figure 5.2). When the rate of entry is greater than the rate of metabolism and/or excretion, storage of the chemical often occurs. Storage or binding sites may not be the sites of toxic action, however. For example, lead is stored primarily in the bone, but acts mainly on the soft tissues of the body. If the storage site is not the site of toxic action, selective sequestration may be a protective mechanism, since only the freely circulating form of the foreign chemical produces harmful effects.

Some chemicals that are stored may remain in the body for a long time without exhibiting direct harmful effects. DDT may be considered as an example. Accumulation or buildup of free chemicals may be prevented, until the storage sites are saturated. Selective storage limits the amount of foreign chemicals to be excreted, however. Since bound or stored toxicants are in equilibrium with their free forms, a chemical will be released from the storage site as it is metabolized or excreted. On the other hand, accumulation may result in illnesses that develop slowly, as exemplified by fluorosis and lead and cadmium poisoning.

METABOLISM OF ENVIRONMENTAL CHEMICALS: BIOTRANSFORMATION

Subsequent to the entry of an environmental chemical into an organism such as a mammal, chemical reactions occur within the body to alter the structure of the chemical. This metabolic conversion process is known as biotransformation and occurs in any of several tissues and organs such the intestine, lung, kidney, skin, and liver.

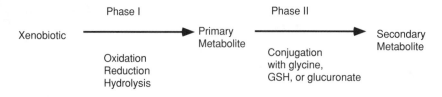

Figure 8.1 The two phases of xenobiotic metabolism.

By far, the largest number of these chemical reactions are carried out in the liver. The liver metabolizes not only drugs but also most of the other foreign chemicals to which the body is exposed. Biotransformation in the liver is thus a critical factor not only in drug therapy but also in the body's defense against the toxic effects of a wide variety of environmental chemicals (Kappas and Alvares 1975). The liver plays a major role in biotransformation because it contains a number of nonspecific enzymes responsible for catalyzing the reactions involved. As a result of the process xenobiotics are converted to more water-soluble and more readily excretable forms. While the purpose of such metabolic processes is probably to reduce the toxicity of chemicals, this does not prove to be always the case. Occasionally the metabolic process converts a xenobiotic to a reactive electrophile that is capable of causing injuries through interaction with liver cell constituents (Reynolds 1977).

TYPES OF BIOTRANSFORMATION

The process of xenobiotic metabolism includes two phases commonly known as Phase I and II. The major reactions involved in Phase I are oxidation, reduction, and hydrolysis, as shown in Figure 8.1. Among the representative oxidation reactions are hydroxylation, dealkylation, deamination, and sulfoxide formation, whereas reduction reactions include azo reduction and addition of hydrogen. Such reactions as splitting of ester and amide bonds are common in hydrolysis. During Phase I, a chemical may acquire a reactive group such as OH, NH_2, COOH, or SH.

Phase II reactions, on the other hand, are synthetic or conjugation reactions. An environmental chemical may combine directly with an endogenous substance, or may be altered by Phase I and then undergo conjugation. The endogenous substances commonly involved in conjugation reactions include glycine, cysteine, glutathione, glucuronic acid, sulfates, or other water-soluble compounds. Many foreign compounds sequentially undergo Phase I and II reactions, whereas others undergo only one of them. Several representative reactions are shown in Figure 8.2.

MECHANISMS OF BIOTRANSFORMATION

In the two phases of reactions shown in Figure 8.1, the lipophilic foreign compound is first oxidized so that a functional group (usually a hydroxyl group) is introduced into the molecule. This functional group is then coupled by conjugating

Phase I reactions

Oxidation

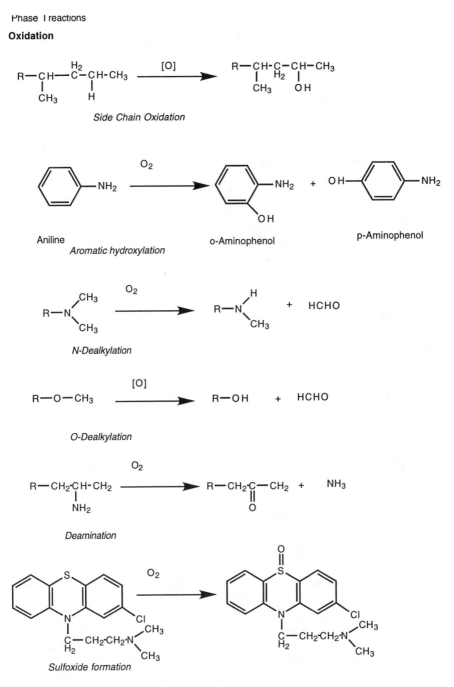

Side Chain Oxidation

Aniline   o-Aminophenol   p-Aminophenol
Aromatic hydroxylation

N-Dealkylation

O-Dealkylation

Deamination

Sulfoxide formation

Figure 8.2 Examples of biotransformation.

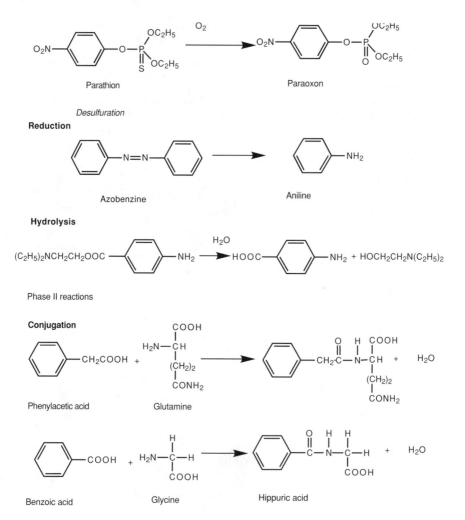

Figure 8.2 (continued).

enzymes to a polar molecule so that the excretion of the foreign chemical is greatly facilitated.

The NADPH-cytochrome P-450 system, commonly known as the mixed-function oxygenase (MFO) system, is the most important enzyme system involved in the Phase I oxidation reactions. Cytochrome P-450 system, localized in the smooth endoplasmic reticulum of cells of most mammalian tissues, is particularly abundant in the liver. This system contains a number of isozymes that are versatile in that they catalyze many types of reactions including aliphatic and aromatic hydroxylations and epoxidations, N-oxidations, sulfoxidations, dealkylations, deaminations, dehalogenations, and others (Wislocki et al. 1980). These isozymes are responsible for the oxidation of different substrates or for different types of oxidation of the same

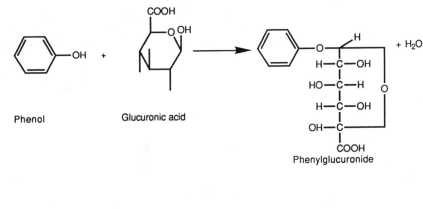

Displacement of aromatic halogens by glutathione

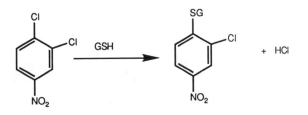

3,4-Dichloronitrobenzene

Figure 8.2 (continued).

substrate. Carbon monoxide binds with the reduced form of the cytochrome, forming a complex with an absorption spectrum peak at 450 nm. This is the origin of the name of the enzyme. As a result of the complex, inhibition of the oxidation process occurs.

At the active site of cytochrome P-450 is an iron atom that, in the oxidized form, binds the substrate (SH) (Figure 8.3). Reduction of this enzyme-substrate complex then occurs, with an electron being transferred from NADPH via NADPH cytochrome P-450 reductase. This reduced (Fe^{2+}) enzyme-substrate complex then binds molecular oxygen in some unknown fashion, and is then reduced further by a second electron, possibly donated by NADH via cytochrome b_5 and NADH cytochrome b_5 reductase. The enzyme-substrate-oxygen complex splits into water, oxidized substrate, and the oxidized form of the enzyme. The overall reaction is therefore:

$$SH + O_2 + NADPH + H^+ \rightarrow SOH + H_2O + NADP^+ \tag{8.1}$$

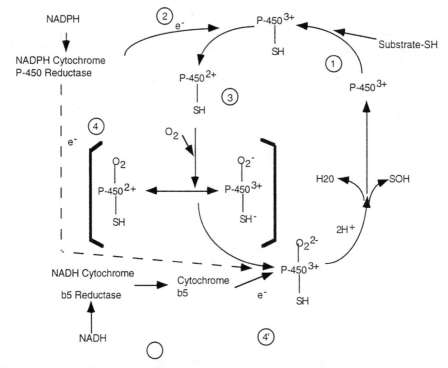

Figure 8.3   The cytochrome P-450 monooxygenase system. P-450^{3+}: cytochrome P-450 with heme iron in oxidized state (Fe^{3+}); P-450^{2+}: cytochrome P-450 with iron in reduced state; SH: substrate; e$^-$: electron. (Adapted from J.A. Timbrell. 1982. *Principles of Biochemical Toxicology.* Taylor and Francis Ltd., London.)

where SH is the substrate. As shown in the above equation, one atom from molecular oxygen is reduced to water and the other is incorporated into the substrate. The requirements for this enzyme system are oxygen, NADPH, and Mg^{2+} ions.

Contrary to the cytochrome P-450 system, most hepatic Phase II enzymes are located in the cytoplasmic matrix. In order for these reactions to occur efficiently, adequate activity of the enzymes involved is essential. In addition, it is clear that adequate intracellular contents of cofactors such as NADPH, NADH, O$_2$, glucuronate, ATP, cysteine, and GSH are required for one or more reactions.

CONSEQUENCE OF BIOTRANSFORMATION

Although hepatic enzymes that catalyze Phase I and II reactions convert the lipid-soluble xenobiotic to a more water soluble metabolite, they also participate in the metabolism or detoxification of endogenous substances. For example, the hormone testosterone is deactivated by cytochrome P-450. The S-methylases detoxify hydrogen sulfide formed by anaerobic bacteria in the intestinal tract. It can be seen, therefore, that chemicals or conditions that influence the activity of the Phase I and II enzymes can affect the normal metabolism of endogenous substances.

As previously mentioned, the biotransformation of lipophilic xenobiotics by Phase I and II reactions might be expected to produce a stable, water-soluble, and readily excretable compound. However, there are examples of hepatic biotransformation mechanisms by which xenobiotics are converted to reactive electrophilic species. Unless detoxified, these reactive electrophiles may interact with a nucleophilic site in a vital cell constituent, leading to cellular damage. There is evidence that many of these reactive substances bind covalently to various macromolecular constituents of liver cells. For example, carbon tetrachloride, known to be hepatotoxic, covalently binds to lipid components of the liver endoplasmic reticulum (Reynolds and Moslen 1980). Some of the reactive electrophiles are carcinogenic as well.

Although liver cells are dependent on the detoxification enzymes for protection against reactive electrophilic species produced during biotransformation, endogenous antioxidants such as vitamins E and C and glutathione also provide protection. As mentioned in Chapter 5, these substances are widely known as free radical scavengers. The main role of vitamin E is to protect the lipid constituents of membranes against free radical-initiated peroxidation reactions. Experimental evidence has shown that livers of animals fed diets deficient in vitamin E were more vulnerable to lipid peroxidation following poisoning with CCl_4 (Reynolds and Moslen 1980). Glutathione, on the other hand, is a tripeptide, and has a nucleophilic sulfhydryl (SH) group that can react with and thus detoxify reactive electrophilic species (Van Bladeren et al. 1980). Glutathione can also donate its sulfhydryl hydrogen to a reactive free radical (GS·). The glutathione radical formed can react with another glutathione radical to form stable oxidized GSSG. The GSSG can then be reduced back to GSH through an NADPH-dependent reaction catalyzed by glutathione reductase. The NADPH is generated in reactions involved in the pentose phosphate pathway.

In addition to vitamins E and C and GSH, there are enzymatic systems that are important in the defense against free radical-mediated cellular damage. These include superoxide dismutase (SOD), catalase, and GSH peroxidase. Figure 8.4 shows the interrelationship between these enzymatic components.

MICROBIAL DEGRADATION

Microbial degradation of xenobiotics is crucial in the prediction of the longevity and, thereby, the long-term effects of the toxicant, and may also be crucial in the actual remediation of a contaminated site. Utilization of the propensity of microorganisms to degrade a wide variety of materials may actually provide an opportunity for environmental toxicologists to not only diagnose and provide a prognosis, but also to prescribe a treatment to assist the ecosystem in the removal of the xenobiotic.

Microbial cell structure is varied with a tremendous diversity in size and shape. Prokaryotic cells typically contain a cell wall, 70s ribosomes, a chromosome that is not membrane bound, various inclusions and vacuoles, and extrachromosomal DNA or plasmids. Eukaryotic microorganisms are equally varied with a variety of forms; many are photosynthetic or harbor photosynthetic symbionts. Many eukaryotic cells

(i) **Superoxide dismutase**

$$2 O_2^- + 2 H^+ \rightarrow H_2O_2 + O_2$$

(ii) **Catalase**

$$2 H_2O_2 \rightarrow 2 H_2O + O_2$$

(iii) **Glutathione peroxidase**

$$ROOH + 2 GSH \rightarrow ROH + GSSG$$

(iv) **Glutathione reductase**

$$GSSG + NADPH + H^+ \rightarrow 2 GSH + NADP^+$$

Figure 8.4 The four cellular antioxidant enzymes. Superoxide dismutase (SOD) catalyzes the dismutation of superoxide free radical (O_2^-) to hydrogen peroxide (H_2O_2). The H_2O_2 formed is converted to H_2O and O_2 by catalase or glutathione peroxidase. Glutathione peroxidase, which requires GSH, also detoxifies ROOH TO ROH. The GSSG formed is then re-reduced to GSH in the presence of NADPH, derived mainly from the pentose phosphate pathway.

contain prokaryotic endosymbionts, some of which contain their own set of plasmids. Given the variety of eukaryotic microorganisms they have been labeled protists, since they are often a mixing of algal and protozoan characteristics within apparently related groups.

Many of these microorganisms have the ability to use xenobiotics as a carbon or other nutrient source. In some instances it may be more appropriate to ascribe this capability to the entire microbial community since often more than one type of organism is responsible for the stages of microbial degradation.

Microorganisms often contain a variety of genetic information. In prokaryotic organisms the chromosome is a closed circular DNA molecule. However, other genetic information is often coded on smaller pieces of closed circular DNA called plasmids. The chromosomal DNA codes the sequences that are responsible for the normal maintenance and growth of the cell. The plasmids or extrachromosomal DNA often code for metal resistance, antibiotic resistance, conjugation processes, and often the degradation of xenobiotics. Plasmids may be obtained through a variety of processes including conjugation, infection, and the absorption of free DNA from the environment (Figure 8.5).

Eucaryotic microorganisms have a typical genome with multiple chromosomes as mixtures of DNA and accompanying proteins. Extrachromosomal DNA also exists within the mitochondria and the chloroplasts that resembles prokaryotic genomes. Many microbials also contain prokaryotic and eukaryotic symbionts that can be essential to the survivorship of the organism. The ciliate protozoan *Paramecium bursaria* contains symbiotic chlorella that can serve as a source of sugar given sufficient light. Several of the members of the widespread species complex, *Paramecium aurelia*, contain symbiotic bacteria that kill paramecium not containing the identical bacteria. Apparently, this killing trait is coded by plasmid DNA contained

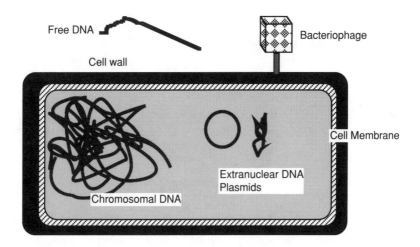

Figure 8.5   Schematic of a typical prokaryote. Genetic information and, thereby, coding for the detoxification and degradation of a xenobiotic may be available from a variety of sources.

within the symbiotic bacteria. Protists generally reproduce by asexual fission but sexual reproduction is available. Often during sexual reproduction an exchange of cytoplasm takes place, allowing cross-infection of symbionts and their associated DNA.

Microorganisms are found in a variety of environments: aquatic, marine, ground water, soil, and even in the Arctic. Many are found in extreme environments, from tundra to the superheated smokers at sites of seafloor spreading. The adaptability of microorganisms extends to the degradation of many types of xenobiotics.

Many organic xenobiotics are completely metabolized under aerobic conditions to carbon dioxide and water. The essential criteria is that the metabolism of the material results in a material able to enter the tricarboxylic acid or TCA cycle. Molecules that are essentially simple chains are readily degraded since they can enter this cycle with relatively little modification. Aromatic compounds are more challenging metabolically. The 3-ketoadipic acid pathway is the generalized pathway for the metabolism of aromatic compounds with the resulting product acetyl-CoA and succinic acid, materials that easily enter into the TCA cycle (Figure 8.6). In this process the aromatic compound is transformed into either catechol or protocatechuic acid. The regulation of the resultant metabolic pathway is dependent upon the group and basic differences exist between bacteria and fungi.

Often the coding process for degradation of a xenobiotic is contained on both the extrachromosomal DNA, the plasmid, and the chromosome. Often the initial steps that lead to the eventual incorporation of the material into the TCA cycle are coded by the plasmid. Of course, two pathways may exist: a chromosomal and a plasmid pathway. Given the proper DNA probes, pieces of DNA with complimentary sequences to the degradation genes, it should be possible to follow the frequency and, thereby, the population genetics of degradative plasmids in procaryotic communities.

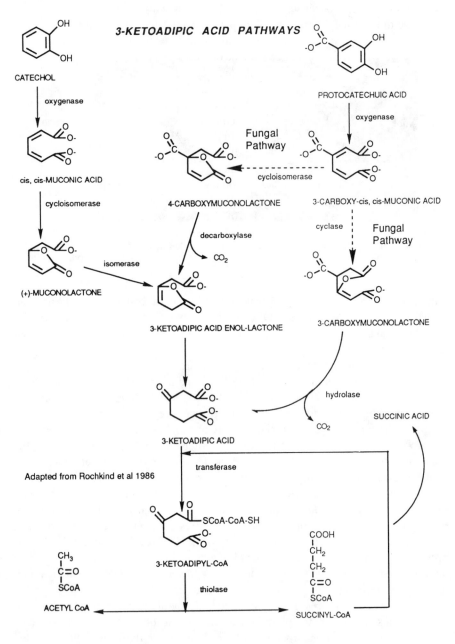

Figure 8.6   The 3-ketoadipic acid pathway.

In procaryotic mechanisms the essential steps allowing an aromatic or substituted aromatic to enter the 3-ketoadipic acid pathway are often, but not always, encoded by plasmid DNA. In some cases both a chromosomal and plasmid pathway are available. Extrachromosomal DNA can be obtained through a variety of mechanisms and can be very infectious. The rapid transmission of extrachromosomal DNA has the potential to enhance genetic recombination and result in rapid evolutionary change. In addition, the availability of the pathways on relatively easy-to-manipulate genetic material enhances our ability to sequence and artificially modify the code and perhaps enhance the degradative capability of microorganisms.

Simple disappearance of a material does not imply that the xenobiotic was biologically degraded. There are two basic methods of assessing the biodegradation of a substance. The first is an examination of the mass balance or materials balance resulting from the degradative process. This is accomplished by the recovery of the original substrate or by the recovery of the labeled substrate and the suspected radiolabeled metabolic products. Mineralization of the substrate is also a means of assessing the degradative process. Production of CO_2, methane, and other common congeners derived from the original substrate can be followed over time. With compounds that have easily identified compounds such as bromide, chloride, or fluoride, these materials can be analyzed to estimate rates of degradation. One of the crucial steps is to compare these rates and process with sterilized media or media containing specific metabolic inhibitors to test whether the processes measured are biological in nature.

Although the specific determination of the fate of a compound is the best means to establish the degradation of a compound, nonspecific methods exist that can be used when it is difficult or impossible to label or analytically detect the substrate. Measurement of oxygen uptake as the substrate is introduced in the culture is a means of confirming the degradation of the toxic material. Biological oxygen demand as determined for wastewater samples can be used, but it is not particularly sensitive. Respirometry with a device such as the Warburg respirometer is more sensitive and can be used to measure the degradation rates of suspected intermediates. Often it is possible to grow the degradative organism using the xenobiotic substrate as the sole carbon source, additionally confirming the degradative process. Controls using sterilized media or inhibitors are again important since microorganisms are able to grow on surprisingly minimal media and with only small amounts of materials that may be present as contaminants.

A wide variety of aromatic organics are degraded by a variety of microorganisms. Table 8.1 provides a compilation from a recent review giving both the compound and the strains that have so far been found that are responsible for the degradation. Only a few examples will be discussed below.

Substituted benzenes are commonly occurring xenobiotics. In Figure 8.7 the biodegradation pathway for toluene is diagrammed. The process begins with the hydroxylation of the toluene. In one case the hydroxylation of the substituent, the methyl group, occurs to form benzyl alcohol. Additional steps result in catechol, a

Table 8.1   Examples of Organic Compounds and Degradative Bacterial Strains

Aniline	*Frateuria sp.* ANA — 18
	Nocardia sp.
	Pseudomonas sp.
	Pseudomonas multivorans AN1
	Rhodococcus sp. AN-117
	Rhodococcus sp. SB3
Anthracene	*Beijerinckia sp.* B836
	Cunninghamella elegans
	Psuedomonas sp.
	Pseudomonas putida 199
Benzene	*Achromobacter sp.*
	Pseudomonas sp.
	Pseudomonas aeruginosa
	Pseudomonas putida
Benzoic acid	*Alcaligenes eutophus*
	Aspergillus niger
	Azotobacter sp.
	Bacillus sp.
	Pseudomonas sp.
	Pseudomonas acidovorans
	Pseudomonas testosteroni
	Pseudomonas sp. strain H1
	Pseudomonas PN-1
	Pseudomonas sp. WR912
	Rhodopseudomonas palustris
	Streptomyces sp.
	By consortia of bacteria
2-Chlorobenzoic acid	*Aspergillus niger*
3-Chlorobenzoic acid	*Acinetobacter calcoaceticus* Bs5 (grown on succinic acid and pyruvic acid)
	Alcaligenes eutrophus B9
	Arthrobacter sp. (grown on benzoic acid)
	Aspergillus niger
	Azotobacter sp. (grown on benzoic acid)
	Bacillus sp. (grown on benzoic acid)
	Pseudomonas aeruginosa B23
	Pseudomonas putida (w/plasmid p AC25)
	Psudomonas sp. B13
	Pseudomonas sp. H1
	Pseudomonas sp. WR912
	By consortia of bacteria
4-Chlorobenzoic acid	*Arthrobacter sp.*
	Arthrobacter globiformis
	Azotobacter sp. (grown on benzoic acid)
	Pseudomonas sp. CBS 3
	Pseudomonas sp. WR912
4-Chloro-3,5-dinitrobenzoic acid	*Chlamydomonas sp.* A2
2,5-Dichlorobenzoic acid	By consortia of bacteria
3,4-Dichlorobenzoic acid	By consortia of bacteria
3,5-Dichlorobenzoic acid	*Pseudomonas sp.* WR912
	By consortia of bacteria
2,3,6-Trichlorobenzoic acid	*Brevibacterium sp.* (grown on benzoic acid)
Biphenyl	*Beijerinckia sp.*
	Beijerinckia sp. B836
	Beijerinckia sp. 199
	Cunninghamella elegans
	Pseudomonas putida
	By consortia of bacteria

Table 8.1 (continued) Examples of Organic Compounds and Degradative Bacterial Strains

Catechol	*Pyrocatechase I*
4-Chlorocatechol	*Achromobacter sp.*
3,5-Dichlorocatechol	*Achromobacter sp.*
Chlorobenzene	*Pseudomonas putida* (grown on toluene)
	Unidentified bacterium, strain WR1306
Chlorocatechol	Pyrocatechases
3,5-Dichlorocatechol	*Achromobacter sp.* (grown on benzoic acid)
Chlorophenol	*Arthrobacter sp.*
2-Chlorophenol	*Alcaligenes eutrophus*
	Nocardia sp. (grown on phenol)
	Pseudomonas sp. B13
3-Chlorophenol	*Nocardia sp.* (grown on phenol)
	Pseudomonas sp. B13
	Rhodotorula glutinis
4-Chlorophenol	*Alcaligenes eutrophus*
	Arthrobacter sp.
	Nocardia sp. (grown on phenol)
	Pseudomonas sp. B13
	Pseudomonas putida
2,4,6-Trichlorophenol	*Arthrobacter* sp.
2,3,4,6-Tetrachlorophenol	*Aspergillus sp.*
	Paecilomyces sp.
	Penicillium sp.
	Scopulariopsis sp.
Chlorotoluene	*Pseudomonas putida* (grown on toluene)
Gentisic acid	*Trichosporon cutaneum*
Guaiacols (*o*-methoxyphenol)	*Arthrobacter sp.*
3,4,5-Trichloroguaiacol	*Arthrobacter sp.* 1395
Homoprotocatechuic acid	*Trichosporon cutaneum*
Naphthalene	*Cunninghamella elegans*
	Oscillatoria sp.
	Pseudomonads
Pentachlorophenol (PCP)	*Arthrobacter sp.*
	Coniophora pueana
	Mycobacterium sp.
	Pseudomonas sp.
	Saprophytic soil corynebacterium
	KC3 isolate
	Mutant ER-47
	Mutant ER-7
	Trichoderma viride
Phenanthrene	*Aeromonas sp.*
	Fluorescent and nonfluorescent pseudomonad groups
	Vibrios
Protocatechuic acid	*Neurospora crassa*
	Trichosporon cutaneum
Sodium pentachlorophenate (Na-PCP)	*Trichoderma sp.*
	Trichoderma virgatum
Tetrachlorohydroquinone	KC3
Toluene	*Achromobacter sp.*
	Pseudomonas sp.
	Pseudomonas aeruginosa
	Pseudomonas putida
4-Amino-3,5-Dichlorobenzoic acid	By consortia of bacteria
2,4,5-Trichlorophenoxyacetic acid	*Psuedomonas cepacia* AC1100

List compiled from: Rochkind, M.L., J.W. Blackburn, and G.S. Sayler. 1986. *Microbial Decomposition of Chlorinated Aromatic Compounds*. EPA160012-861090, pp. 45-98.

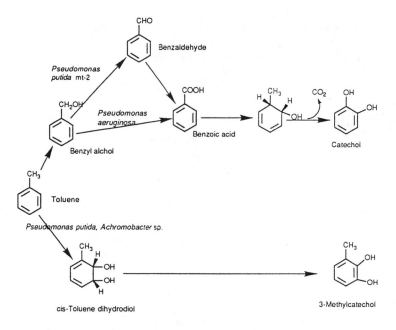

Figure 8.7 Alternate pathways for the degradation of a substituted benzene, toluene.

material readily incorporated into the 3-ketoadipic acid pathway. Another set of species hydrolyze the ring itself producing a substituted catechol as the end process.

The degradation mechanism of materials such as naphthalene by fungi has been found comparable in a broad sense to the detoxification mechanisms found in the liver in vertebrates. Fungi use a monooxygenase system that incorporates an atom of oxygen into the ring as the other atom is incorporated to water (Figure 8.8). The resulting epoxide can be further hydrolyzed to form an intermediate ultimately ending with a transhydroxy compound. The epoxide can also isomerize to form a variety of phenols. Both of these mechanisms occur in the degradation of naphthalene by the fungus *Cunninghamella elagans*.

A particularly widespread environmental contaminant is the pesticide pentachlorophenol (PCP). PCP has been used as a bactericide, insecticide, fungicide, herbicide, and molluscicide in order to protect a variety of materials from decomposition. Although it has bactericidal properties, PCP has been found to be degraded in a variety of environments by both bacteria and fungi. In some instances degradation occurs with PCP being used as an energy source.

A proposed pathway for the degradation of PCP by two bacterial strains is represented in Figure 8.9. Cultures of Pseudomonas were found to transform PCP into tetrachlorocatechol and tetrachlorohydroquinone (TeCHQ). These materials are then metabolized and radiolabeled carbon can be found in the amino acids of the degradative bacteria. Mycobacterium methylates PCP to pentachloroanisole but does not use PCP as an energy source. Fungi also metabolize PCP to a less toxic metabolite.

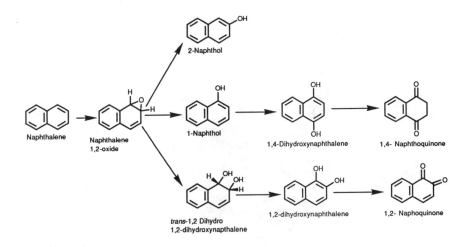

Figure 8.8   Biodegradation of naphthalene by *Cunninghamella elagans*.

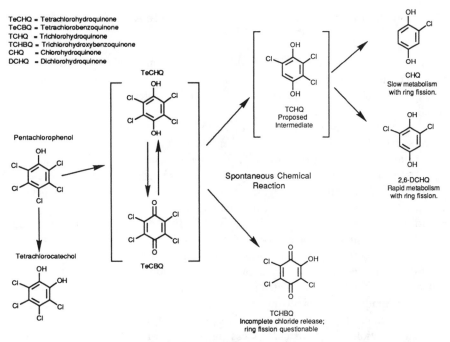

Figure 8.9   Possible mechanisms for the degradation of pentachlorophenol by *Pseudomonas* sp.

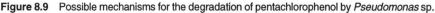

BIOREMEDIATION

Given the ability of many organisms to degrade toxic materials within the environment, a practical application would be to use these degradative capabilities in the removal of xenobiotics from the environment. In the broadest sense this might

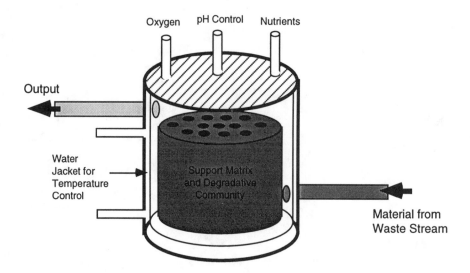

Figure 8.10 Schematic of a bioreactor for the detoxification of a waste stream or for inclusion in a pump and water treatment process.

entail the introduction of a specifically designed organism into the polluted environment to ensure the degradation of a known pollutant. Other examples of attempts at using biodegradation for remediation are the addition of fertilizers to enhance degradation of oil spills and the construction of biological reactors, bioreactors, through which contaminated water or a soil slurry can be passed. In some instances these attempts have appeared successful, in others the data are not so clear.

The most important design criteria for attempting bioremediation is the complexity of the environment and the complexity and concentration of the toxicants. Controlled and carefully defined waste streams such as those derived from a specific synthesis at a manufacturing plant may be especially amenable to degradation. A reactor such as the one schematically depicted in Figure 8.10, could be developed using a specific strain of bacteria or protist that has been established on a substrate. Nutrients, temperature, oxygen concentration, and toxicant concentration can be carefully controlled to offer a maximum rate of degradation. As the complexity of the effluent or the site to be remediated increase, a consortia of several organisms or of an entire degradative community may be necessary. Consortia, groups of co-occurring organisms, can also be established in a bioreactor-type setting.

Concentration of the toxicant is essential in determining the success of the bioremediation attempt. As shown in Figure 8.11, too low a concentration will not stimulate growth of the degradative organism. At too high a concentration the toxic effects become apparent and the culture dies. The shape of the curve is dependent not only upon the degradative system of the organism, but also upon the availability of nutrients, temperature, and the other factors essential for microbial growth. One of the advantages of the bioreactor system is that all of these factors can be carefully

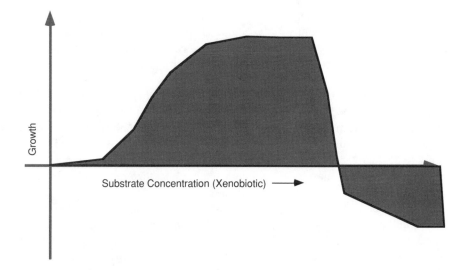

Figure 8.11 Degradative growth curve. At low concentrations, degradation may not occur due to the lack of nutritive content of the xenobiotic as substrate. Eventually a maximal rate of degradation and also growth may occur with a plateau. Eventually the concentration of the toxic material overwhelms the ability of the organism to detoxify the material and death ensues.

controlled. In a situation where it may be necessary to attempt the *in situ* remediation of a toxicant these factors are more difficult to control. Biotic factors, such as competitors and predators, also become important as the process is taken out of the bioreactor and placed in a more typical environment. Not only do the degradative organisms have to be able to degrade the toxicant, they must be able to compete effectively with other microflora and escape predation.

To enhance degradation frequent plowing and fertilization of a terrestrial site may be done to ensure proper aeration of the soil. Groundwater is often nutrient- and oxygen-limited and both of these materials can be introduced. Often hydrogen peroxide is pumped into groundwater as an effective means of delivering oxygen as the hydrogen peroxide decomposes.

ISOLATION AND ENGINEERING OF DEGRADATIVE ORGANISMS

The basic scheme of isolating degradative organisms is relatively straightforward. Samples from a site likely to contain degradative bacteria are collected. If the degradation of oil products is sought, soils and sediments near pumping stations or other sites likely to be contaminated with the materials of interest are sampled. PCP has been widely used as a preservative, so old wood processing plants may be appropriate.

The next step is to enhance the selection process for the ability to degrade the toxicant by using increasing concentrations of the material. This process can be accomplished into two related ways. First, the toxicant and sample are mixed in a chemostat. A chemostat maintains the culture at specific conditions, adds nutrients, and often has a mixing apparatus. At an initial low concentration, samples are taken in order to determine whether or not the xenobiotic has been degraded. It may take many months for the evolution of the degradative ability in the original microbial community. As degradation is observed, successively higher concentrations of the toxicant can be added to the chemostat to further strengthen the ability of the selection to degrade the toxicant. At very high concentrations only a few bacterial or fungal species may survive. These survivors can then be plated and examined for the ability to degrade the toxicant. The researcher must be prepared for the possibility that no one organism may be able to completely mineralize the xenobiotic and a consortia of several organisms may be required.

A similar process can be accomplished without access to a chemostat. Samples from a culture of an initial concentration of xenobiotic can be placed in other containers with successively higher concentrations of the toxicant, achieving the same selective pressures as found in the chemostat (Figure 8.12). Again, it may take long periods for evolution of a degradative organism or community to arise.

As the degradative organism or consortia is isolated, further studies may actually isolate a particular plasmid or even genes responsible for the degradation. It may be possible to construct organisms with several of these plasmids or the genes may be inserted into the hosts chromosome. If the desire is to place the organisms into a field situation, basic survival traits must also be maintained.

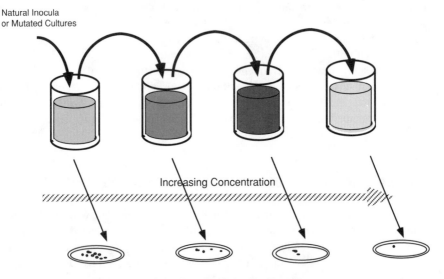

Figure 8.12   Selection protocol for the isolation of degradative microorganisms.

THE GENETICS OF DEGRADATIVE ELEMENTS

Once formed, a degradative element can suffer a number of fates (Figure 8.13). Using an organophosphate degradative or *opd* gene as an example, a number of recombination and other genetic events can occur that affect the reproduction and expression of the gene.

First, the gene exists on a plasmid within the host cell. The plasmid can replicate, increasing the copy number of the plasmid that is the host of the degradative genetic element. In some instances, the plasmid can be incorporated into the host chromosome through a recombination event. The entire plasmid or sections can enter the host genome. Expression of the genes contained in the plasmid may or may not occur. Occasionally, the genetic elements can be excised from the host and again reproduce as an independent plasmid. This scenario is similar to that for the life cycle of λ phage.

At a conjugation event, the plasmid may be passed on in its entirety and the new host translating the genetic code into a viable degradative enzyme. However, a mistake in replication or a mismatch with the new protein-generating machinery of the new host may result in the plasmid being passed on but the activity of the gene product not being realized. In some cases a protein may be manufactured, but the degradative activity lost through mutation.

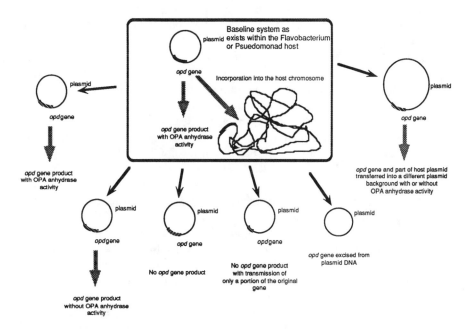

Figure 8.13 Outcomes in the evolution of a degradative element in a prokaryote.

Deletions also may occur that result in only part of the degradative element remaining on the plasmid. If only a portion of the original gene is being transmitted, an inactive protein may result. If the deletion is in the base sequence that is recognized by the transcription machinery of the cell, no mRNA and derivative protein will be produced.

A deletion event may also excise the degradative element from the plasmid, resulting in a loss of the information from the resulting host cells. In this case, the ability to degrade a xenobiotic is lost, and will probably not be recovered unless recombination with a plasmid containing the degradative element occurs.

Of course, many procaryotes contain more than one plasmid. Recombination between the plasmid containing the degradative gene and a plasmid of the same neighborhood can pass the degradative gene to a new host.

AN EXAMPLE OF A DETOXIFICATION ENZYME — THE OPA ANHYDROLASES

The examples provided above give only a brief overview of the variety of enzymatic functions that alter, biotransform, and biodegrade xenobiotics. In many instances numerous enzymes are known, as in the case of the mixed function oxidases. In order to provide a concrete example of a system of detoxification enzymes that is widely distributed we have chosen the organophosphate acid anhydrolases. Enzymes that may aid in the understanding of organophosphate intoxication may also provide a means for the detoxification and bioremediation of these materials.

An interesting example of a series of enzymes able to hydrolyze a variety of organophosphates are the organophosphorous acid anhydrolases (OPA anhydrolases). OPA anhydrolases are a wide-ranging group of enzymes. As will be shown below, there are often several distinguishable enzymes within an organism. The ability to hydrolyze a particular substrate varies tremendously. Inhibitors have been found and cations seem to be important for activity. The enzymatic mechanism has been described for the *opd* gene product, but is still unknown for the remaining OPA anhydrolases. Currently, the natural role of these enzymes is unknown, although suggestions have recently been made that the OPA anhydrolases evolved for the degradation of naturally occurring organophosphates and halogenated organics (Haley and Landis 1988; Chester et al. 1988; Landis et al. 1989).

Two categories of organofluorophosphate OPA anhydrolases have been recognized in the literature (Hoskin et al. 1984). Typically, the Mazur type is characterized as being stimulated by Mn^{2+}, hydrolyzing soman faster than DFP, intolerant of ammonium sulfate precipitation, and is usually found to be dimeric with a molecular weight of approximately 62,000 D (Storkebaum and Witzel 1975), and is competitively or reversibly inhibited by Mipafox (Hoskin 1985). Mipafox is a structural analog to DFP (Figure 8.14). The Mazur-type OPA anhydrase demonstrates a stereospecificity in the

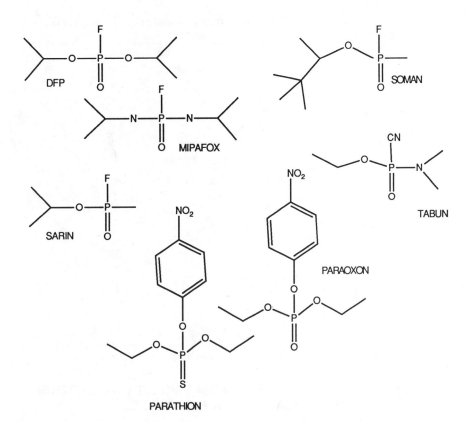

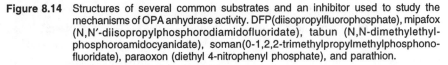

Figure 8.14 Structures of several common substrates and an inhibitor used to study the mechanisms of OPA anhydrase activity. DFP(diisopropylfluorophosphate), mipafox (N,N'-diisopropylphosphorodiamidofluoridate), tabun (N,N-dimethylethyl-phosphoroamidocyanidate), soman(0-1,2,2-trimethylpropylmethylphosphono-fluoridate), paraoxon (diethyl 4-nitrophenyl phosphate), and parathion.

hydrolysis of tabun (Hoskin and Trick 1955) and soman. The archetypal Mazur-type OPA anhydrase can be found in hog kidney. Typically, squid-type OPA anhydrase (Hoskin et al. 1984) hydrolyzes DFP faster than soman, is stable, can be purified using ammonium sulfate, has a molecular weight of approximately 26,000 D, is usually unaffected or slightly inhibited by Mn^{2+}, experiences no inhibition of DFP hydrolysis by mipafox (Hoskin et al. 1984), and does not demonstrate stereospecificity towards the hydrolysis of soman. Squid-type OPA anhydrase is present in nerve (optic ganglia, giant nerve axon), hepatopancreas, and salivary gland of cephalopods (Hoskin et al. 1984). Cephalopods also contain OPA anhydrase resembling the Mazur type in other tissues. Table 8.2 lists the characteristics of several of the different OPA anhydrolases studied to date.

Table 8.2   Comparison of Several Aquatic OPA Anhydrase Activities with Typical Squid and Mazur-Type OPA Anhydrolases.

Characteristic activity	Substrate hydrolysis			
	Mol wt (KD)	Soman/ DFP ratio	Mn^{2+} stimulation	Mipafox inhibition
T. thermophila				
Tt DFPase-1	80	1.12	2.5–4.0	+
Tt DFPase-2	75	1.26	2.0	+
Tt DFPase-3	72	0.71	1.7–2.5	+
Tt DFPase-4	96	1.95	17–30	nt
R. cuneata				
Rc OPA-1	19–35	nt	1	—
Rc OPA-3	82–138	nt	nt	Hydrolyzes mipafox
Thermophile isolate OT (JD.100)	84	nt	+	—
Halophile isolate JD6.5				
OPAA I	98	nt	nt	
OPAA II	62^4	0.5	3–5	nt
opd Gene product (parathion hydrolase)	60–65 (35,418 subunits)	nt	+	
Squid-type OPA anhydrase (*Loligo pealei*)	23–30	0.25	1	—
Mazur-type OPA anhydrase (hog kidney)	62–66 (30,000 subunits)	6.5	2	+

Note: The enzymes vary in molecular weight, reaction to ions, and in the Soman/DFP ratios.

CHARACTERISTICS OF THE *OPD* GENE PRODUCT AND OTHER BACTERIAL OPA ANHYDROLASES

Currently under intense scrutiny, the protein product of the *opd* gene of *Pseudomonas diminuta* is perhaps the best studied of the bacterial OPA anhydrolases. It has recently been shown that the *opd* OPA anhydrase (also called phosphotriesterase) has the capability to hydrolyze DFP and perhaps other organofluorophosphates (Dumas et al. 1989; Donarski et al. 1988). Until recently, this activity was labeled as a phosphotriesterase and was characterized by the capability to hydrolyze materials such as paraoxon and parathion. Although not strictly aquatic, this OPA anhydrase is apparently widely distributed among bacteria and the genetic code has been sequenced and the mechanism of hydrolysis elucidated.

The *opd* OPA anhydrase is coded by a plasmid-borne gene of 1079 base pairs in length (McDaniel et al. 1988). The gene sequence is identical in both *Flavobacterium* and *P. diminuta* although the plasmids bearing this gene are not. Crude preparations of bacteria containing the *opd* gene have been demonstrated to have the ability to hydrolyze a variety of phosphotriesters, such as paraoxon, fensulfothion, *O*-ethyl *O*-*p*-nitrophenyl phenylphosphothioate (EPN) and chlorofenvinophos (Brown 1980; Chiang et al. 1985; McDaniel 1985). However, in at least the case of malathion hydrolysis, the active agent is not the *opd* OPA anhydrase. Activity that can degrade malathion exists even in *P. diminuta* cured of the plasmid containing the *opd* gene

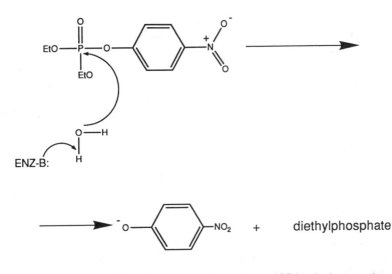

Figure 8.15 Mechanism of hydrolysis of parathion by the *opd* OPA anhydrase as determined by Lewis et al. The reaction is a single displacement using a base at the active site to activate a water molecule. The activated water attacks the phosphorus, producing diethyl phosphate and 4-nitrophenol. The same active site is able to hydrolyze DFP (Dumas et al. 1989) and related organofluorophosphates (Dumas et al. 1990). (Modified from Lewis, V. E., W. J. Donarski, J. R. Wild, and F. M. Raushel. 1988. *Biochemistry* 27:1591-1597. With permission.)

(Wild and Raushel 1988); 80 to 90% of the OPA anhydrase activity apparently is associated with the pseudomonad membrane. The *opd* OPA anhydrase is insensitive to ammonium sulfate (Dumas et al. 1989). Molecular weight as determined by analysis of the gene sequence is 35,418 D (McDaniel et al. 1988). However, disassociated from the membrane using a Triton-X-100 or Tween 20, the apparent molecular weight is estimated to be 60,000 to 65,000 D. These data raise the possibility that the active enzyme is dimeric.

In an elegant series of experiments, the mechanism of the *opd* OPA anhydrase was elucidated (Lewis et al. 1988). Using oxygen-18-containing water and the (=) and (–) enantiomers of O-ethyl phenylphosphonothioic acid, it was determined that the reaction was a single in-line displacement by an activated water molecule at the phosphorus center of the substrate (Figure 8.15). It is significant that this same enzyme was also able to hydrolyze DFP and other related organofluorophosphates.

Attaway et al. (1987) have screened a number of bacterial isolates for OPA anhydrase activity including strains of *Pseudomonas diminuta*, *P. aeruginosa*, *P. putida*, *Vibrio alginolyticus*, *V. parahaemolyticus*, *Escherichia coli*, and *Flavobacterium* sp. m. Chettur et al. (1988) and Hoskin et al. (1989) have recently published findings on the OPA anhydrase activities of the obligate thermophile (OT) organism, also known as JD.100, from the DeFrank collection. The thermophilic bacteria were isolated by J. DeFrank from soil samples from the Edgewood Area of the Aberdeen Proving Ground, MD. OT has been identified as a strain of *Bacillus stearothermophilus*. The OPA anhydrase activity was purified using a Pharmacia G-

100 column followed by a DEAE ion exchange column. A five- to ten-fold purification was accomplished. Estimated molecular weight was 84,000 D. The OT OPA anhydrase hydrolyzed soman, sarin, and dimebu (3,3-dimethylbutyl methylphosphonfluoridate) but not DFP. The catalysis was markedly stimulated by Mn^{2+}. Dimebu hydrolysis was also stimulated but less stimulation by Mn^{2+} is apparent. Sarin hydrolysis followed the pattern of dimebu. Mipafox was not inhibitory. DFP was reported to be a weak noncompetitive inhibitor of soman hydrolysis. A suggestion was made in this report that since hydrolysis and the reduction of acetylcholinesterase inhibition coincide, the OT OPA anhydrase activity hydrolyzed all four isomers simultaneously, similar to the squid-type OPA anhydrase.

Several halophilic isolates that exhibit OPA anhydrase activity have been collected and studied by DeFrank (1988). One isolate, designated JD6.5, was obtained from Grantsville Warm Springs, UT. Two OPA anhydrase activities were present; however, 90% of the activity was represented by one of the enzymes, OPA-2. According to SDS-PAGE electrophoresis and gel permeation chromatography the molecular weight has been estimated at approximately 62,000 D. OPA-2 is stimulated by Mn^{2+} and hydrolyzes soman and NPEPP. Optimum pH was approximately 7.2. Attempts at purification using Sepharose CL-4B indicate that the enzyme may be very hydrophobic. Isolate JD30.3 was isolated from Wilson Hot Springs, UT and also contained OPA anhydrase activity able to hydrolyze DFP and soman. The purified activity was stimulated by divalent cations with Mg^{2+} being the best. Molecular weight was approximately 76,000 D as determined by gel molecular sieve chromatography. The OPA anhydrases from JD30.3 were insensitive to ammonium sulfate.

Zech and Wigand (1975) demonstrated that the DFP hydrolyzing and paraoxon hydrolyzing activities in at least one strain of *Escherichia coli*, $K_{12}sr$, were distinct. Separated by gel filtration, the activities showed no overlap. Two peaks of DFP hydrolyzing activity were found using gel filtration and four peaks were found at isoelectric points of 5.3, 5.7, 6.1, and 7.8. Three isoelectric points at 5.3, 5.6, and 6.2 were found for the paraoxon hydrolyzing activity. The optimal pH for DFP hydrolysis was found to be 8.3, and for paraoxon hydrolysis it was 9.3.

EUCARYOTIC OPA ANHYDROLASES

The ability of crude extracts of the protozoan *Tetrahymena thermophila* to hydrolyze the organophosphate DFP was discovered by Landis et al. (1985). Purification of the Tetrahymena material was conducted using a Sephacryl S-200 and S-300 molecular sizing column using a fraction volume of approximately half of that used in previous studies in order to increase resolution. Three repeatable peaks capable of the hydrolysis of DFP immediately became apparent. Upon the addition of Mn^{2+}, a fourth peak appeared. The activities were identified as *Tt* DFPase-1, *Tt* DFPase-2 ...*Tt* DFPase-5, and their characteristics can be found in Table 8.3. Molecular weights of the Tetrahymena OPA anhydrases range from 67,000 to

Table 8.3   Comparison of DFP and Soman Hydrolysis Ratio and Stimulation by Mn^{2+} for Aquatic Organisms

Enzyme source	Soman/DFP ratio		Stimulation by Mn^{2+}	
	Mn^{2+}	no Mn^{2+}	DFP	Soman
Proteus vulgaris	19	22	1.5	1.3
Saccharomyces cerevisae	8	4	0.5	1.0
Homarus (lobster) nerve	11	9.1	2.9	3.4
Spisula (surf clam) nerve	6.1	3.0	1.0	2.0
Electrophorus electricus (torpedo fish) liver	16	14	1.7	1.8
T. thermophila (crude extract)	20	10	1.0	2.0

Note: Interestingly, *S. cerevisae*, *Spisula*, and *T. thermophila* do not show a stimulation in DFP hydrolysis with Mn^{2+}. Perhaps, like *T. thermophila*, the other two species have at least two enzymes, one of which hydrolyzes soman and is stimulated by Mn^{2+}.

From Hoskin, F.C.G., M.A. Kirkish, and K. E. Steinman, 1984. *Fund. Appl. Tox.*, 4:5165-5172. With permission.

96,000 D. The activity of DFPase-4 is stimulated 17 to 30-fold with Mn^{2+}. *Tt DFPase-1*, *Tt* DFPase-2, and *Tt* DFPase-3 are only stimulated two- to four-fold, and part of this increase may be due to contamination by the higher molecular weight *Tt* DFPase-4. Soman-to-DFP ratios are approximately 1:1 for the Tetrahymena OPA anhydrases.

Mipafox is reversible and competitively inhibits *Tt* DFPase-1, *Tt* DFPase-2, and *Tt* DFPase-3 (Landis et al. 1989). Hydrolysis of the mipafox by partially purified Tetrahymena extract was only 13% of the rate of DFP.

Of all the conventionally recognized OPA anhydrases, the squid-type as found in *Loligo pealei* is perhaps the best studied. The distribution of the squid-type OPA anhydrase is relatively narrow, being found in only the nervous tissue, saliva, and hepatopancreas of cephalopods. The molecular weight of the squid-type OPA anhydrase is approximately 23,000 to 30,000 D. The term "squid type" is specific to the activities found in these tissues. At times, more than one peak is apparent upon molecular sizing chromatography at this molecular weight range (Steinmann 1988). It has been estimated that the squid-type OPA anhydrase constitutes approximately 0.002% of the intracellular protein (Hoskin 1989). Squid-type OPA anhydrase does hydrolyze soman although at a rate of only about 0.25 that of DFP. However, squid-type OPA anhydrase apparently hydrolyzes all of the four stereoisomers of soman with some stereospecificity in rates.

Mipafox is not inhibitory to the squid-type OPA anhydrase. As reported by Gay and Hoskin (1979), the active site prefers an isopropyl side chain compared to an ethyl or methyl group.

Although the primary investigation into the OPA anhydrases of squid tissue has been of the squid-type OPA anhydrase, squid contains the more widespread Mazur-type OPA anhydrase. Gill, heart, mantle, and blood tissues all exhibit OPA anhydrase activities that are Mn^{2+} stimulated and hydrolyze soman faster than DFP (Hoskin et al. 1984).

CHARACTERISTICS OF OTHER INVERTEBRATE
METAZOAN ACTIVITIES

Nervous tissue of a variety of invertebrates has been screened for OPA anhydrase activity. Other mollusks have been reported to contain OPA anhydrases, notably *Octopus*, *Anisodoris* (sea-lemon), *Aplysia* (sea-hare), and *Sepia* (cuttlefish) (Hoskin and Long 1972). *Sepia* and *Octopus* hydrolyze DFP faster than tabun, a squid-type OPA anhydrase characteristic employed at that time, and now replaced by the DFP/soman ratio. Conversely, *Aplysia*, *Spisula*, and *Homarus* (lobster) hydrolyze tabun faster than DFP (Hoskin and Brande 1973), a typically Mazur characteristic. Soman-to-DFP ratios and Mn^{2+} stimulation for several species are shown in Table 8.3. *Homarus* and *Spisula* were further examined by Hoskin et al. (1984). These organisms were found to have activities broadly defined as Mazur type, using Mn^{2+} stimulation and DFP/soman ratio as criteria. In *Spisula* (surf clam), DFP hydrolysis was not stimulated by Mn^{2+} although soman hydrolysis was doubled. In light of research conducted since then, this result may indicate that more than one OPA anhydrase system may be present.

Anderson et al. (1988) discovered an OPA anhydrase activity in the estuarine clam *Rangia cuneata*. The clams were collected from Chesapeake Bay sediment. Of the tissues examined, OPA anhydrase activity was highest in the digestive gland and lowest in the foot muscle (Anderson et al. 1988). Soman was hydrolyzed faster than DFP. Exogenous Mn^{2+} did not increase the rate of DFP hydrolysis although soman hydrolysis was increased by 40% in the presence of 1 mM Mn^{2+}. The temperature range was determined to be from 15 to 50°C. Initial estimate of molecular weight was 22,000 D for the digestive gland as determined by molecular sieve chromatography. Interestingly, the molecular weight for the OPA anhydrase from the visceral mass was higher, implying a different enzyme and some tissue specificity. Except for molecular weight, the clam activity appeared to more closely resemble Mazur-type OPA anhydrase.

CHARACTERISTICS OF THE FISH ACTIVITIES

Hogan and Knowles (1968) examined the OPA anhydrases of liver homogenates from the bluegill sunfish, *Lepomis macrochirus*, and the channel catfish, *Ictalarus punctatus*. Initially, a 1.5% (w/v) homogenate of the excised livers from each species was determined to hydrolyze 10^{-2} M concentrations of DFP and 2,2-dichlorovinyl dimethyl phosphate (cochlorvos); 90% of the activity was found in the supernatant after a 1 h centrifugation at 100,000 G. For both species a Mn^{2+} concentration ranging from 0.3 to 1.0 mM was found to promote hydrolysis. Co^{2+} was optimal at a concentration of 0.1 mM, but was inhibitory at concentrations greater than 1.0 mM. Mg^{2+} and Ca^{2+} had no detectable effect. For studies using other organophosphates a 1 mM Mn^{2+} concentration was included in the reaction system.

Bluegill and catfish were both able to hydrolyze DFP, dichlorvos, and dimethyl 2,2,2-trichloro-1-n-butyryloxyethyl phosphonate (butonate). Catfish enzymes were also able to hydrolyze paraoxon and methyl 3-hydroxy-alpha-cronate, dimethyl phosphate (mevinphos) although at a very slow rate. K_ms calculated for the enzymes of both species indicated that each had a greater affinity for DFP than dichorvos. Sulfhydryl reagents and Cu^{2+} were found to inhibit the enzymatic activity of both organisms. Paraoxon had no effect. Cleavage products were identified as dimethyl phosphate and 2,2-dichloroacetaldehyde from dichlorovos hydrolysis and diisopropyl phosphate from the hydrolysis of DFP.

The fish *Electrophorus* was examined by Hoskin et al. (1984) and found to have an activity that hydrolyzes soman faster than DFP and to be stimulated by Mn^{2+}. This activity may be similar to those of catfish and bluegill.

COMPARISON OF THE OPA ANHYDRASES

It is natural to wish to impose a classification scheme upon the OPA anhydrases that would imply a set of phylogenetic relationships. The classification scheme of squid-type and Mazur-type anhydrases has proven useful in that it was quickly possible to differentiate the squid-type OPA anhydrase from the other forms. As will be seen below, many of the OPA anhydrase activities lie somewhere in between.

The multiple activities in *T. thermophila* share some of the characteristics of both the squid-type OPA anhydrase and classical Mazur-type OPA anhydrase found in hog kidney. In crude preparations, the OPA anhydrase activity has the characteristics of the hog kidney OPA anhydrase in that it hydrolyzes soman faster than DFP, is stimulated by Mn^{2+}, and is inhibited by mipafox. Further purification has revealed that the hydrolysis of soman and the stimulation of this hydrolysis by Mn^{2+} is principally due to the *Tt* DFPase-4. The *Tt* DFPase-1, *Tt* DFPase-2, and *Tt* DFPase-3 hydrolyze soman and DFP at approximately the same rates and demonstrate only moderate stimulation of soman hydrolysis by Mn^{2+}, and yet are inhibited by mipafox. The Tetrahymena OPA anhydrases fall within a narrow range: from 96,000 to 67,000 D. However, this range of molecular weights is larger than typically ascribed to the Mazur-type enzymes. The Tetrahymena OPA anhydrases can be purified by ammonium sulfate precipitation, like the squid-type OPA anhydrase.

Although possessing very similar kinetics and characteristics in homogenate form (Table 8.3), the OPA anhydrase activities of *T. thermophila* and *R. cuneata* are markedly different after even a simple purification. *R. cuneata* has a low molecular weight activity, *Rc* OPA-1, that is not inhibited by mipafox, and has a molecular weight close to that of the squid type OPA anhydrase. The clam also has a mipafox-hydrolyzing activity that hydrolyzes mipafox faster than DFP.

The bacterial activities again point to the diversity of the OPA anhydrases. The OT strain JD.100 is able to degrade soman, sarin, and dimebu, but not DFP. The bacterial activities reported to date all seem insensitive to ammonium sulfate inhibitions and have molecular weights above that of the hog kidney OPA anhydrase.

The *opd* OPA anhydrase is smaller than the bacterial OPA anhydrase studied to date and has an apparent molecular weight of 60,000 to 65,000 D with 35,000 D subunits. To date, the other bacterial OPA anhydrases have not been tested using paraoxon as a substrate although JD6.5 hydrolyzes the related compound NPEPP.

Even though they are a diverse set of enzymes, some generalizations on the OPA anhydrases can be reached. Generally, the substrate range of the OPA anhydrases is quite broad. Sensitivity to ammonium sulfate is a characteristic found in only a few cases and not in those OPA anhydrases so far examined from aquatic organisms. Subunits have been demonstrated in the case of hog kidney and the *opd* OPA anhydrase and may exist in the larger enzymes in Tetrahymena. A variety of OPA anhydrases seem to exist within an organism, be it a squid, Tetrahymena, clam, or bacteria. Differentiation among OPA anhydrases of various tissues has also been demonstrated.

To date, the active site of the OPA anhydrases has not been mapped by X-ray crystallography, yet some indications of the topography can be made. The size of the leaving group does not seem to be important. Enzymes from Tetrahymena, the *opd* gene, and *R. cuneata* can hydrolyze compounds with both fluoride and nitrophenol leaving groups. It is as if the leaving group is perpendicular to the surface of the enzyme with the remainder of the molecule inserted into the active site. If the mechanism for the *opd* OPA anhydrase can be generalized as an attack at the phosphorus by an activated water, the configuration may be important to catalytic activity. Indeed, small changes in side chains apparently make a tremendous difference: NPEPP is readily hydrolyzed by the Tetrahymena OPA anhydrases, but its close analog NPIPP is not. The squid-type OPA anhydrase does not hydrolyze either the NPEPP or NPIPP. The squid type OPA anhydrase does hydrolyze the four isomers of soman at roughly comparable rates, showing a substrate tolerance of a different sort.

NATURAL ROLE OF THE OPA ANHYDRASES

An enzymatic activity that phylogenetically is as widespread as that of the OPA anhydrases must be important to the cellular metabolism and the survival of the organism. The strength of the selective pressure for the *opd* OPA anhydrase is evident: divergent plasmids in *Pseudomonas* and *Flavobacterium* share identical *opd* gene sequences. The widespread nature of the OPA anhydrases also argues for a strong selective pressure over a much longer period than the last 45 years. However, the natural substrate and role(s) of the OPA anhydrases are unknown. Correlations to isethionate, pyruvate, and squid neurotoxin (Hoskin et al. 1984; Hoskin 1971; Hoskin and Brande 1973) exist but no cause-effect relationship has been found. Generally unrecognized, however, is that many types of naturally occurring organophosphates have been identified from a variety of sources (Rosenburg 1964; Kitteridge and Roberts 1969; Rouser et al. 1963; Simon and Rouser 1967; Quin and Shelburn

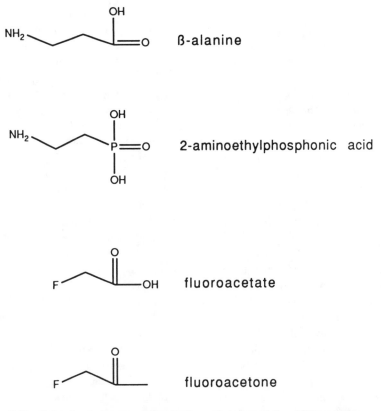

Figure 8.16 Natural substances similar to the substrates of the OPA anhydrases. AEP, naturally synthesized, is an organophosphate analog to the amino acid β-alanine. Several naturally synthesized fluorometabolites are known, fluoroacetate, and fluorocitrate are two examples. OPA anhydrases may be involved in the metabolism of these or similar compounds.

1969; Neidleman and Geigert 1986). The alanine amino acid analog, 2-aminoethylphosphonic acid (AEP) (Figure 8.16), is synthesized by Tetrahymena (Rosenburg 1964). Other types of phosphonates are found free in cells, incorporated into glycerophosphonolipids, sphingophosphonolipids, and phosphonoproteins. The linkage of AEP to phosphonolipids appears to be covalent, although this has not been conclusively demonstrated (Rosenburg 1964; Kitteridge and Roberts 1969; Rouser et al. 1963; Simon and Rouser 1967; Quin and Shelburn 1969; Neidleman and Geigert 1986). It has been previously suggested that the OPA anhydrases are parts of a metabolic system handling the various organophosphonates incorporated into the cellular matrix and encountered in food sources (Landis et al. 1986; Landis et al. 1987; Landis et al. 1989). That hypothesis must be expanded as some OPA anhydrases may also be important in dehalogenation of naturally occurring halogenated organic compounds.

A wide variety of halogenated organics are also naturally occurring. Neidleman and Geigert (1986) reviewed the variety of halometabolites that naturally occur. Chlorotetracycline and chloramphenicol are two important chlorinated halometabolites. Fungi produce a variety of ringed and aromatic chlorinated organics. The richest known source of halometabolites are the marine algae with approximately 20% of the extractable material being halogenated organics. Freshwater blue-green algae also produce halogenated molecules. The variety of halogenated molecules is amazing. The production of halogenated molecules is not restricted to microorganisms or plants since marine animals also produce a variety of bromo-, chloro-, and iodosometabolites. Fluorometabolites are not as common but do occur, especially in higher plants. Fluoroacetate and fluorocitrate are synthesized by a number of plants (Figure 8.16). Fluorinated fatty acids are found in the seeds of *Dichapetalum toxicarium*. The fungi *Streptomyces calvus* produces the fluorinated antibiotic nucleocidin, a adenosine analog. The number of fluorinated organics may even be larger than those currently identified because of the difficulty of distinguishing a C–F bond from a C–H bond (Neidleman and Geigert 1986). With the use of F-electrodes, mass spectrometry, and ion chromatography, the list of fluorinated organics and their degradation products is certain to grow.

Neidleman and Geigert (1986) also review the evidence that halometabolites are used as chemical defense in marine and perhaps other organisms. These organisms range from a green algae, *Avrainvillea longicalulis*, to the Nudibranch mollusk, *Diaulula sandiegensis*. One of the more interesting speculations of Neidleman and Geigert is the role that toxic halogenated compounds may play in prey-predator interactions. Perhaps the synthesis of active halogenated compounds is sufficiently damaging to a predator to reduce the efficiency of the predation or to kill the predator. Competitive relationships among microorganisms may also be mediated by the production of halogenated organics. Detection of the very low concentrations of these molecules appears to be the major stumbling block in further elucidating the role of halogenated organics in predator-prey and competitive relationships.

REFERENCES AND SUGGESTED READINGS

Aldridge, W.N. 1989. A-esterases and B-esterases in perspective. In *Enzymes Hydrolysing Organophosphorus Compounds*. E. Reiner, W.N. Aldridge, and F.C.G. Hoskin, Eds., Ellis Horwood, Chichester, England, pp. 1-14.

Aldridge, W.N., F.C.G. Hoskin, E. Reiner, and C.H. Walker. 1989. Suggestions for a nomenclature and classification of enzymes hydrolyzing organophosphorus compounds. In *Enzymes Hydrolysing Organophosphorus Compounds*. E. Reiner, W.N. Aldridge, and F.C.G. Hoskin, Eds., Ellis Horwood, Chichester, England, pp. 246-253.

Anderson, R.S., H.D. Durst, and W.G. Landis. 1988. Organofluorophosphate-hydrolyzing activity in an estuarine clam *Rangia cuneata*. *Comp. Biochem. Physiol.* 91C:575-578.

Attaway, H., J.O. Nelson, A.M. Baya, M.J. Voll, W.E. White, D.J. Grimes, and R.R. Colwell. 1987. Bacterial detoxification of diisopropylfluorophosphate. *Appl. Env. Micro.* 53:1685-1689.

Awasthi, Y.C., D.D. Dao, and R.P. Saneto. 1980. Interrelationships between anionic and cationic forms of glutathione s-transferases of human liver. *Biochem. J.* 191:1-10.

Bianchi, M.A., R.J. Portier, K. Fujisaki, C.B. Henry, P.H. Templet, and J.E. Matthews. 1988. Determination of optimal toxicant loading for biological closure of a hazardous waste site. In *Aquatic Toxicology and Hazard Assessment,* 10th Volume, ASTM STP 971. W.J. Adams, G.A. Chapman, and W.G. Landis, Eds., American Society for Testing and Materials, Philadelphia, PA, pp. 503-516.

Brown, E.J., J.J. Pignatello, M.M Martinson, and R.L. Crawford. 1986. Pentachlorophenol degradation — a pure bacterial culture and an epilithic microbial consortium. *Appl. Environ. Microbiol.,* 52:92-97.

Brown, K.A. 1980. Phosphotriesterases of flavobacterium sp. *Soil Biol. Biochem.* 12:105-112.

Chemnitus, J., H. Losch, K. Losch, and R. Zech. 1983. Organophosphate detoxicating hydrolases in different vertebrate species. *Comp. Biochem. Physiol.* 76C:85-93.

Chester, N.A., R.S. Anderson, and W.G. Landis. 1988. Mipafox as a substrate for Rangia-DFPase. CRDEC-TR-88153.

Chettur, G., J.J. DeFrank, B.J. Gallo, F.C.G. Hoskin, S. Mainer, F.M. Robbins, K.E. Steinmann, and J.E. Walker. 1988. Soman-hydrolyzing and -deoxifying properties of an enzyme from a therophilic bacterium. *Fund. Appl. Tox.* 11:373-380.

Chiang, T., M.C. Dean, and C.S. McDaniel. 1985. A fruit-fly bioassay with phosphotriesterase for detection of certain organophosphorus insecticide residues. *Bull Environ. Contam. Toxicol.* 34:809-814.

Crawford, R.L. and W.W. Mohn. 1985. Microbiological removal of pentachlorophenol from soil using a flavobacterium. *Enzyme Microbiol Technol.,* 7:617-620.

DeFrank, J.J., personal communication, December 1988.

Donarski, W.J., D.P. Dumas, D.P. Heitmeyer, V.E. Lewis, and F.M. Raushel, 1988. Structure-activity relationships in the hydrolysis of substrates by phosphotriesterase from *Pseudomonas diminuta. Biochemistry* 28:4650-4655.

Dumas, D.P., H.D. Durst, W.G. Landis, F.M. Raushel, and J.R. Wild. 1990. Inactivation of organophosphorus nerve agents by the phosphotriesterase from *Pseudomonas diminuta. Arch. Biochem. Biophy.,* 277:155-159.

Dumas, D.P., J.R. Wild, and F.M. Raushel. 1989. Diisopropylfluorophosphate hydrolysis by a phosphotriesterase from *Pseudomonas diminuta. J. Appl. Biotech.* 11:235-243.

Garden, J.M., S.K. Hause, F.C.G. Hoskin, and A.H. Roush. 1975. Comparison of DFP-hydrolyzing enzyme purified from head ganglion and hepatopancreas of squid (*Loligo pealei*) by means of isoelectric focusing. *Comp. Biochem. Physiol.* 52C:95-98.

Gay, D.D. and F.C.G. Hoskin. 1979. Stereospecificity and active-site requirements in a diisopropylphorofluoridate-hydrolyzing enzyme. *Biochem. Pharmacol.,* 128:1259-1261.

Haley, M.V. and W.G. Landis. 1988. Confirmation of multiple organofluorophate hydrolyzing activities in the protozoan *Tetrahymena thermophila.* CRDEC-TR-88009, pp. 1-11.

Harper, B.G., L.P. Midgley, I.G. Resnick, and W.G. Landis. 1986. Scale-up production and purification of diisopropylfluorophosphotase from *Tetrahymena thermophila. Proceedings of the 1986 Army Science Conference.*

Heymann, E. 1989. A proposal to overcome some general problems of the nomenclature esterases. In *Enzymes Hydrolysing Organophosphorus Compounds.* E. Reiner, N. Aldridge, and F.C.G. Hoskin, Eds., Ellis Horwood, Chichester, England, 226-235.

Hogan, J.W. and C.O. Knowles. 1968. Some enzymatic properties of brain acetylcholinesterase from bluegill and channel catfish. *J. Fish. Res. Bd. Can.,* 25:1571-1579.

Hoskin, F.C.G. 1971. Diisoprophylphosphorofluoridate and tabun enzymatic-hydrolysis and nerve function. *Science* 172, p. 1243.

Hoskin, F.C.G. 1976. Distribution of diisopropylphosphorofluoridate hydrolyzing enzyme between sheath and axoplasm of squid giant-axon. *J. Nerochem.* 26:1043-1045.

Hoskin, F.C.G. 1985. Inhibition of soman- and di-isopropylphosphorofluoridate (DFP)-detoxifying enzyme by mipafox. *Biochem. Pharmacol.* 34:2069-2072.

Hoskin, F.C.G. 1989. An organophosphorus detoxifying enzyme unique to cephalopods. In *Squid as Experimental Animals.* D.L. Gilbert, W.J. Adelman, Jr., and J.M. Arnold, Eds., Plenum Press, New York, pp. 469-480.

Hoskin, F.C.G. and M. Brande. 1973. An improved sulfur assay applied to a problem of isethione metabolism in squid axon and other nerves. *J. Neurochem.* 20:1317-1327.

Hoskin, F.C.G., G. Chettur, S. Mainer, and K.E. Steinmann. 1989. Soman hydrolysis and detoxification by a thermophilic bacterial enzyme. In *Enzymes Hydrolyzing Organophosphorus Compounds.* E. Reiner, F.C.G. Hoskin, and N.W. Aldridge, Eds., Ellis Horwood, Chichester, England, pp. 53-64.

Hoskin, F.C.G., M.A. Kirkish, and K.E. Steinman. 1984. Two enzymes for the detoxification of organophosphorus compounds — sources, similarities, and significance. *Fund. Appl. Toxicol,* 4:5165-5172.

Hoskin, F.C.G. and R.J. Long. 1972. Purification of DFP-hydrolyzing enzyme from squid head ganglion. *Arch Biochem. Biophys.* 150:548-555.

Hoskin, F.C.G., P. Rosenberg, and M. Brzin. 1966. Re-examination of the effect of DFP and electrical cholinesterase activity of squid giant nerve axon. *Proc. Natl. Acad. Sci. USA* 55:1231-1235.

Hoskin, F.C.G. and A.H. Rouch. 1982. Hydrolysis of nerve gas by squid type diisopropylphosphorofluoridate hydrolyzing enzyme on agarose beads. *Science,* 215:1255-1257.

Hoskin, F.C.G. and G.S. Trick. 1955. Stereospecificity in the enzymatic hydrolysis of tabun and acetyl-B-methylcholine chloride. *Can. J. Biochem. Physiol.* 3:963-969.

Jakoby, W. B. 1978. The glutathione s-transferases: a group of multifunctional detoxification proteins. *Adv. Enzymol.* 46:383-414.

Kappas, A. and A.P. Alvares. 1975. How the liver metabolizes foreign substances. *Scientific Amer.* 232(6)22-31.

Kitteridge, J.S. and E. Roberts. 1969. A carbon-phosphorus bond in nature. *Science* 164:37-42.

Landis, W.G., R.S. Anderson, N.A. Chester, H.D. Durst, M.V. Haley, D.W. Johnson, and R.M. Tauber. 1989. The organophosphate acid anhydrases of the protozoan, *Tetrahymena thermophila,* and the clam, *Rangia cuneata.* In *Aquatic Toxicology and Environmental Fate:* 12th Volume. ASTM STP 1027. U.M. Cowgill and L.R. Williams, Eds., American Society for Testing and Materials, Philadelphia, PA, pp. 74-81.

Landis, W.G., N.A. Chester and R.S. Anderson. 1989. Identification and comparison of the organophosphate acid anhydrase activities of the clam *Rangia cuneata. Comp. Biochem. Physiol.* 94C:365-371.

Landis, W. G., N.A. Chester, M.V. Haley, D.W. Johnson, R.M. Tauber, and H.D. Durst. 1989. Alternative substrates and an inhibitor of the organophosphate acid anhydrase activities of the protozoan *Tetrahymena thermophila. Comp. Biochem. Phys.* 94C:365-371.

Landis, W.G., H.D. Durst, R.E. Savage, Jr., D.M. Haley, M.V. Haley, and D.W. Johnson. 1987. Discovery of multiple organofluorophosphate hydrolyzing activities in the protozoan *Tetrahymena thermophila. J. Appl. Toxicol,* 7:35-41.

Landis, W.G., M.V. Haley, and D.W. Johnson. 1986. Kinetics of the DFPase activity in *Tetrahymena thermophila. J. Protozool.* 32:216-218.

Landis, W.G., R.E. Savage, Jr., and F.C.G. Hoskin. 1985. An organofluorophosphate hydro-lyzing activity in *Tetrahymena thermophila*. *J. Protozool*. 32:517-519.

LeBlanc, G.A. and B.J. Cochrane. 1985. Modulation of substrate-specific glutathione s-transferase activity in *Daphnia magna* with concomitant effects on toxicity tolerance. *Comp. Biochem. Physiol*. 82C:37.

Lech, J.J. and M.J. Vodicnik. 1985. Biotransformation. In *Fundamentals of Aquatic Toxicology*. G.M. Rand and S.R. Petrocelli, Eds., Hemisphere Publishing Corporation, New York, p. 526.

Lee, M.D., J.M. Thomas, R.C. Borden, P.B. Bedient, C.H. Ward, and J.T. Wilson. 1988. Biorestoration of aquifers contaminated with organic-compounds. *Crit. Rev. Environ. Contr*. 18:29-89.

Lewis, V.E., W.J. Donarski, J.R. Wild, and F.M. Raushel. 1988. Mechanism and stereochemical course at phosphorus of the reaction catalyzed by a bacterial phosphotriesterase. *Biochemistry*, 27:1591-1597.

Lippmann, M. and R.B. Schlesinger. 1979. *Chemical Contamination in the Human Environment*. Oxford University Press, New York.

Mackness, M. I., H.M. Thompson, A.R. Hardy, and C.H. Walker. 1987. Distinction between 'A'-esterases and arylesterases. *Biochem, J*. 245:293-296.

Mazur, A. 1946. An enzyme in animal tissue capable of hydrolyzing phosphorus-fluorine bond alkyl fluorophosphates. *J. Biol. Chem*. 164:271-289.

McDaniel, C.S. 1985. Plasmid-mediated Degradation of Organophosphate Pesticides. Ph.D. dissertation, Texas A&M University.

McDaniel, C.S., L.L. Harper, and J.R. Wild. 1988. Cloning and sequencing of a plasmid-borne gene (opd) encoding a phosphotriesterase. *J. Bacteriol*. 170:2306-2311.

Mottet, N.K., Ed. 1985. *Environmental Pathology*. Oxford University Press, New York.

Mounter, L.A. 1963. Metabolism of organophosphorus and anticholinesterase agents. In *Hanbuch de Experimentellen Pharmakologie: Cholinesterases and Anticholinesterase Agents*. G.B. Kolle, Ed., Springer-Verlag, Berlin, pp. 486-504.

Mounter, L.A., R.F. Baxter, and A. Chanutin. 1955. Dialkylfluorophosphates of microorganisms. *J. Biol. Chem*. 215:699-704.

Neidleman, S.L. and J. Geigert. 1986. *Biohalogenation: Principals, Basic Roles and Applications,* Ellis Horwood, Chichester, England, pp. 128-130.

Noellgen, R. 1992. Characterization of the Organophosphate Acid Anhydrase in the Marine Mussel *Mytilus Edulis*, M.S. thesis, Western Washington University.

Omenn, G.S. 1988. *Environmental Biotechnology: Reducing Risks from Environmental Chemicals through Biotechnology*. Basic Life Sciences, Vol. 45, Plenum Press, New York.

Omenn, G.S., and A. Hollaender. 1984. *Genetic Control of Environmental Pollutants*. Basic Life Sciences, Vol. 28, Plenum Press, New York.

Peterson and Holtzman. 1980. In *Extrahepatic Metabolism of Drugs and Other Foreign Compounds*. T.E. Gram, Ed., Spectrum Publications, Jamaica, New York.

Picardi, A., P. Johnston, and R. Stringer. 1991. *Alternative Technologies for the Detoxification of Chemical Weapons: An Information Document,* Greenpeace International, Washington D.C.

Portier, R. and K. Fujisaki. 1988. Enhanced biotransformation and biodegradation of polychlorinated biphenyls in the presence of aminopolysaccharides. In *Aquatic Toxicology and Hazard Assessment,* 10th Volume. ASTM STP 971. W.J. Adams, G.A. Chapman, and W.G. Landis, Eds., American Society for Testing and Materials, Philadelphia, PA, pp. 517-527.

Quin, L.D. and F.A. Shelburn. 1969. An examination of marine animals for the presence of carbon bound phosphorus. *J. Marine Res.,* 27:73-84.

Reiner, E., F.C.G. Hoskin, and N.W. Aldridge. 1989. *Enzymes Hydrolyzing Organophosphorus Compounds,* Ellis Horwood, Chichester, England.

Reynolds, E.S., and M.T. Moslen. 1980. Environmental liver injury: Halogenated hydrocarbons. In *Toxic injury of the liver.* E. Farber and M.F. Fisher, Eds., Marcel Dekker, New York, pp. 541-596.

Rochkind, M.L., J.W. Blackburn, and G.S. Sayler. 1986. *Microbial Decomposition of Chlorinated Aromatic Compounds.* EPA/600/2-86/090.

Rosenburg, H. 1964. Distribution and fate of 2-aminoethylphosphonic acid in Tetrahymena. *Nature* (London), 203:299-300.

Rouser, G., G. Kritchevsky, D. Heller, and E. Lieber. 1963. Lipid composition of beef brain, beef liver, and the sea anemone: two approaches to quantitative fraction of complex lipid mixtures *J. Am. Oil Chem. Soc.* 40:425-545.

Saber, D.L. and R.L. Crawford. 1985. Isolation and characterization of flavobacterium strains that degrade pentachlorophenol. *Appl. Environ. Microbiol.,* 50:1512-1518.

Simon, G. and G. Rouser. 1967. Phospholipids of the sea anemone: quantitative distribution; absence of carbon-phosphorus linkages in glycerol phospholipids; structural elucidation of ceramide aminoethylphosphonate. *Lipids* 2:55.

Steiert, J.G. and R.L. Crawford. 1985. Microbial degradation of chlorinated phenols. *Trends Biotechnol.,* 3:300-305.

Steinmann, K.E. 1988. Personal Communication.

Stenersen, J. and N. Oien. 1981. Glutathione s-transferases in earthworms (Lubricidae) substrate-specificity, tissue and species distribution, and molecular weight. *Comp. Biochem. Physiol.* 69:243-252.

Storkebaum, W. and H. Witzel. 1975. Study on the enzyme catalyzed splitting of triphosphates. Forschungsber. *Landes Nordrhein-Westfalen* No. 2523:1-22.

Van Bladeren, P.J., D.D. Briemer, G.M.T. Rottiveel-Smijs, R.A.W. deJong, W. Buijs, A. Van der Gen, and G.R. Mohn. 1980. The role of glutathione conjugation in the mutagenicity of 1,2-dibromoethane. *Biochem. Pharmacol.,* 29: 2975-2982.

Walker, C.H. 1989. The development of an improved system of nomenclature and classification of esterases. in *Enzymes Hydrolysing Organophosphorus Compounds.* E. Reiner, W.N. Aldridge, and F.C.G. Hoskin, Eds., Ellis Horwood, Chichester, England, pp. 236-245.

Walker, C.H. and M.I. Mackness. 1983. Esterases: Problems of identification and classification. *Biochem. Pharmocol.,* 32(22):3265-3269.

Walls, L.H. 1987. Isolation and purification of a diisopropyl phosphorofluoridate hydrolase from thermophilic bacteria. CRDEC-CR-87072.

Wild, J.R. and F.M. Raushel. 1988. Personal Communication.

Wislocki, P.G., G.T. Mirva, and A.Y.H. Yu. 1980. Reactions catalyzed by the cytochrome P-450 system. In *Enzymatic basis of detoxication,* Vol. I., W.B. Jakoby, Ed., Academic Press, New York, pp. 135-182.

Zech, R. and K.D. Wigand. 1975. Organophosphate-detoxifying enzymes in *E. coli.* Gel filtration and isoelectric focusing of DFPase, paraoxonase, and unspecified phosphohydrolases. *Experientia.,* 31:157-158.

STUDY QUESTIONS

1. What is biotransformation of an environmental chemical? Where does it occur?
2. What are Phases I and II in the process of xenobiotic metabolism?
3. Describe the NADPH-cytochrome P-450 system.
4. Hepatic enzymes that catalyze Phase I and II reactions perform what functions in addition to detoxifying xenobiotics?
5. Discuss the conversion of xenobiotics to reactive electrophilic species by hepatic biotransformation mechanisms.
6. Many microorganisms have the ability to use xenobiotics for what use?
7. Discuss the genetic information contained in microorganisms, including their functions and origins.
8. Discuss the aerobic metabolism of organic xenobiotics.
9. How could the degradative capability of microorganisms be enhanced?
10. How is biodegradation of a substance measured? What nonspecific methods can be used as alternatives?
11. Describe the degradation of PCP by bacteria and fungi.
12. How can biodegradation be used for remediation?
13. Explain the use of a bioreactor as a bioremediation tool. What factors determine the success of the bioremediation attempt?
14. Discuss the isolation and engineering of degradative organisms.
15. What are OPA anhydrolases?
16. Summarize the hypothesized natural role of the OPA anhydrolases.

CHAPTER 9

Measurement and Evaluation of the Ecological Effects of Toxicants

INTRODUCTION

This chapter deals with perhaps the most difficult topic in environmental toxicology, how to measure and then evaluate the impact of toxicants at ecological levels of organization. The chapter starts with an evaluation of methods and ends with a discussion of the responses of ecosystems to chemical stressors.

MEASUREMENT OF ECOLOGICAL EFFECTS AT VARIOUS LEVELS OF BIOLOGICAL ORGANIZATION

Biomonitoring is a term that implies a biological system is used in some way for the evaluation of the current status of an ecosystem. Validation as to the predictions and protections derived from the elaborate series of tests and our understanding presented in previous chapters can only be done by effective monitoring of ecosystems (Landis 1991). In general, biomonitoring programs fall into two categories, exposure and effects. Many of the traditional monitoring programs involve the analytical measurement of a target compound with the tissue of a sampled organism. The examination of pesticide residues in fish tissues, or PCBs in terrestrial mammals and birds are examples of this application of biomonitoring. Effects monitoring looks at various levels of biological organization to evaluate the status of the biological community. Generically, effects monitoring allows a toxicologist to perform an evaluation without an analytical determination of any particular chemical concentration. Synergistic and antagonistic interactions within complex mixtures are integrated into the biomonitoring response.

In the biomonitoring process, there is the problem of balancing specificity with the reliability of seeing an impact (Figure 9.1). Specificity is important since it is crucial to know and understand the causal relationships in order to set management

Biomonitoring Tug of War

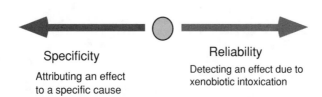

Specificity

Attributing an effect
to a specific cause

Reliability

Detecting an effect due to
xenobiotic intoxication

Figure 9.1   The tug-of-war in biomonitoring. An organismal or community structure monitoring
system may pick up a variety of effects but lack the ability to determine the precise
cause. On the other hand, a specific test, such as looking at the inhibition of a
particular enzyme system, may be very specific but completely miss other modes
of action.

or clean up strategies. However, an increase in specificity generally results in a focus
on one particular class of causal agent and effects, and in many cases chemicals are
added to ecosystems as mixtures. Emphasis upon a particular causal agent may mean
that effects due to other materials can be missed. A tug-of-war exists between
specificity and reliability.

There is a continuum of monitoring points along the path that an effect on an
ecosystem takes from introduction of a xenobiotic to the biosphere to the final series
of effects (Chapter 2). Techniques are available for monitoring at each level, although they are not uniform for each class of toxicant. It is possible to outline the
current organizational levels of biomonitoring:

- Bioaccumulation/biotransformation/biodegradation
- Biochemical monitoring
- Physiological and behavioral
- Population parameters
- Community parameters
- Ecosystem effects

A graphical representation of the methods used to examine each of these levels are
depicted in Figure 9.2.

Many of these levels of effects can be examined using organisms native to the
particular environment, or exotics planted or introduced by the researcher. There is
an interesting trade-off for which species to use. The naturally occurring organism
represents the population and the ecological community that is under surveillance.
There is no control over the genetic background of the observed population and little
is usually known about the native species from a toxicological viewpoint. Introduced
organisms, either placed by the researcher or enticed by the creation of a habitat, have
the advantage of a database and some control over the source. Questions dealing with

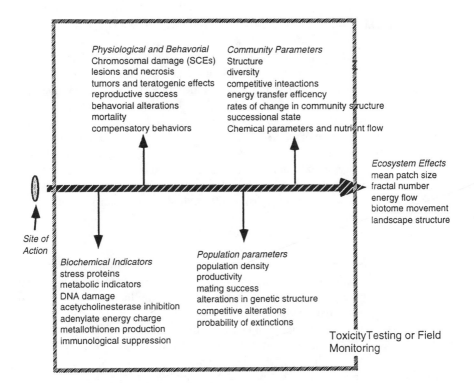

Figure 9.2 Methods and measurements used in biomonitoring for ecological effects. A number of methods are used both in a laboratory situation and in the field to attempt to classify the effects of xenobiotics upon ecological systems. Toxicity tests can be used to examine effects at several levels of biological organization and can be performed with species introduced as monitors for a particular environment.

the realism of the situation and the alteration of the habitat to support the introduced species can be raised.

It may also prove useful to consider a measure of biomonitoring efficacy as a means to judge biomonitoring. Such a relationship may be expressed in the terms of a safety factor as

$$E = \frac{U_i}{B_i} \qquad (9.1)$$

Where E is the efficacy of the biomonitoring methodology, U_i is the concentration at which undesirable effects upon the population or ecosystem in system i occur, and B_i is the concentration at which the biomonitoring methods can predict the undesirable effect or effects in system i. The usefulness of such an idea is that it measures the ability to predict a more general effect. Methods that can predict effects rather than observe detrimental impacts are under development. Several of the methods discussed below are developments that may have a high efficacy factor.

BIOACCUMULATION/BIOTRANSFORMATION/BIODEGRADATION

Much can happen to the introduced pesticide or other xenobiotic from its intro-duction to the environment to its interaction at the site of action. Bioaccumulation often occurs with lipophilic materials. Tissues or the entire organism can be analyzed for the presence of compounds such as PCBs and halogenated organic pesticides. Often the biotransformation and degradation products can be detected. For example, DDE is often an indication of past exposure to DDT. With the advent of DNA probes it may even be possible to use the presence of certain degradative plasmids and specific gene sequences as indications of past and current exposure to toxic xenobiotics. Biosensors are new developments that also may hold promise as new analytical tools. In this new class of sensors a biological entity such as the receptor molecule or an antibody for a particular xenobiotic is bound to an appropriate electronic sensor. A signal can then be produced as the material bound to the chip interacts with the toxicant.

One of the great advantages to the analytical determination of the presence of a compound in the tissue of an organism is the ability to estimate exposure of the material. Although exposure cannot necessarily be tied to effects at the population and community levels, it can assist in confirming that the changes seen at these levels are due to anthropogenic impacts and are not natural alterations. The difficulties in these methods lay in the fact that it is impossible to measure all compounds. Therefore, it is necessary to limit the scope of the investigation to suspect compounds or to those required by regulation. Compounds in mixtures can be at low levels, even those not detected by analytical means, yet in combination can produce ecological impacts. It should always be noted that analytical chemistry does not measure toxicity. Although there is a correspondence, materials easily detected analytically may not be bioavailable, and conversely, compounds difficult to measure may have dramatic effects.

MOLECULAR AND PHYSIOLOGICAL INDICATORS
OF CHEMICAL STRESS — BIOMARKERS

A great deal of research has been done recently on the development of a variety of molecular and physiological tests to be used as indicators and perhaps eventually as predictors of the effects of toxicants. McCarthy and Shugart (1990) have recently published a book reviewing in detail a number of biomarkers and their use in terrestrial and aquatic environments. The collective term, biomarkers, has been given to these measurements, although they are a diversified set of measurements ranging from DNA damage to physiological and even behavioral indices. To date, biomarkers have not proven to be predictive of effects at the population, community, or ecosys-tem levels of organization. However, these measurements have demonstrated some usefulness as measures of exposure and can provide clinical evidence of causative agent. The predictive power of biomarkers is currently a topic of research interest.

Biomarkers have been demonstrated to act as indicators of exposure (Fairbrother 1989). Often specific enzyme systems are inhibited by only a few classes of materials. Conversely, induction of certain detoxification mechanisms, such as specific mixed function oxidases, can be used as indications of the exposure of the organism to specific agents, even if the agent is currently below detectable levels. Additionally, the presence of certain enzymes in the blood plasma, which is generally contained in a specific organ system, can be a useful indication of lesions or other damage to that specific organ. These uses justify biomarkers as a monitoring tool even if the predictive power of these techniques has not been demonstrated. The following discussion is a brief summary of the biomarkers, currently under investigation.

ENZYMATIC AND BIOCHEMICAL PROCESSES

The inhibition of specific enzymes such as acetylcholinesterase has proven to be a popular biomarker and with justification. The observation is at the most basic level of toxicant-active site interaction. Measurement of acetylcholinesterase activity has been investigated for a number of vertebrates, from fish to birds to man. It is also possible to examine cholinesterase inhibition without the destruction of the organism. Blood plasma acetyl and butyl cholinesterase can be readily measured. The drawbacks to using blood samples are the intrinsic variability of the cholinesterase activity in the blood due to hormonal cycles and other causes. Brain cholinesterase is a more direct measure, but requires sacrifice of the animal. Agents exist that can enhance the recovery of acetylcholinesterase from inhibition by typical organophosphates, providing a measure of protection due to an organophosphate agent.

Not only are enzyme activities inhibited, but they can also be induced by a toxicant agent. Quantitative measures exist for a broad variety of these enzymes. Mixed-function oxidases are perhaps the best studied with approximately 100 now identified from a variety of organisms. Activity can be measured or the synthesis of new mixed-function oxidases may be identified using antibody techniques. DNA repair enzymes can also be measured and their induction is an indication of DNA damage and associated genotoxic effects.

Not all proteins induced by a toxicant are detoxification enzymes. Stress proteins are a group of molecules that have gathered a great deal of attention in the last several years as indicators of toxicant stress. Stress proteins are involved in the protection of other enzymes and structure from the effects of a variety of stressors (Bradley 1990). A specialized group, the heat shock proteins (hsps) are a varied set of proteins with four basic ranges of molecular weights 90, 70, 58 to 60, and 20 to 30 kD. A related protein, ubiquitin, has an extremely small molecular weight, 7 kD. Although termed heat shock proteins, stressors other than heat are known to induce their formation. The exact mechanism is not known. Other groups of stress-related proteins are also known. The glucose regulated proteins are 100 to 75 kD and form another group of proteins that respond to a variety of stressors.

The stress-related proteins discussed above are induced by a variety of stressors. However, other groups of proteins are induced by specific materials. Metallothioneins are proteins that are crucial in reducing the effects of many heavy metals. Originally evolved as important players in metal regulation, these proteins sequester heavy metals and thereby reduce the toxic effects. Metallothioneins are induced and like many proteins can be identified using current immunological techniques.

At an even more fundamental level there are several measurements that can be made to examine damage at the level of DNA and the associated chromosomal material (Shugart 1990; Powell and Kocan 1990). DNA strand breakage, unwinding of the helix, and even damage to the chromosomal structure can be detected. Formation of micronuclei as remnants of chromosomal damage can be observed. Some toxics bind directly to the DNA causing an adduct to form. Classical mutagens can actually change the sequence of the nucleotides, cause deletions or other types of damage.

Immunological endpoints can provide evidence of a subtle, but crucial indication of a chronic impact to an organism or its associated population (Anderson 1975; Anderson et al. 1981). Most organisms have cells that perform immunological functions, and perhaps the most common are the many types of macrophages. Toxicants can either enhance or inhibit the action of macrophages in their response to bacterial challenges. Rates of phagocytosis in the uptake of labeled particles can be used as an indicator of immune activation or suppression. The passage of macrophages, recently obtained from the organisms under examination, can be examined as they pass through microscopic pores as they are attracted to a bacterial or other immunological stimulus. Macrophage immunological response is widespread and is an important indicator of the susceptibility of the test organisms to disease challenges.

Birds and mammals have additional immunological mechanisms and can produce antibodies. Rates of antibody production, the existence of antibodies against specific challenges, and other measures of antibody-mediated immunological responses should prove useful in these organisms.

PHYSIOLOGICAL AND HISTOLOGICAL INDICATORS

Physiological and behavioral indicators of impact within a population are the classical means by which the health of populations are assessed. The major drawback has been the extrapolation of these factors based upon the health of an individual organism, attributing the damage to a particular pollutant, and extrapolating this to the population level.

As described in earlier chapters, toxicants can cause a great deal of apparent damage that can be observed at the organismal level. Animals often exhibit deformations in bone structure, damage to the liver and other organs, and alterations in bone structure at the histological and morphological levels. Changes in biomass and overall morphology can also be easily observed. Alterations to the skin and rashes are often indicators of exposure to an irritating material. Plants also exhibit readily observed damage that may be linked to toxicant impact. Plants can exhibit chlorosis, a fading of green color due to the lack of production or destruction of chlorophyll. Necrotic tissues can also be found on plants and are often an indicator of airborne pollutants. Histological indicators for both plants and animals include various lesions, especially due to irritants or materials that denature living tissue. Cirrhosis is often an indication of a variety of stresses. Parasitism at abnormally high levels in plants or animals also indicate an organism under stress.

Lesions and necrosis in tissues have been the cornerstone of much environmental pathology. Gills are sensitive tissues and often reflect the presence of irritant materials. In addition, damage to the gills has an obvious and direct impact upon the health of the organism. Related to the detection of lesions are those that are tumorogenic. Tumors in fish, especially flatfish, have been extensively studied as indicators of oncogenic materials in marine sediments. Oncogenesis has also been extensively studied in Medaka and trout as a means of determining the pathways responsible for tumor development. Development of tumors in fish more commonly found in natural communities should follow similar mechanisms. As with many indicators used in the process of biomonitoring, relating the effect of tumor development to the health and reproduction of a wild population has not been as closely examined as the endpoint.

Blood samples and general hematology are additional indicators of organisms with organ damage or metabolic alterations. Anemia can be due to a lack of iron or an inhibition of hemoglobin synthesis. Abnormal levels of various salts, sodium, potassium, or metals such as calcium, iron, copper, or lead can give direct evidence as to the causative agent.

Perhaps most promising in a clinical sense is the ability to detect enzymes present is the blood plasma due to the damage and subsequent lesion of organs. Several enzymes such as the LDHs are specific as to the tissue. Presence of an enzyme not normally associated with the blood plasma can provide specific evidence for organ system damaged and perhaps an understanding of the toxicant.

Cytogenetic examination of miotic and mitotic cells can reveal damage to genetic components of the organism. Chromosomal breakage, micronuclei, and various trisomies can be detected microscopically. Few organisms, however, have the requisite chromosomal maps to accurately score more subtle types of damage. Properly developed, cytogenetic examinations may prove to be powerful and sensitive indicators of environmental contamination for certain classes of materials.

Molecular and physiological indicators offer specific advantages in monitoring an environment for toxicant stressors. Many enzymes are induced or inhibited at low concentrations. In addition, the host organism samples the environment in an ecologically relevant manner for that particular species. Biotransformation and detoxification processes are included within the test organism, providing a realistic metabolic pathway that is difficult to accurately simulate in laboratory toxicity tests used for biomonitoring. If particular enzyme systems are inhibited, it is possible to set a lower limit for environmental concentration when the kinetics of site of action/toxicant interaction are known. The difficulties with molecular markers, however, must be understood. In the case of stress proteins and their relatives they are induced by a variety of anthropogenic and natural stressors. It is essential that the interpretation is made with as much detailed knowledge of the normal cycles and natural history of the environment as possible. Likewise, immunological systems are affected by numerous environmental factors that are not toxicant related. Comparisons to populations at similar but relatively clean reference sites are essential to distinguish natural from anthropogenic stressors. Shugart has long maintained that a variety of molecular markers be sampled, thereby increasing the opportunities to observe effects and examine patterns that may tell a more complete story.

Perhaps the best recent example of using a suite of biomarkers is the investigation by Theodorakis et al. (1992) using bluegill sunfish and contaminated sediments. Numerous biomarkers were used, including stress proteins, EROD (ethoxyresorufin-O-deethylase activity), liver and spleen somatic indexes, and DNA adducts and strand breaks. Importantly, patterns of the biomarkers were similar in the laboratory bluegills to the native fish taken from contaminated areas. Some of the biomarkers responded immediately, such as the ATPase activities of intestine and gill. Others were very time dependent, such as EROD and DNA adducts. These patterns should be considered when attempting to extrapolate to population or higher level responses.

Currently, it is not possible to accurately transform data gathered from molecular markers to predict effects at the population and community levels of organization. Certainly, behavioral alterations caused by acetylcholinesterase inhibitors may cause an increase in predation or increase the tendency of a parent to abandon a brood, but the long-term populational effects are difficult to estimate. In the estimation and classification of potential effects it may be the pattern of indicators rather than the simple occurrence of one that is important.

TOXICITY TESTS AND POPULATION LEVEL INDICATORS

Perhaps the most widely employed method of assessing potential impacts upon ecological systems has been the array of effluent toxicity tests used in conjunction with National Pollution Discharge and Elimination System (NPDES) permits. These tests are now being required by a number of states as a means of measuring the toxicity of discharges into receiving waters. Often the requirements include an invertebrate such as Ceriodaphnia acute or chronic tests, toxicity tests using a variety of fish, and in the case of marine discharges, echinoderm species. These tests are a means of directly testing the toxicity of the effluent although specific impacts in the discharge area have been difficult to correlate. Since the tests require a sample of effluent and take several days to perform, continuous monitoring has not proven successful using this approach.

Although not biomonitoring in the sense of sampling organisms from a particular habitat, the use of the cough response and ventilatory rate of fish has been a promising system for the prevention of environmental contamination (van der Schalie 1986). Pioneered at Virginia Polytechnic Institute and State University, the measurement of the ventilatory rate of fish using electrodes to pick up the muscular contractions of the operculum has been brought to a very high stage of refinement. It is now possible to continually monitor water quality as perceived by the test organisms with a desktop computer analysis system at relatively low cost. Although the method has now been available for a number of years it is not yet in widespread use.

This reaction of the fish to a toxicant has promise over conventional biomonitoring schemes in that the method can prevent toxic discharges into the receiving environment. Samples of the effluent can be taken to confirm toxicity using conventional methods. Analytical processes can also be incorporated to attempt to identify the toxic component of the effluent. Drawbacks include the maintenance of the fish facility, manpower requirements for the culture of the test organisms, and the costs of false positives. The ecological relevance of such subtle physiological markers can be questioned; however, sensitive measures such as the cough response have proven successful in several applications.

An ongoing trend in the use of toxicity tests designed for the monitoring of effluents and receiving waters has been in the area of toxicity identification evaluation and toxicity reduction evaluations (TIE/TRE). TIE/TRE programs have as their goal the reduction of toxicity of an effluent by the identification of the toxic component and subsequent alteration of the manufacturing or the waste-treatment process to reduce the toxic load. Generally, an effluent is fractioned into several components by a variety of methods. Even gross separations into particulate and liquid phase can be used as the first step to the identification of the toxic material. Each component of the effluent is then tested using a toxicity test to attempt to measure the fraction generating the toxicity. The toxicity test is actually being used as a bioassay or a measure using biological processes of the concentration of the toxic material in the effluent. Once the toxicity of the effluent has been characterized, changes in the manufacturing process can then proceed to reduce the toxicity. The effects of these changes can then be tested using a new set of fractionations and toxicity tests. In some cases simply reducing ammonia levels or adjusting ion concentrations can significantly reduce toxicity. In other cases, biodegradation processes may be important in reducing the concentrations of toxicants. Again questions as to the type of toxicity tests to be used and the overall success in reducing impacts to the receiving ecosystem exist; however, as a means for reducing the toxicant burden, this approach is useful.

In addition to monitoring effluents, toxicity tests have also been proven useful in the mapping of toxicity in a variety of aquatic and terrestrial contaminated sites. Sediments of both freshwater and marine systems are often examined for toxicity using a variety of invertebrates. Water samples may be taken from suspected sites and tested for toxicity using the methods adopted for effluent monitoring. Terrestrial sites are often tested using a variety of plant and animal toxicity tests. Soils elutriates can be tested using species such as the fathead minnow. Earthworms are popular test organisms for soils and have proven to be straightforward test organisms.

The advantages to the above methods are that they measure toxicity and are rather comparable in design to the traditional laboratory toxicity test. Many of the controls possible with laboratory tests and the opportunity to run positive and negative references can assist in the evaluation of the data. However, there are some basic drawbacks to the utility of these methods. As with the typical NPDES monitoring tests, the samples project only a brief snapshot of the spatial and temporal distribution of the toxicant. Soils, sediments, and water are mixed with media that may change the toxicant availability or nutritional state of the test organism. Non-native species typically are used since the development of culture media and methods is a time-consuming and expensive process. A preferable method may be the introduction of free ranging or foraging organisms that can be closely monitored for the assessment of the actual exposure and the concomitant effects upon the biota of a given site.

SENTINEL ORGANISMS OR *IN SITU* BIOMONITORING

In many instances, monitoring of an ecosystem has been attempted by the sampling of organisms from a particular environment. Another approach has been the introduction of organisms that can be readily recovered. Upon recovery, these organisms can be measured and subjected to a battery of biochemical, physiological, and histological tests. Lower and Kendall (1990) have recently published a book about these methods for terrestrial systems.

Reproductive success is certainly another measure of the health of an organism and is the principal indicator of the Darwinian fitness. In a laboratory situation, it certainly is possible to measure fecundity and the success of offspring in their maturation. In nature, these parameters may be very difficult to measure accurately. Sampling of even relatively large vertebrates is difficult and mark-recapture methods have a large degree of uncertainty associated with them. Radio-telemetry of organisms with radio collars is perhaps the preferred way of collecting life-history data on organisms within a population. Plants are certainly easier to mark and it is easier to make note of life span, growth, disease, and fecundity in number of seeds or shoots produced. In many aquatic environments, the macrophytes and large kelp can be examined. Large plants form an important structural as well as functional component of systems, yet relatively little data exist for the adult forms.

It is sometimes possible to introduce organisms into the environment and confine them so that recapture is possible. The resultant examinations are used to measure organismal and populational level factors. This type of approach has been in wide-spread use. Mussels, *Mytilus edulis*, have been placed in plastic trays and suspended in the water column at various depths to examine the effects of suspected pollutants upon the rate of growth of the organism (Nelson 1990; Stickle et al. 1985). Sessile organisms or those easily contained in an enclosure have a tremendous advantage over free ranging organisms. A difficulty in such enclosure-type experiments is maintaining the same type of nutrients as the reference site so that effects due to habitat differences other than toxicant concentration can be eliminated.

The introduction of sentinel organisms has also been accomplished with terrestrial organisms. Starling boxes have been used by Kendall and others and are set up in areas of suspected contamination so that nesting birds would occupy the area.

Exposure to the toxicant is difficult to accurately gauge since the adults are free to range and may limit their exposure to the contaminated site during foraging. However, exposure to airborne or gaseous toxicants may be measurable given these methods.

Birds contained in large enclosures in a suspected contaminated site or a site dosed with a compound of interest may have certain advantages. In a study conducted by Matz, Bennett, and Landis (Matz 1992), bobwhite quail chicks were imprinted upon chicken hens. Both the hens and the chicks were placed in pens with the adult chicken constrained within a shelter so that the chicks were free to forage. The quail chicks foraged throughout the penned area and returned to the hen in the evening, making counts and sampling straightforward. It was found that the chicks were exposed to chemicals by all routes and that the method holds promise as a means of estimating risks due to pesticide applications and a means of examining the toxicity of contaminated sites.

Many factors other than pollution can lead to poor reproductive success. Secondary effects, such as the impact of habitat loss on zooplankton populations essential for fry feeding, will be seen in the depression or elimination of the young age classes.

Mortality is certainly easy to assay on the individual organism; however, it is of little use as a monitoring tool. Macroinvertebrates, such as bivalves and cnidaria, can be examined and as they are relatively sessile, the mortality can be attributed to a factor in the immediate environment. Fish, being mobile, can die due to exposure kilometers away or due to multiple intoxications during their migrations. Also, by the time the fish are dying, the other levels of the ecosystem are in a depleted state.

In summary, sentinel species have several distinct advantages. These organisms can be used to demonstrate the bioavailablity of xenobiotics since they are exposed in a realistic fashion. If the organisms can be maintained in the field for long periods, indications of the impacts of the contamination upon the growth and population dynamics of the system can be documented. Organisms that are free to roam within the site of interest can serve to integrate, in a realistic fashion, the spatial and temporal heterogeneity of the system. Sentinel organisms are also available for residue measurements, can be assayed for the molecular, physiological, and behavioral changes due to chemical stress, and can serve as a genetic baseline so that effects in a variety of environments can be normalized. Introduced organisms are not generally full participants in the structure and dynamics of an ecosystem and assessments of the native populations should be conducted.

POPULATION PARAMETERS

A variety of endpoints have been used to characterize the stress upon populations. Population numbers or density have been widely used for plant, animal, and microbial populations in spite of the problems in mark recapture and other sampling strategies. Since younger life stages are considered to be more sensitive to a variety of pollutants, shifts in age structure to an older population may indicate stress. Unfortunately, as populations mature, older organisms tend to dominate, comparisons become difficult. In addition, cycles in age structure and population size occur due to the inherent properties of the age structure of the population and predator-prey interactions. Crashes in populations such as that of the stripped bass in the Chesapeake Bay do occur and certainly are observed. A crash often does not lend itself to an easy cause-effect relationship, making mitigation strategies difficult to create.

The determination of alterations in genetic structure, i.e., the frequency of certain marker alleles, has become increasingly popular. The technology of gel electrophoresis has made this a seemingly easy procedure. Population geneticists have long used this method to observe alterations in gene frequencies in populations of bacteria, protozoa, plants, various vertebrates, and the famous Drosophilla. The largest drawback in this method is ascribing differential sensitivities to the genotypes in question. Usually, a marker is used that demonstrates heterogeneity within a particular species. Toxicity tests can be performed to provide relative sensitivities. However, the genes that have been looked at to date are not genes controlling the xenobiotic metabolism, but are genes that have some other physiological function and act as a marker for the remainder of the genes within a particular linkage group. Although with some problems, this method promises to provide both populational and biochemical data that may prove useful in certain circumstances.

Alterations in the competitive abilities of organisms can be an indication of pollution. Obviously, bacteria that can use a xenobiotic as a carbon or other nutrient source or that can detoxify a material have a competitive advantage, all other factors being equal. Xenobiotics may also enhance species diversity if a particularly competitive species is more sensitive to a particular toxicant. These effects may lead to an increase in plant or algal diversity after the application of a toxicant.

ASSEMBLAGE AND COMMUNITY PARAMETERS

The structure of biological communities has always been a commonly used indicator of stress in a biological community. Early studies on cultural eutrophication emphasized the impacts of pollution as they altered the species composition and energy flow of aquatic ecosystems. Various biological indices have been developed to judge the status of ecosystems by measuring aspects of the invertebrate, fish, or plant populations. Perhaps the largest drawback is the effort necessary to accurately determine the structure of ecosystems and to distinguish pollution-induced effects from normal successional changes. There is also the temptation to reduce the data to a single index or other parameter that eliminates the dynamics and stochastic properties of the community.

One of the most widely used indexes of community structure has been species diversity. Many measures for diversity are used, from such elementary forms as species number to measures based on information theory. A decrease in species diversity is usually taken as an indication of stress or impact upon a particular

Table 9.1   Index of Biological Integrity for Fish Communities

Metrics	Rating of metric		
	5	3	1
Species richness and composition			
1. Total number of fish species[a] (native fish species)[b]	Expectations for metrics		
2. Number and identity of darter species (benthic species)	1-5 vary with stream		
3. Number and identity of sunfish species (water-column species)	size and region.		
4. Number and identity of sucker species (long-lived species)			
5. Number and identity of intolerant species			
6. Percentage of individuals as green sunfish (tolerant species)	<5	5–20	>20
Trophic composition			
7. Percentage of individuals as omnivores	<20	20–45	>45
8. Percentage of individuals as insectivorous cyprinids (insectivores)	>45	45–20	<20
9. Percentage of individuals as piscivores (top carnivores)	>5	5-1	<1
Fish abundance and condition			
10. Number of individuals in sample	Expectations for metric 10 vary with stream size and other factors.		
11. Percentage of individuals as hybrids (exotics or simple lithophils)	0	>0–1	>1
12. Percentage of individuals with disease, tumors, fin damage, and skeletal anomalies	0–2	>2–5	>5

[a] Original IBI metrics for midwestern United States.

[b] Generalized IBI metrics (see Miller and Levy 1989).

Modified from Karr, J. R. 1991. *Ecol. Appl.* 1:66-84.

ecosystem. However, measures of species diversity are just that — measures of diversity and not measures of toxicant impacts. Indices that have been designed to measure the alteration in the structure and function of ecosystems have been derived, and among the most widely used has been the index of biological integrity (IBI) developed by Karr (1991) for fish assemblages and communities.

An index such as the IBI is a means of rating the structure of a community from a one-time set of samples. Standard methods can be used in the procedures set to produce the IBI and the resulting numbers can be used in the establishment of management programs. The setting of an IBI does require prior detailed knowledge of the assemblage or community under study so that comparisons can be made to normal communities. The rankings require expert judgment so that components such as stream or lake type, seasonal components, and natural variation in assemblage composition can be accounted for. The components and rankings of the IBI for fish communities are presented in Tables 9.1 and 9.2.

The utility of a measure such as the IBI is that it is transferable with modifications to other fish assemblages and to other types of organisms. Given adequate modification the basic premise should be broadly transferable to even terrestrial communities. Dickson et al. (1992) have reported a relationship between measurements such as the IBI and biomonitoring toxicity tests. Another advantage of the index approach is that a great deal of information is condensed to a single number, which also is a disadvantage.

Table 9.2   Index of Biological Integrity Scores with Attributes

Total IBI score (sum of the 12 metric ratings)[a]	Integrity class of site	Attributes
58–60	Excellent	Comparable to the best situations without human disturbance; all regionally expected species for the habitat and stream size, including the most intolerant forms, are present with a full array of age (size) classes; balanced trophic structure.
48–52	Good	Species richness somewhat below expectation, especially due to the loss of the most intolerant forms; some species are present with less than optimal abundances or size distributions; trophic structure shows some signs of stress.
40–44	Fair	Signs of additional deterioration include loss of intolerant forms, fewer species, and highly skewed trophic structure (e.g., increasing frequency of omnivores and green sunfish or other tolerant species); older age classes of top predators may be rare.
28–34	Poor	Dominated by omnivores, tolerant forms, and habitat generalists; few top carnivores; growth rates and condition factors commonly depressed; hybrids and diseased fish often present.
12–22	Very poor	Few fish present, mostly introduced or tolerant forms; hybrids common; disease, parasites, fin damage, and other anomalies regular.
[b]	No fish	Repeated sampling finds no fish.

[a] Sites with values between classes assigned to appropriate integrity class following careful consideration of individual criteria/metrics by informed biologists.

[b] No score can be calculated where no fish were found.

Modified from Karr, J. R. 1991. *Ecol. Appl* 1:66-84.

All indexes collapse in a somewhat arbitrary fashion the numerous dimensions that comprise them into a single number that is treated as an accurate measurement of the condition of the area or environment sampled. Of course, the variables that comprise the index, and indeed the values assigned to the components, are often based upon professional judgment. Indexes can be fooled, and quite different systems can result in indexes of comparable scores. Interpretation of such scores should be taken with the above caveats.

Perhaps the most commonly measured aspects of communities has been the number of species, evenness of the composition, and diversity based on information theory. These measures are not measures of toxicant stress, but do describe the communities. Prior judgment as to the depletion of diversity relative to a reference site due to anthropogenic causes is not warranted unless any other factors that control these community level impacts are understood. Among the factors that can naturally alter these types of measures relative to a so-called reference site are history of the colonization of that habitat, catastrophic events, gene pool, colonization area, stability of the substrate and the environment, and stochastic events. All of these factors can alter community structure in ways that may mimic toxicant impacts.

INTERPRETATION OF EFFECTS AT THE POPULATION, COMMUNITY, AND ECOSYSTEM LEVELS OF ORGANIZATION

Related to diversity is the notion of static and dynamic stability in ecosystems. Traditional dogma stated that diverse ecosystems were more stable and, therefore, healthier than less rich ecosystems. May's work in the early 1970s did much to test these, at the time, almost unquestionable assumptions about properties of ecosystems. Biological diversity is important, but diversity itself may be an indication of the longevity and size of the habitat rather than the inherent properties of the ecosystem. Rarely are basic principles such as island biogeography and evolutionary time incorporated into comparisons of species diversity when assessments of community status are made. Diversity should be examined closely as to its worth in determining xenobiotic impacts upon biological communities.

The impacts of toxicants upon the structure of communities has been investigated using the resource competition models of Tilman. Species diversity may be decreased or increased and a rationale for studying indirect effects emerges.

RESOURCE COMPETITION AS A MODEL OF THE DIRECT AND INDIRECT EFFECTS OF POLLUTANTS

Resource competition as modeled by David Tilman and adopted for toxicological purposes by Landis may assist in putting into a theoretical framework the varied effects of toxicants on biological systems. Detailed derivations and proof can be found in Tilman's excellent monograph. This brief review is to demonstrate the utility of resource competition to the prediction, or at least explanation, of community level impacts.

The basis for the description of resource competition is the differential uptake and utilization of resources by species. The use of the resource, whether it is space, nutrients, solar radiation, or prey species can be described by using growth curves with the rate of growth plotted against resource concentration or amount.

Figure 9.3 illustrates growth curves for species A and B as plotted against the concentration of resource 1. At a point for each species, the rate of growth exceeds mortality at a certain concentration of resource 1. Above this concentration the population grows; below this concentration extinction occurs. A different zero net growth point, the point along the resource concentration where the population is at break-even, differs for the two species unless differential predation forces coincide. These curves, at least for nutrients, are easily constructed in a laboratory setting.

In order to describe the uptake of the toxicant by the organism, a resource consumption vector is constructed. Figure 9.4 diagrams a consumption vector for the two-species case. This vector is the sum of the consumption vectors for each of the resources and the slope is the ratio of the individual resource vectors. Although it is certainly possible that the consumption vector can change according to resource concentration, it is assumed in this discussion to be constant unless altered by a toxicant.

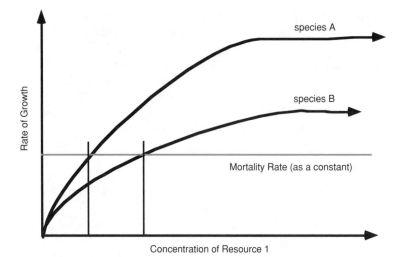

Figure 9.3   Rate of growth and resource supply. As the supply of resource increases so does the reproductive rate of an organism until a maximum is reached. At one point the rate of growth exceeds the rate of mortality and the population increases. As long as the resource concentration exceeds this amount the population grows; below this amount, extinction will occur.

Figure 9.4   Consumption vector. Consumption vector for species A, $\vec{C}_A$, is the sum of the vectors for the rate of consumption of resource 1 and resource 2. The consumption vector determines the path of the concentrations of resources as it moves through the resource space. In the single-species case, the eventual equilibrium of resources occurs where the sum of the utilization vectors and the $\vec{C}_A$ is zero.

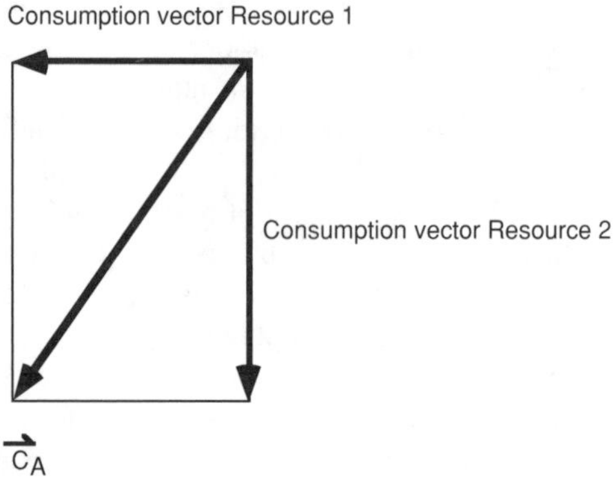

The zero net growth point expanded to the two-dimensional resource space produces a zero net growth isocline (ZNGI) as illustrated in Figure 9.5. At the ZNGI, the rate of reproduction and the mortality rates are equal, resulting in no net growth of the population. In the shaded region the concentration or availability of the resource results in an increase in the population. In the clear area, the population declines and ultimately becomes extinct.

The shape of the ZNGI is determined by the utilization of the resource by the organism. If the resources are essential to the survivorship of the organism, then the

Figure 9.5   Zero net growth isocline (ZNGI). The ZNGI is the line in the resource space that represents the lowest concentration of resources that can support a species. In an equilibrium situation, the equilibrium will eventually be drawn to a point along the ZNGI. In the shaded area of the resource space, the population will grow. In the unshaded area extinction will eventually occur.

shape is as drawn. Eight different types of resources have been classified according to the ZNGI.

The eventual goal in the single-species case is the prediction of where the equilibrium point on the ZNGI will be with an initial concentration of resources. A supply vector, U1, can be derived that describes the rate of proportion of supply from the resource supply point. At equilibrium in a single-species case, the resources in a habitat will be at a point along the ZNGI where $C_A + U_{1.2} = 0$. Tilman has shown that this point exists and is stable. Metaphorically speaking, the $\bar{C}$ 1 pulls the equilibrium point along the ZNGI until the consumption of the two resources is directly offset by the rate and proportion of the supply of the resources. Although the description is for two essential resources, the same holds true for other resource types.

The two-species case can be represented by the addition of a new ZNGI and $\bar{C}$ to the graph of the resource space. In the case of essential resources, six regions are defined (Figure 9.6). Region 1 is the area in which the supply of resources is too low for the existence of either species. In region 2, only species A can survive since the resource concentration is too low for the existence of species B. In region 3, coexistence is possible for a time but eventually species A can drive the resources

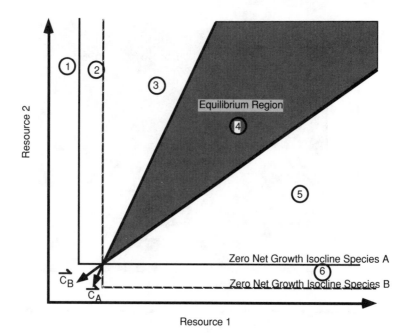

Figure 9.6. Two species graph. The $\vec{C}_A$ and ZNGI for each species is incorporated into the graph. Six regions of the resource space are created. In region 1, neither species can exist; in region 2, only species A can survive; in region 3, species A and species B can survive but B is driven to extinction; region 4 is the equilibrium region; in region 5 both species A and species B can survive but A is driven to extinction; and in region 6, only species B can survive. In the case illustrated, if the original resource point, S1,S2 lies within the shaded equilibrium region, both species will exist.

below the ZNGI for species B. Region 4 is the area in which an equilibrium is possible and the consumption vectors will drive the environment to the equilibrium point. The equilibrium point lies at the intersections of the two ZNGI. In region 5, coexistence is possible for some period, but eventually species B can drive the resources below the ZNGI for species A. Finally, within region 6, only species B can survive.

An unstable equilibrium can exist if the consumption vectors are transposed. However, since any perturbation would result in the extinction of one species, this situation in unlikely to be persistent.

The basic assumptions made in order to model the impacts of toxicants on the competitive interactions discussed above are (1) the toxicant affects the metabolic pathways used in the consumption of a resource and (2) this alteration of the metabolism affects the growth rate vs. resource curve. In the terms of resource competition, the consumption vector is changed and the shape and placement of the ZNGI is altered. In the following discussions the implications of these changes on examples using essential resources are depicted.

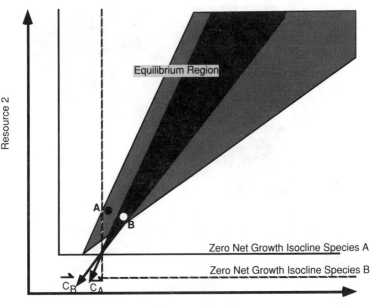

Figure 9.7 Case 1: toxicant impacts on species B. The introduction of a toxicant alters the ability of species B to use resource 1. The slope of the consumption vector is altered and the ZNGI shifts compared to the initial condition. The equilibrium point moves and the equilibrium region shifts and shrinks. With a smaller equilibrium region, the probability of coexistence of the two species also is decreased.

CASE 1

In the first example, the initial conditions are the same as used to illustrate the two-species resource competition model with essential resources (Figure 9.7). The toxicant alters the ability of species B to use resource 1. The slope of $\vec{C}_B$ increases and the ZNGI and the $\vec{C}_B$ shift the equilibrium point and reduce the area of the equilibrium region. The resource supply point A, which was part of the original equilibrium region, is now in an area that will lead to the eventual extinction of species B. Conversely, point B is now contained within the equilibrium region. However, the overall reduction of the size of the equilibrium region will make the likelihood of a competitive equilibrium smaller.

CASE 2

In this example the toxicant affects species A, increasing the slope of the $\vec{C}_A$ as the ability of species A to use resource 1 is altered. In Figure 9.8A the toxicant has forced the $ZNGI_A$ to a near overlap with the $ZNGI_B$ in the utilization of resource 1. In only a small region can species A drive species B to extinction. As the $ZNGI_A$ and

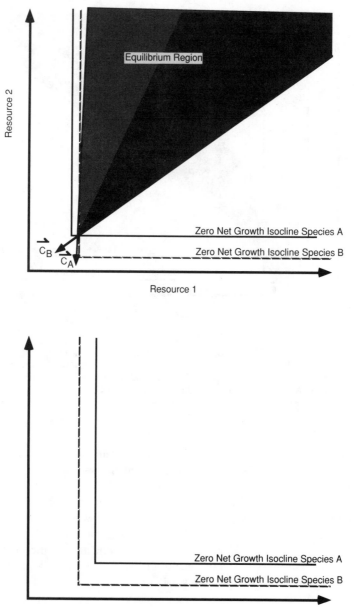

Figure 9.8 Case 2: toxicant impacts on species A. The delivery of the toxicant impacts upon the ability of species A to use resource 1. In this case, the equilibrium point has not moved but the equilibrium region has greatly increased thus increasing the opportunities for a coexistence of the two species (A). However, an increase in the equilibrium and an increase in species diversity does not mean that the system is less stressed. (B) the addition of a toxicant has forced the $ZNGI_A$ inside the $ZNGI_B$, resulting in the eventual extinction of species A.

$ZNGI_B$ overlap in regards to resource 1, the equilibrium region would be at a maximum. The addition of more toxicant would drive the $ZNGI_A$ inside the $ZNGI_B$, and in all regions of the resource space, species A can drive B to extinction. Coexistence over any protracted time is now impossible (Figure 9.8B). Interestingly, the situation that produces the greatest likelihood of a competitive equilibrium also borders on extinction.

In the examples presented above, resource heterogeneity was not incorporated. Resources in nature are variable in regards to supply over both time and space and this does much to explain the coexistence of competing species. Tilman represents this by projecting a 95% bivariate confidence interval, a circle, upon the resource space (Figure 9.9). In this case, the dynamics of the competitive interactions between the two species change depending upon the resource availability. In part of the confidence interval, a competitive equilibrium is possible. In other parts of the confidence interval, competitive displacement of species A is possible.

The significance of this cannot be missed. If the confidence interval is based on time, competitive relationships differ on a seasonal basis and the lack of a species at certain times may not be due to an increase or decrease in pollutants, but may be attributable to yearly changes in resource availability. Seasonal changes in species composition are expected and the limitations of one-time sampling are well known. However, the confidence interval can also be expressed over space as well. Slight differences in resources ratios that are part of the normal variation within a stream, lake, or forest can result in different species compositions unrelated to toxicant inputs.

Conversely, toxicants that do not directly affect the competing species but instead alter the availability of resources can alter the species composition of the community. In Figure 9.10, the case of the moving resource confidence interval is presented. In this case, the ratio of resource 2 has been increased relative to resource 1. This could be the alteration in microbial cycling of nutrients or the alteration in relative proportions of prey species for a predator, to name two examples. The confidence region is now outside the equilibrium region and species B becomes extinct.

Even more subtle differences in populations may occur. The genetic variation within a population can be rather substantial. The two-dimensional ZNGIs can be expanded to demonstrate the fact that the ability of organisms to consume and use resources is not a point but a continuum dictated by the genetic variation of the population. Figure 9.11 illustrates this idea.

The lines representing the ZNGIs have become bars and the equilibrium point has now been transformed into a confidence region. Depending upon the amount of variation within a population relating to the physiological parameter impacted by the toxicant, resource competition could also occur between the various phenotypes within the population. Guttman and colleagues have attempted to document these changes by following changes in allelic frequencies in polluted and so-called reference sites. The approach may have promise, but the difficulty of sorting natural variation from toxicant-induced selection can be daunting.

The use of resource competition models also leads to a classification or a flow diagram describing the potential impacts of toxicants upon competitive interactions

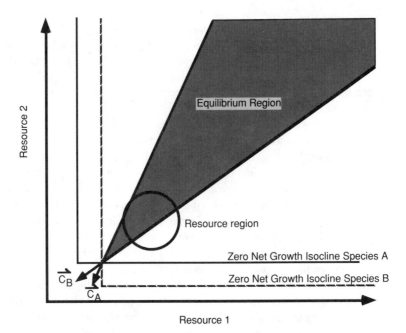

Figure 9.9   Resource heterogeneity. The heterogeneity of the resource can be represented by two-dimensional 95% intervals projected upon the graph. The placement of the circle can help to predict the dynamics of the system and describe the occasional extinction of one species and the coexistence of the two.

(Figure 9.12). The toxicant can directly or indirectly alter every aspect of the competitive interaction except the nonspecific or density-independent mortality.

Genetics — The effects of the toxicant can be both long-lasting and severe. Since the genome ultimately controls the biochemical, physiological, and behavioral aspects of the organism that set the consumption vector and the ZNGI, alterations can have a major impact.

Predation — Often a toxicant affects more than one species. Perhaps the predators, disease organisms, or herbivores that crop a food resource are affected by the toxicant. Predation is an important aspect of mortality.

Reproduction — Teratogenicity and the reduction of reproductive capacity are well-known effects of toxicants, especially in vertebrate systems.

Mortality — An increase in mortality moves the minimal amount of resource necessary to maintain a population. The combination of mortality and reproduction determines the ZNGI for that population.

Consumption vectors — The consumption vectors express the relative efficiencies of the uptake and utilization of resources. An alteration in the metabolic activity of even one resource will shift the slope of the vector. In conjunction with the ZNGI, the consumption vector fixes the equilibrium region within the resource space.

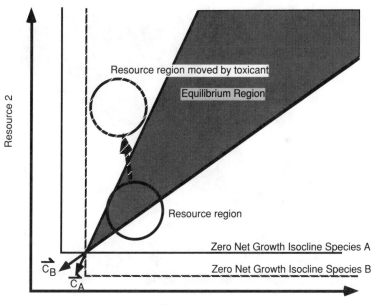

Figure 9.10 Shifting of the confidence interval of resources. The addition of a toxicant that impacts organisms that act as resources for other organisms can have dramatic effects without any direct impact upon the consumers. A shift in the resource region due to a shift in competitive interactions at other energetic levels can alter the competitive relationships of the consumers. Structure of the community is then altered even more dramatically. In this case, a situation with a general competitive equilibrium is shifted so that species A can be driven to extinction with the movement of the resource area.

Biotic components of the resource region — The confidence regions describing the supply of resources are dependent on the biotic components in both the temporal and spatial variability. The organisms that compose the resources can be affected as presented above. A population boom or bust can shift the confidence interval of the resource supply. Excessive production of a resource can affect other resources. An algal bloom can lead to oxygen depletion during darkness.

Since the organisms that are competing at one level are resources for other trophic levels, the effects can reverberate throughout the system. Therefore these models have the potential for describing a variety of interactions in a community.

One of the major implications of these models is the importance of resources and initial conditions in the determination of the outcome of a toxicant stressor. Depending upon the resource ratio, three different outcomes are possible given the same stressor. History of the system, therefore, plays a large part in determining the response of a community to a stressor.

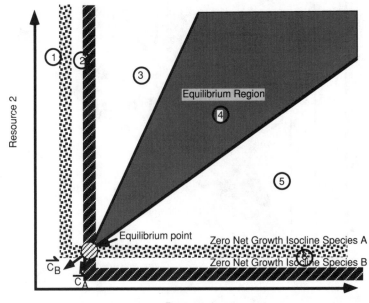

Figure 9.11   Genetic diversity. The genetic diversity of a population will alter the sharp lines of the ZNGIs into bars representing 95% confidence intervals. The consumption vectors can be similarly altered although for this diagram they are still conventionally represented. The equilibrium point and equilibrium region then become probabilistic.

MODELING OF POPULATIONS USING AGE STRUCTURE AND SURVIVORSHIP MODELS

Barnthouse and colleagues (Barnthouse 1993; Barnthouse et al. 1990, 1989) have explored the use of conventional population models to study the interactions among toxicity, predation, and harvesting pressure for fish populations. These studies are excellent illustrations of the use of population models in the estimation of toxicant impacts.

Distinguishing between the change in population or community structure due to a toxicant input or the natural variation is difficult. The use of resource competition models can aid in determining the factors that lead to alterations in competitive dynamics and the ultimate structure of a community. A great deal of knowledge about the system is required and an indication of exposure is necessary to differentiate natural changes from anthropogenic effects. This categorization may be even more difficult due to the inherent dynamics of populations and ecosystems.

POPULATION BIOLOGY, NONLINEAR SYSTEMS, AND CHAOS

Simple models for the description of populations of organisms with nonoverlapping generations can lead to a variety of interesting dynamics. May (1974), and May and

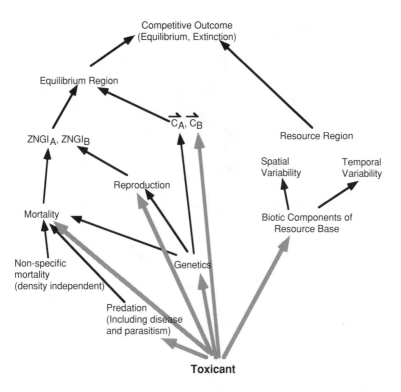

Figure 9.12   Impacts of toxicants upon the components of resource competition. The relationships among the factors incorporated into resource competition models can be affected in several ways by a toxicant. Only the density-independent factors governing mortality escape.

Oster (1976) demonstrated that the use of difference equations such as that for population growth:

$$N_{t+1} = N[1 + r[1 - N/K]] \tag{9.2}$$

where N = population size at time t; N_{t+1} = population size at the next time interval; K = carrying capacity of the environment; and r = intrinsic rate of increase over the time interval can yield a variety of dynamics. At different sets of initial r values, populations can reach an equilibrium, fluctuate in a stable fashion around the carrying capacity, or exhibit dynamics that have no readily discernible pattern; i.e., they appear chaotic.

The investigation of chaotic dynamics has also spread to weather forecasting and the physical sciences. An excellent popularization by Gleick (1987) reviews the discovery of the phenomena, from the butterflies of Lorenz in the modeling of weather to complexity theory. What follows is only a brief introduction.

Figure 9.13 compares two outcomes. In the first instance, r is set at 2.0, the carrying capacity at 10,000, and N at 2500. Within ten time intervals, the population

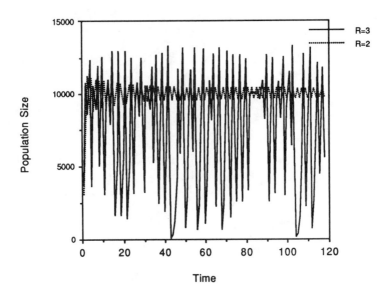

Figure 9.13 Comparison of the population dynamics of two systems that begin at the same initial conditions but with different rates of increase.

is oscillating around the carrying capacity in a regular fashion. It is as if the carrying capacity is attracting the system, and the system slowly but perceptively falls toward the attractor. The width of the oscillations slowly shrinks. In stark contrast is the system that is identical, except that the r value is 3.0. The system initially climbs toward the carrying capacity, but soon exhibits a complex dynamics that does not repeat itself. The system oscillates in an apparently random fashion, but is bounded. In this instance it is bounded by 13,000 and 0. The apparently stochastic pattern, is however, completely derived by Equation 9.2. The system is deterministic and not stochastic. When this occurs the system is defined as chaotic, a deterministic system that exhibits dynamics that cannot be typically determined as different from a stochastic process.

One of the characteristics of nonlinear systems and chaotic dynamics is the dependence upon the initial conditions. Slight differences can produce very different outcomes. In Equation 9.2, there are specific values of r that determine the types of oscillations around the carrying capacity. At a specific finite value of r, the system becomes chaotic. Different initial values of the population also produce different sets of dynamics. Figure 9.14 provides an example. Using Equation 9.2 the initial N in Figure 9.14A is 9999 with a carrying capacity of 10,000. Overlaid on this figure in Figure 9.14B is the dynamic of a population whose initial N = 10,001. Notice that after ten time intervals the two systems have dramatically diverged from each other. An error of 1/10,000 in determining the initial conditions would have provided an incorrect prediction of the behavior of these populations. Chaotic systems are very dependent upon initial conditions. Slight differences in initial conditions can give rise to the maximum differences allowed by the system.

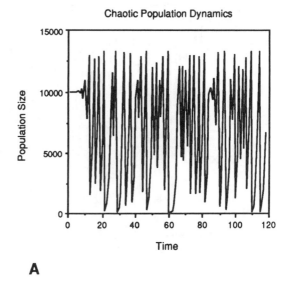

A

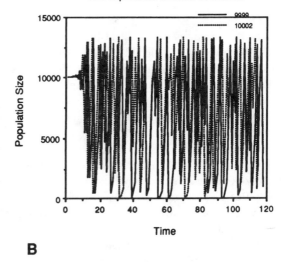

Figure 9.14 The importance of initial conditions. Although the equations governing the populations are identical, a slight 1/10,000 difference in the initial conditions results in very different dynamics. (A) The dynamics are set from an initial condition of N = 9999. (B) An overlay of the dynamics from an initial starting point of N = 10,001 is presented. Notice that after only a few interactions that the two lines do not overlap.

B

Can chaotic systems be differentiated from random fluctuations? Yes. Even though the dynamics are complex and resemble a stochastic system, they can be differentiated from a truly stochastic dynamic. Figure 9.15 compares the plots of N = 10,001 and a selection of points chosen randomly from 13,000 to 0. Note that after approximately ten time intervals the dynamics of both are quite wild and would be difficult to distinguish one from another as far as one is deterministic and the other chaotic. However, there is a simple way to differentiate these two alternatives: the phase space plot.

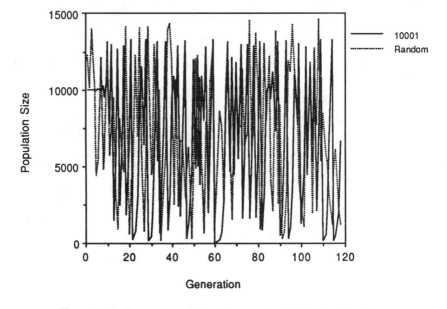

Figure 9.15   Comparison of chaotic vs. random population dynamics.

Figure 9.16A is the phase space plot for the N = 10,001 graph. In this plot N vs. the N at an arbitrary yet constant time interval are plotted against each other. For these illustrations, N is plotted vs. N_{t+1}. Notice that the points fit along a simple arch; this pattern is unique to the equation and is in fact somewhat conserved despite the initial conditions. In Figure 9.16B, the phase space plot of the randomly generated plot, no such pattern is apparent. The phase space plot resembles a shotgun blast upon a target. This pattern is typical of a randomly generated pattern and is quite distinct from the chaotic yet deterministic pattern.

The importance of these findings is still under much debate in the biological sciences. A search for chaotic dynamics in population biology was undertaken by a variety of researchers, notably Hassell et al. (1991); Schaffer (1985); Schaffer and Kot (1985); and recently Tilman and Weldin (1991). Chaotic dynamics certainly are not universal, but have been found in several ecological and epidemiological contexts as described in Table 9.3.

As can be seen, chaotic dynamics can be found in a variety of systems. Even in the classical population dynamics of the Canadian Lynx, the results were demonstrably chaotic in nature. Perhaps one of the most recent studies that has particular relevance to environmental toxicology is the demonstration that grass populations studied by Tilman and Weldin (1991) became chaotic over the period of the extended study. They hypothesize that the increase in plant litter in the experimental plots pushed the system toward the chaotic dynamics.

The implications for population ecology and the interpretation of field data are important. First, these dynamics exist in nonequilibrium states. Since many of the tenets of ecological theory depend on an assumption of equilibrium, they may be misleading. Schaffer and Kot (1985) make a stronger statement, "Our own opinion

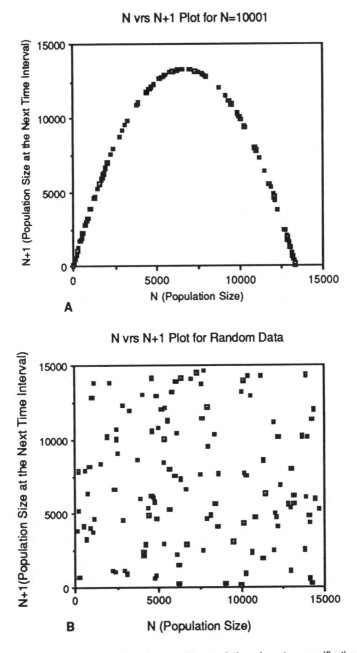

Figure 9.16   Plots of the population size vs. the population size at a specific time interval reveals the structure of a chaotic system. (A) Derived from the deterministic yet stochastic looking dynamics, a pattern readily forms that is characteristic of the underlying equation. (B) A shotgun blast or random pattern is revealed.

Table 9.3   Examples of Chaotic Dynamics in Ecological Systems

Organism (as compiled by Schaffer. 1985)	Chaotic dynamics observed
Mammals	
Canadian Lynx	Yes
Muskrat	No
Insects	
Thrips	Yes
Leucoptera caffeina	Yes
L. meyricki	Yes
Blowflies	Yes
Human Diseases	
Chickenpox-New York City	No
Chickenpox-Copenhagen	No
Measles-New York City	Yes
Measles-Baltimore	Yes
Measles-Copenhagen	Yes
Mumps-New York City	No
Mumps-Copenhagen	Yes
Rubella-Copenhagen	Yes
Scarlet fever-Copenhagen	No
Whooping cough-Copenhagen	No
Sugihara, G., B. Grenfell, and R. M. May (1990)	
Measles City by City (U.K.)	Yes
Measles (Country wide)	No — Noisy 2-year cycle
Tilman, D. and D. Weldin (1991)	
Perennial grass *Agrostis*	Yes

is that what passes for fundamental concepts in ecology is as mist before the fury of the storm — in this case, a full, nonlinear storm". One of the crucial recommendations of this paper is the importance of understanding the current dynamic status of the ecological system. Only then can perturbation experiments designed to elucidate interactions be considered valid.

The implications of nonlinear dynamics in environmental toxicology have recently been discussed by Landis et al. (1993a, 1993b). First, if ecological systems are nonequilibrium systems, then attempts to measure stability or resilience may have no basis. In fact, it may be impossible to go back to the original state, or after a perturbation to the state of the reference state. Second, the dynamics of the system will not allow a return to the reference state. Nonlinear systems are very sensitive to original conditions and record a history of previous alterations within the dynamics of their structure. Third, historical events give rise to unique dynamics that are likely unique for each situation. As stated by Schaffer and Kot (1985), unless the initial dynamics are understood, perturbation experiments, either accidental or deliberate, are impossible to interpret. Fourth, the future cannot be predicted beyond the ability to measure initial conditions. Since nonlinear systems are so sensitive to initial conditions, predictions can only be accurate for short periods of time.

The replicability of field studies can also be seen as impossible beyond certain limits. That is not to say that patterns of impacts cannot be reproduced, but reproducibility in the dynamics of individual species is unlikely unless the initial conditions of the experiments can be made identical.

As interesting and powerful as the development of the understanding of nonlinear systems has been, it is only part of the study of system complexity. Nicolis and Prigogine (1989) have produced an excellent introduction, and the understanding of complexity theory promises to have a major impact on ecology and environmental toxicology.

COMMUNITY AND ECOSYSTEM EFFECTS

The difficulty of measuring community and ecosystem effects has been extensively discussed in the literature (Suter 1993). Ecological systems can be perceived as mechanisms for energy flow, materials cycling, and as assemblages of species. Ecosystem properties may also be examined. In each case the system as described is multidimensional in nature, and the tools for analysis should represent this fact.

Ecosystems are multidimensional constructs and they have been seen in that fashion for a number of years. For example, the Hutchinsonian idea of organisms and populations residing in a n-dimensional hypervolume is the basis of current niche theory (Hutchinson 1959). The n-dimensional hypervolume is the ecosystem with all its components as perceived by the population. The variability of these parameters over time as well as the quantity and quality of nutrient inputs to the system are used to account for the diversity of species within this system (Hutchinson 1961; Richerson et al. 1970; Tilman and Kilham 1976). An accurate description of an ecosystem should in some fashion correspond to its multidimensional nature.

Often impacts are quantified using a reference site as a negative control for comparison to other sites under question. Similarly, multispecies toxicity tests and microcosms and mesocosms, attempt to detect differences between the control treatment and the dosed treatment groups.

A number of methods have been developed to attempt to measure these differences. Analysis of variance is the classical method to examine single-variable differences from the control groups. However, problems with Type II error and the difficulty of graphically representing the data set have been problematic. Conquest and Taub (1989) developed a method to overcome some of the problems of classical ANOVA, intervals of nonsignificant difference. This method corrects for the likelihood of a Type II error and produces intervals that are readily graphed to ease examination. This method is routinely used in the examination of data derived from the SAM and is applicable to other data sets. The major drawback to these methods is again the examination of only one variable at a time over the course of the experiment. In many instances the interactions may not be as straightforward as the classical predator-prey or nutrient limitation dynamics usually picked as examples of community level interactions. Ideally, multivariate statistical tests used for evaluating complex data sets will have the following characteristics:

- It does not combine counts from dissimilar taxa by means of sums of squares or other mathematical techniques.
- It does not require transformations of the data, such as normalizing the variance.

- It works without modification on incomplete data sets.
- It can work without further assumptions on different data types (e.g., species counts or presence/absence data).
- Significance of a taxon to the analysis is not dependent on the absolute size of its count, so that taxa having a small total variance, such as rare taxa, can compete in importance with common taxa, and taxa with a large, random variance will not automatically be selected, to the exclusion of others.
- It provides an integral measure of "how good" the clustering is, i.e. whether the data set differs from a random collection of points.
- It can, in some cases, identify a subset of the taxa that serve as reliable indicators of the physical environment.

The remainder of this section details the potential application of multivariate methods in the selection of endpoints and in the evaluation of exposure and effects of stressors in ecosystems. Particular reference is made to the application of these methods to the current framework for ecological risk assessment. Examples of the use of multivariate methods in detecting effects and in selecting important measurement variables is covered using both field surveys and multispecies toxicity tests.

APPLICATION AND OF MULTIVARIATE TECHNIQUES

The application of these methods have been examined in a series of field studies and multispecies toxicity tests. These examinations have demonstrated the power and usefulness of multivariate techniques in elucidating patterns in biological communities of varying complexity. Several researchers have attempted to employ multivariate methods to the description of ecosystems and the impacts of chemical stressors. Perhaps the best developed approaches have been those of Kersting (1984, 1988), Johnson (1988a), and a new approach developed by Matthews et al. (1991a, 1991b).

NORMALIZED ECOSYSTEM STRAIN

Normalized ecosystem strain (NES) was developed by Kersting (1984, 1988) as a means of describing the impacts of several materials to the three-compartment microecosystems containing an autotrophic, herbivore, and decomposer subsystems. These variables in the unperturbed control systems are used to calculate the normal operating range (NOR) of the microecosystem. The NOR is the 95% confidence ellipsoid of the unperturbed state of a system. The center of the NOR is defined as the reference point for the calculation of the NES. The NES is calculated as the quotient of the Euclidean distance from a state to the reference state divided by the distance from the reference state to the 95% confidence (also called tolerance) ellipsoid, along the vector that connects the reference state to the newly defined state (Figure 9.17). A value of 1 or less indicates that the new state is within the 95%

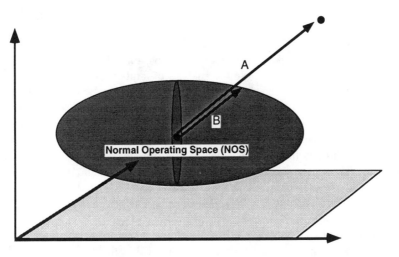

Figure 9.17 Normalized ecosystem strain.

confidence ellipsoid; values greater than 1 indicate that the system is outside this confidence region.

Originally limited to ellipsoids, the use of Mahalonobis distances allows the use of more variables because the confidence ellipsoid can be transformed to a confidence or tolerance hypersphere. These ideas were examined using the microecosystem test method developed by Kersting for the examination of multispecies systems. These three-compartment microecosystems are comprised of an autotrophic, herbivore, and decomposer subsystems that are connected by tubing and pumps. Although relatively simple and small, these systems are operable over a number of years.

Several variable measurements are obtained weekly for these experiments:

- Algal biomass in the autotrophic and herbivore systems
- Number of *Daphnia magna*
- pH of the autotrophic and herbivore subsystems
- Molybdate-reactive phosphorus in the autotrophic subsystem and in the return flow between the decomposer and autotrophic subsystems.

These variables allow for the determination of the NOR and after dosing with a toxicant, the NES. In some instances impacts that are not significant using univariate analysis are detectable using NES. The sensitivity of the NES increased as the number of variables used to describe the system increased (Kersting 1988). Another interesting observation was the increasing distance from the normal space of the system after a perturbation as measured by NES as time increased. This increasing distance indicates that the perturbed system is drifting from its original state. Kersting hypothesized that the system may even shift to a different equilibrium state or domain and that the system would remain there even after the release of the stressor.

STATE SPACE OF ECOSYSTEMS

Apparently as an independent development, Johnson (1988a) proposed the idea of using a multivariate approach to the analysis of multispecies toxicity tests. This state space analysis is based upon the common representation of complex and dynamic systems as an n-dimensional vector. In other words, the system is described at a specific moment in time as a representation of the values of the measurement variables in an n-dimensional space. A vector can be assigned to describe the motion of the system through this n-dimensional space to represent successional changes, evolutionary events, or anthropogenic stressors. The direction and position information form the trajectory of the state space and this can be plotted over time.

In the n-dimensional hypervolume that describes the placement and trajectory of the ecosystem it is possible to compare the positions of systems at a specified time. This displacement can be measured by literally computing the distance from the systems and this displacement vector can be regarded as the displacement of these systems in space. The displacement vectors can be easily calculated and compared. Using the data generated by Giddings et al. (1980) in a series of classic experiments comparing results of the impacts of synthetic oil on aquarium and small pond multispecies systems, Johnson (1988a) was able to plot dose-response curves using the mean separation of the replicate systems. These plots are very reminiscent of dose-response curves from typical acute and chronic toxicity tests.

As summarized by Johnson, the strengths of this methodology are the objectivity for quantifying the behavior of the stressed ecosystem and the power of this methodology to summarize large amounts of data. As with the work of Kersting, this methodology allows the investigator to examine the dynamics of the ecosystem and the eventual fate of the system relative to the control treatment.

Another important application proposed by Johnson (1988b) was the use of multivariate analysis to identify diagnostic variables that can be applied in the monitoring of ecosystems. Diagnostic variables, if reliable in differentiating anthropogenically stressed systems from control systems, would be extremely valuable in monitoring for compliance and in determining clean up standards. In a follow-up publication Johnson (1988b) detailed the derivation and use of these diagnostic variables. The use of such variables is justified due to the fact that decisions often have to be made with incomplete data sets due to technical difficulties, cost, and a general lack of knowledge. Techniques proposed for the determination of these variables included linear regression, discriminant analysis, and visual inspection of graphed data. Johnson conducted a cost-benefit analysis using an ecosystem model that demonstrated under the condition of that model, the benefits of diagnostic variables. In the discussion, Johnson proposes simulation modeling to attempt to find generalized diagnostic variables that best describe the state space and trajectory of an ecosystem.

One of the difficulties in the past of using multivariate methodologies such as those proposed by Johnson and Kersting was the computational effort required. Computational requirements are not the limiting factor that they may have once been, even for large data sets.

The major difficulty with the methods detailed above is the reliance on conventional metric statistics. Vector distances in an n-dimensional space including such disparate variables as pH, cell counts, and nutrient concentrations are difficult to compare from one experiment to another. Another consideration is the fact that many of the variables may be compilations of others. Algal biomass is often calculated by multiplying cell counts by an appropriate constant for each species. Species diversity and many indices of ecosystem health are similarly composited variables. As discussed in the previous sections, the use of metric methods with nonmetric clustering may prove a useful combination.

The attempt by Johnson to derive diagnostic variables is an interesting approach. However, current research indicates that the variables that contribute the most to separating control treatment from dosed treatment groups change from sampling period to sampling period. The variables change in the multispecies toxicity test experiments no doubt in response to the successional trajectory of the system as nutrients become depleted. As nutrients become limiting and the ability of the system to exhibit large differences in community structure become less, the metric measures do not exhibit the same magnitudes of separation.

NONMETRIC CLUSTERING

Multivariate methods have proved promising as a method of incorporating all of the dimensions of an ecosystem. Both of the methods presented above have the advantage of examining the multispecies test systems as a whole and can track such process as succession, recovery, and the deviation of a system due to an anthropogenic input. The disadvantage to these systems and to conventional multivariate techniques is that all of the data are generally incorporated without regard to the metric (unit of measurement) or the contribution of a variable to the separation of the clusters. It can be difficult to reconcile variables such as pH with a 0 to 14 metric to the numbers of bacterial cells per milliliter, where low numbers are in the 10^6 range. Random data indiscriminately incorporated with large metrics may overwhelm important variables with a different metric. Recently developed for the analysis of ecological data is a multivariate derivative of artificial intelligence research, nonmetric clustering, that has the potential of circumventing many of the problems of conventional multivariate analysis.

Unlike the more conventional multivariate statistics, nonmetric clustering is an outgrowth of artificial intelligence and a tradition of conceptual clustering. In this approach, an accurate description of the data is only part of the goal of the statistical analysis technique. Equally important is the intuitive clarity of the resulting statistics. For example, a linear discriminant function to distinguish between groups might be a complex function of dozens of variables, combined with delicately balanced factors. While the accuracy of the discriminant may be quite good, use of the discriminant for evaluation purposes is limited because humans cannot perceive hyperplanes in highly dimensional space. By contrast, a conceptual clustering will attempt to distinguish groups using as few variables as possible, and by making simple use of each one. Rather than combining variables in a linear function, for example, conjunctions of elementary "yes-no" questions could be combined: species A greater than 5, species B less than 2, and species C between 10 and 20. Numerous examples throughout the artificial intelligence literature have proven over and over again that such conceptual statistical analysis of the data provides much more useful insight into the patterns in the data, and, indeed, are often more accurate and robust. Delicate linear discriminants and other traditional techniques chronically suffer from overfitting, particularly in highly dimensioned spaces. Conceptual statistical analysis attempts to fit the data, but not at the expense of a simple, intuitive result.

Before we can determine whether a toxin has affected a group of organisms or the dynamics of an ecological community, we must first determine what types of changes would occur that are independent of the toxin. In field situations, this is usually attempted by using a reference site, monitoring the changes that occur at that site, and comparing these with the changes that occur in organisms at the dosed site.

However, one of the most difficult analytical challenges in ecology is to identify patterns of change in large ecological data sets. Often these data are not linear, they rarely conform to parametric assumptions, they have incommensurable units (e.g., length, concentration, frequency, etc.), and they are incomplete (due to both sample loss and sampling design whereby different parameters are collected at different frequencies). These difficulties exist regardless of whether there are toxicants present; the only difference is that with the presence of a toxicant, we must try to separate the response to the toxicant from the other changes that occur at the site(s).

The following examples demonstrate the usefulness of multivariate methods in the evaluation of field ecological data and laboratory multispecies toxicity tests. In each of the examples several multivariate techniques were used; generally, Euclidean and cosine distances, principal components, and nonmetric clustering and association analysis. Cosine and vector distances are diagrammatically presented in Figure 9.18.

Matthews et al. (1991a, 1991b) have compared several types of multivariate techniques to evaluate two types of ecological data: a limnological data set that included spatial and temporal changes in water chemistry and phytoplankton populations, and a stream data set that included spatial (longitudinal) and temporal changes in benthic macroinvertebrate species assemblages. Their objective was to see whether the multivariate tests could identify obvious patterns involving the influences of stratification in the lake and the effects of substrate and water quality changes on stream macroinvertebrates. We used principal components analysis, hierarchical clustering (k-means with squared Euclidean or cosine of vectors distance measures), correspondence analysis, and nonmetric clustering to look for patterns in the data.

In both studies, nonmetric clustering outperformed the metric tests, although both principal components analysis and correspondence analysis yielded some additional insight on large-scale patterns that was not provided by the nonmetric clustering results. However, nonmetric clustering provided information without the use of inappropriate assumptions, data transformations, or other data set manipulations that usually accompany the use of multivariate metric statistics. The success of these

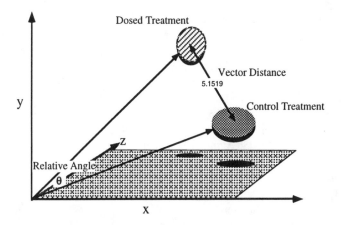

Figure 9.18   Measures of distance between clusters. Two of the commonly used measures of separation of clusters in an n-dimensional space are the cosine of the angle and the vector distance. Each method has advantages and disadvantages. In order to visualize the data as accurately as possible, several measures should be employed.

studies and techniques led to the examination of community dynamics in a series of two multispecies toxicity tests.

The multivariate methods described above have recently been used to examine a series of multispecies toxicity tests. Described below are the data analyses from two recently published tests using methodology derived from the standardized aquatic microcosm (SAM) (ASTM E1366-91). The method is described in some detail in Chapter 4.

In the first example, the riot control material 1,4-dibenz oxazepine (CR) was degraded using the patented organism *Alcaligenes denitrificans denitrificans* CR-1 (*A. denitrificans* CR-1) (Landis et al. 1993). *A. denitrificans* CR-1 was obtained using a natural inoculum set in an environment containing the microcosm medium T82MV containing the toxicant CR. After demonstrating the ability of the organism to degrade the toxicant CR, a microcosm experiment was set up to investigate the ability of the microorganisms to degrade CR in an environment resembling a typical freshwater environment. Toxicity tests of the riot control material demonstrated that although *A. denitrificans* CR-1 eliminated the toxicity of a CR solution towards algae, toxicity did remain to *Daphnia magna*.

The SAM experiment was set up with a control group without the toxicant or *A. denitrificans* CR-1, a second group with only CR, a third group with only *A. denitrificans* CR-1, and the fourth group containing both the toxicant CR and the bacterium *A. denitrificans* CR-1. Conventional analysis demonstrated that the major impact was the increase in algal populations since both CR and the degradative products of the toxicant inhibited the growth of the major herbivore, *D. magna*. The control group and the microcosms inoculated initially with *A. denitrificans* CR-1 were not distinguishable using conventional analysis.

As a first test of the use of multivariate analysis in the interpretation of multispecies toxicity tests, the data set used to analyze the CR microcosm experiment was presented in a blind fashion for analysis. Neither the purpose nor the experimental setup was provided to the analysis team. Nonmetric clustering was used to rank variables in terms of contribution and to set clusters. Surprisingly, the analysis resulted in only two clusters being recognized, control and *A. denitrificans* CR-1 treatments, and the CR and CR plus *A. denitrificans* CR-1 treatments. Variables important in assigning clusters were *D. magna, Ankistrodesmus, Scenedesmus,* and NO_2. Obviously, the inclusion of the principal algal species in these experiments and the daphnia was not a surprise, but NO_2 had not been demonstrated as a significant factor in previous analysis. However, the species *A. denitrificans denitrificans* is classified for its denitrification ability (Matthews and Matthews 1991).

The second major application of nonmetric clustering to the analysis of SAM data has been the investigation of the impact of the complex Jet-A (Landis et al. 1993). The major modification to the SAM protocol was the means of toxicant delivery. Test material was added on day 7 by stirring each microcosm, removing 450 ml from each container, and then adding appropriate amounts of the water soluble fraction (WSF) of Jet-A to produce concentrations of 0, 1, 5, and 15% WSF. After toxicant addition the final volume was adjusted to 3l.

All of the multivariate tests (cosine distance, vector distance, and nonmetric clustering) agree that a significant difference between treatment groups was observed through day 25. From day 28 to day 39, the effect diminished until there were no significant effects observable. However, significant effects were again observable from day 46 through day 56, after which they again disappeared for days 60 and 63.

In Figure 9.19, the average cosine distances within the control group and between the control group and each of the three treatment groups are plotted on a log scale. The initial, strong effect, from day 11 to day 25, is easily seen as a large distance from the treatment 1 (control) and treatment 2, together, to both treatment groups 3 and 4, initially, but then treatment 3 moves closer to the control. The period of no significant difference, from day 35 to day 46, is also clear. During the second period of significant difference, from day 49 to day 59, a perfect dose response for all three treatments is seen, with higher doses becoming more distant from the control. This dose-response relationship is consistently maintained over a period of 11 days for four sampling dates: days 49, 53, 56, and 59. A dose-response relationship like this was not observed earlier, although the magnitude of the distance was considerably greater.

Also of interest are the variables that best described the clusters and the stability of the importance of the variables during the course of the experiment (Table 9.4). In this experiment the variables that are important early are those dealing with the algae and their predators, the daphnia. Towards the end of the experiment bactivorus organisms like ostracodes and the rotifer *Philodina* increase in importance. In general, the number of variables that were important was larger during the start of the test and lower at the end. In addition, a great deal of variability in rankings is apparent during the course of the SAM. As shown in Table 9.5, some of the variables are more important at some times than at other times. However, when a variable such as the

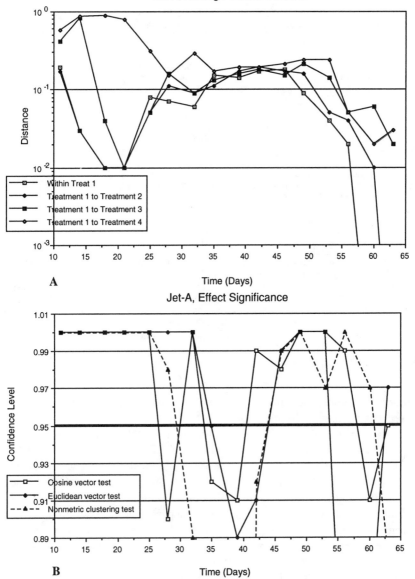

A

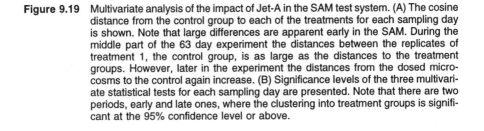

B

Figure 9.19   Multivariate analysis of the impact of Jet-A in the SAM test system. (A) The cosine distance from the control group to each of the treatments for each sampling day is shown. Note that large differences are apparent early in the SAM. During the middle part of the 63 day experiment the distances between the replicates of treatment 1, the control group, is as large as the distances to the treatment groups. However, later in the experiment the distances from the dosed micro-cosms to the control again increase. (B) Significance levels of the three multivariate statistical tests for each sampling day are presented. Note that there are two periods, early and late ones, where the clustering into treatment groups is significant at the 95% confidence level or above.

Table 9.4   Important Variables Ranked by Nonmetric Clustering for Each Sampling Date for the Jet-A SAM Toxicity Test

Day	Important variables in determining clusters in rank order
11	M. Daphnia, Chlorella, Chlamydamonas, Ulothrix, S. Daphnia, Selanstrum, Scenedesmus
14	S. Daphnia, M. Daphnia-Selenastrum,[a] Chlamydamonas, Chlorella, L. Daphnia, *Ankistrodesmus*
18	*Ankistrodesmus*, S. Daphnia, Chlorella, Chlamydamonas, Selanstrum, L. Daphnia
21	*Ankistrodesmus*, S. Daphnia, L. Daphnia-M. Daphnia, Scenedesmus
25	Scenedesmus, S. Daphnia, L. Daphnia, Chlorella, Philodina-M. Daphnia
28	*Ankistrodesmus*, L. Daphnia, Scenedesmus
32	S. Daphnia, M. Daphnia, *Ankistrodesmus*, Chlorella
35	*Ankistrodesmus*
39	M. Daphnia-Selenastrum, Ostracod-Ankistrodesmus
42	M. Daphnia, Ostracod, Scenedesmus
46	Scenedesmus, *Ankistrodesmus*, S. Daphnia. M. Daphnia
49	Chlorella, *Philodina, Ankistrodesmus*, Lyngbya
53	*Ankistrodesmus, Ostracod,* Chlorella
56	M. Daphnia-Scenedesmus, *Ankistrodesmus*, Lyngbya
60	Lyngbya, M. Daphnia, *Philodina*, Chlorella
63	Chlorella, *Ankistrodesmus, Philodina, Ostracod*

Note: Some variables such as *Ankistrodesmus* were consistently important in determining group clusters throughout the experiment. Some of the variables such as Ostracod and Philodina were more important in the latter stages of the experiment. The order of importance of the variables often changed from day to day, with no one variable being common to each sampling date. The variables used as part of the overall analysis were Anabaena, Ankistrodesmus, Chlamydomonas, Chlorella, Daphnia (Ephipia, Small Daphnia, Medium Daphnia, Large Daphnia), Hypotricha, Lyngbya, Miscellaneous sp., Ostracod (Cyprinotus), Philodina (Rotifer), Scenedesmus, Selenastrum, Stigeoclonium, and Ulothrix.

[a] Hyphen between variables denotes equal rank.

number of *Philodina* are important, they are crucial in determining the various experimental groupings.

Conventional analysis using such techniques as the IND plot (Conquest and Taub 1989) was unable to detect the second oscillation. The only leads were statistically significant deviations from the control for one sampling date for the variables pH and the photosynthesis-to-respiration ratio. These deviations were considered cases of Type II error until confirmation of effects using multivariate analysis.

Analysis of the toxicant concentration using purge-and-trap gas chromatography indicated that few of the constituents of the WSF were present in the water column at the end of the SAM experiment.

Examination of individual parameters provided only a limited, and somewhat distorted view of the SAM microcosm response to Jet-A. The univariate data analysis did indeed show that there were some significant responses to the toxicant by individual taxa and chemistry; however, the responses were scattered over time, and did not present a logical, coherent pattern. Furthermore, the individual responses detected were typified by wild swings in a taxon's population density over time.

The repeated oscillation of the dosed replicates compared to the controls can be accounted for in two basic ways:

Table 9.5   Variable According to Success
in Determining Clusters as
Defined by Nonmetric Clustering
in the Jet-A SAM Experiments

Variable	Ranked
Ankistrodesmus	12
M. Daphnia	11
Chlorella	9
Scenedesmus	7
S. Daphnia	6
L. Daphnia	5
Ostracod	4
Philodina	4
Selenastrum	4
Lyngbya	3
Ulothrix	1

Note: Variables such as Ankistrodesmus and the Daphnia classes were important in the course of this study. Reliance on even these two variables would have been misleading in the determination of the second oscillation.

1. A reflection of the functioning of the community best described by parameters not directly sampled by the SAM protocol
2. A repeated fluctuation in community structure initiated by the initial stress and that is visible as an undampened movement in the systems

Until more data can be obtained, the cause-effect of the second oscillation cannot be determined. However, the use of multivariate analysis detected an unexpected result, one providing a new insight into the dynamics of even the relatively simple laboratory microcosm.

However, the search for diagnostic measures to indicate the displacement of an ecosystem may not be fruitless. Although the relative importance of the variables in the SAM experiments may change, there are often variables that are more critical during the earlier stages of the development of the microcosm and those that are more crucial in the latter stages. The variable Ostracods is generally more important in the earlier half of the experimental series than in the latter stages. The crucial aspect is that the clustering algorithm is able to select ecosystem attributes that are the best in differentiating stressed vs. nonstressed systems. Although expert judgment may be able to predict in some cases variables that could be considered important to measure, the clustering approach is rapid, consistent, and not biased.

INTERPRETATION OF ECOSYSTEM LEVEL IMPACTS — STABILITY AND NON-EQUILIBRIUM ECOSYSTEM DYNAMICS

The measurement of the current status of an ecosystem and the assumption that recovery is the likely outcome once the stressor is removed may not hold up to careful scrutiny given new developments in the study of population dynamics and ecosystems. First, it is crucial to know the dynamical aspects of the systems we are studying and second, as with the weather, it may prove inherently impossible to predict the futures of ecosystems.

First, the apparent recovery or movement of a dosed system toward the reference case may be an artifact of our measurement systems that allow the n-dimensional data to be represented in a two-dimensional system. In an n-dimensional sense, the systems may be moving in opposite directions and simply pass by similar coordinates during certain time intervals. Positions can be similar but the n-dimensional vectors describing the movements of the systems can be very different. One-time sampling indexes are likely to miss these movements or incorrectly plot the system in an arbitrary coordinate system.

The apparent recoveries and divergences may also be artifacts of our attempt to choose the best means of collapsing and representing n-dimensional data into a two- or three-dimensional representation. In order to represent such data, it is necessary to project n-dimensional data into three or less dimensions. As information is lost when the shadow of a cube is projected upon a two-dimensional screen, a similar loss of information can occur in our attempt to represent n-dimensional data. The possible illusion of recovery based on this type of projection is diagrammatically represented in Figure 9.20. In Figure 9.20A the dosed and the reference systems appear to converge, i.e., recovery has occurred. However, this may be an illusion created by the perspective chosen to describe and measure the system. Figure 9.20B is the same system but viewed from the "top". When a new point of view is taken, divergence of the systems occurs throughout the observed time period. As the various groups separate, the divergence may be seen as a separate event. In fact, this separation is a continuation of the dynamics initiated earlier upon one aspect of the community. Eventually, the illusion of recovery may simply be the divergence of the replicates within each treatment group becoming large enough, with enough inherent variation, so that even the multivariate analysis cannot distinguish treatment group similarities. Not every divergence from the control treatment may have a causal effect related to it in time; differentiating these events from those due to degradation products or other perturbations will be challenging.

Not only may system recovery be an illusion, but there are strong theoretical reasons that seem to indicate that recovery to a reference system may be impossible or at least unlikely. Systems that differ only marginally in their initial conditions and at levels probably impossible to measure are likely to diverge in unpredictable manners. May and Oster (1976), in a particularly seminal paper, investigated the likelihood that many of the dynamics seen in ecosystems that are generally attributed as chance or stochastic events are in fact deterministic. Simple deterministic models of populations can give rise to complicated behaviors. Using equations resembling those used in population biology, bifurcations occur resulting with several distinct outcomes. Eventually, given the proper parameters, the system appears chaotic in nature although the underlying mechanisms are completely deterministic. Biological systems have limits,

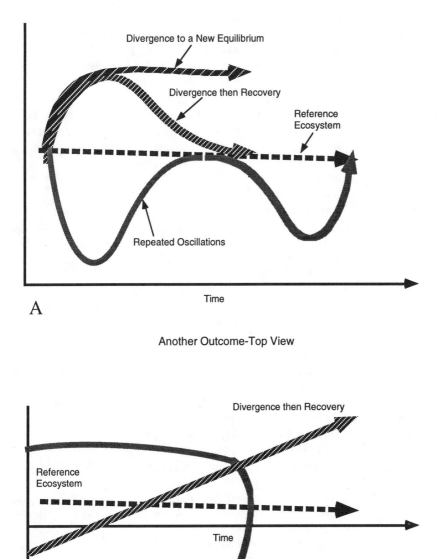

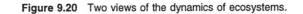

Figure 9.20   Two views of the dynamics of ecosystems.

extinction being perhaps the most obvious and best recorded. Another ramification is that the noise in ecosystems and in sampling may not be the result of a stochastic process but the result of an underlying deterministic chaotic relationship.

These principals also apply to spatial distributions of populations as recently reported by Hassell et al. (1991). In a study using host-parasite interactions as the model, a variety of spatial patterns were developed using the Nicholson-Bailey model. Host-parasite interactions demonstrated patterns ranging from static 'crystal lattice' patterns, spiral waves, chaotic variation, or extinction with the appropriate variation of only three parameters within the same set of equations. The deterministically determined patterns could be extremely complex and not distinguishable from stochastic environmental changes.

Given the possible chaotic nature of populations, it may not be possible to predict species presence, population interactions, or structural and functional attributes. Katz et al. (1987) examined the spatial and temporal variability in zooplankton data from a series of five lakes in North America. Much of the analysis was based on limnological data collected by Brige and Juday from 1925 to 1942. Copepods and cladocera, except *Bosmina*, exhibited larger variability between lakes than between years in the same lake. Some taxa showed consistent patterns among the study lakes. They concluded that the controlling factors for these taxa operated uniformly in each of the study sites. However, in regards to the depth of maximal abundance for calanoid copepods and *Bosmina*, the data obtained from one lake had little predictive power for application to other lakes. Part of this uncertainty was attributed to the intrinsic rate of increase of the invertebrates with the variability increasing with a corresponding increase in r_{max}. A high r_{max} should enable the populations to accurately track changes in the environment. Katz et al. (1987) suggest that these taxa be used to track changes in the environment. Unfortunately, in the context of environmental toxicology, the inability to use one lake to predict the nondosed population dynamics of these organisms in another eliminates comparisons of the two systems as measures of anthropogenic impacts.

A better strategy may be to let the data and a clustering protocol identify the important parameters in determining the dynamics of and impacts to ecological systems. This approach has been recently suggested independently by Dickson et al. (1992) and Matthews et al. (1991a, 1991b). This approach is in direct contrast to the more usual means of assessing anthropogenic impacts. One classical approach is to use the presence or absence of so-called indicator species. This assumes that the tolerance to a variety of toxicants is known and that chaotic or stochastic influences are minimized. A second approach is to use hypothesis testing to differentiate metrics from the systems in question. This second approach assumes that the investigators know *a priori* the important parameters. Given that, at least in our relatively simple SAM systems, the important parameters in differentiating nondosed from dosed systems change from sampling period to sampling period, this assumption cannot be made. Classification approaches such as nonmetric clustering or the canonical correlation methodology developed by Dickson et al. (1992) eliminates these assumptions.

The results presented in this report and the others reviewed above, and the implications of chaotic dynamics suggest that reliance upon any one variable or an index of variables may be an operational convenience that may provide a misleading representation of pollutant effects and the associated risks. The use of indices such as diversity and the index of biological integrity have the effect of collapsing the dimensions of the descriptive hypervolume in a relatively arbitrary fashion. Indices, since they are composited variables, are not true endpoints. The collapse of the dimensions that are composited tends to eliminate crucial information, such as the variability in the importance of variables. The mere presence or absence and the frequency of these events can be analyzed using techniques such as nonmetric clustering that preserve the nature of the data set. A useful function was certainly served by the application of indices. The new methods of data compilation, analysis, and representation derived from the Artificial Intelligence tradition can now replace these approaches and illuminate the underlying structure and dynamic nature of ecological systems.

The implications are important. Currently, only small sections of ecosystems are monitored or a heavy reliance is placed upon so-called indicator species. These data suggest that to do so is dangerous, and may produce misleading interpretations resulting in costly error in management and regulatory judgments. Much larger toxicological test systems are currently analyzed using conventional statistical methods on the limit of acceptable statistical power. Interpretation of the results has proven to be difficult.

The importance of viewpoint and the apparent chaotic nature of ecological systems makes discussion of such parameters as ecosystem stability difficult to accurately determine. In Figure 9.21 a system that hits a perturbation is depicted. Although the distances that each have traveled are the same in a two-dimensional picture, from the viewpoint of the observer one system moves farther than the other, and by some definitions is less stable. Conversely, if the chaotic nature of systems prevents a return to the original state, recovery cannot be considered an inherent property of the system.

The dynamics in the research discussed above make a metaphor such as ecosystem health inappropriate and misleading. In a recent critical evaluation, Suter (1993) dismissed ecosystem health as a misrepresentation of ecological science. Ecosystems are not organisms with the patterns of homeostasis determined by a central genetic core. Since ecosystems are not organismal in nature, health is a property that cannot describe the state of such a system. The urge to represent such a state as health has lead to the compilation of variables with different metrics, characteristics, and casual relationships. Suter suggests a better alternative would be to evaluate the array of ecosystem processes of interest, with an underlying understanding that the fundamental nature of these systems are quite different than those of organisms.

The outcome of the new developments may result in a new set of rules for the evaluation of ecosystems after an anthropogenic stressor. They can be listed as:

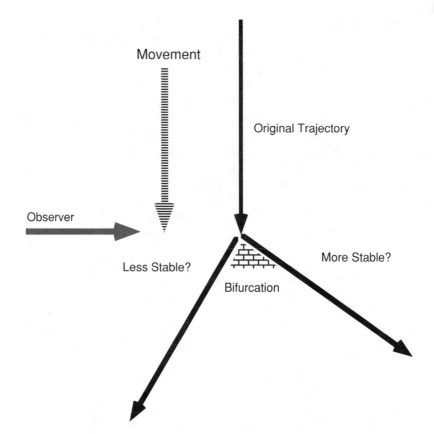

Figure 9.21   Apparent change in an ecosystem depends upon the point of view of the observer.

1. You can never go home again. Ecosystems do not recover to the previous state and they cannot be expected to. The history of the disturbance has changed the initial conditions of the system resulting in one that may be superficially similar but that is different down to its genetic make up. If no two systems are the same then prediction becomes problematical.
2. You cannot even try to go home again. If the underlying dynamics are chaotic the system is unlikely to recur. Even if regular cycles are possible the chance of the systems being in phase may be low. Even if major efforts, such a fertilization or selective colonization force the system into a final outcome, the trajectories, the road getting there, are likely to be different.
3. History is important. Reasons discussed in 1 and 2 above outline some of the reasons. Evolution and the history of speciation points to a chaotic system both in the numbers of species generated and in the rate of speciation. How do we know that many of the current species are in periods of long term decline or increase even without the input of man? History of colonization and the differing interactions that are caused by these events are to a large degree stochastic. Stochastic events overlap chaotic ones.

4. You cannot predict the future no matter how much you know. If the dynamics are chaotic and 1, 2, and 3 occur, then predicting the future and, hence, the impact of the system may be impossible beyond a certain range. The prediction of another chaotic event, weather formation has proven recalcitrant even with the massive resources put into research and data collection. It may be that ecosystems and ability to predict impacts may bear a similar fate.

APPENDIX: MULTIVARIATE TECHNIQUES — NONMETRIC CLUSTERING

In the research described above, three multivariate significance tests were used. Two of them were based on the ratio of multivariate metric distances within treatment groups vs. between treatment groups. One of these is calculated using Euclidean distance and the other with cosine of vectors distance (Good 1982; Smith et al. 1990). The third test used nonmetric clustering and association analysis (Matthews et al. 1990). In the microcosm tests there were four treatment groups with six replicates, giving a total of 24. This example is used to illustrate the applications in the derivations that follow.

Treating a sample on a given day as a vector of values, $\vec{x} = \langle x_1 \dots x_n \rangle$, with one value for each of the measured biotic parameters, allows multivariate distance functions to be computed. Euclidean distance between two sample points $\vec{x}$ and $\vec{y}$ is computed as

$$\sqrt{\sum_i (x_i - y_i)^2}$$

The cosine of the vector distance between the points $\vec{x}$ and $\vec{y}$ is computed as

$$1 - \frac{\sum_i x_i y_i}{\sqrt{\sum_i x_i^2 \sum_i y_i^2}}$$

Subtracting the cosine from one yields a distance measure, rather than a similarity measure, with the measure increasing as the points get farther from each other.

The within-between ratio test used a complete matrix of point-to-point distance (either Euclidean or cosine) values. For each sampling date, one sample point $\vec{x}$ was obtained from each of six replicates in the four treatment groups, giving a 24 × 24 matrix of distances. After the distances were computed, the ratio of the average within-group metric (W) to the average between-group metric (B) was computed (W/B). If the points in a given treatment group are closer to each other, on average, than they are to points in a different treatment group, then this ratio will be small. The significance of the ratio is estimated with an approximate randomization test. This test is based on the fact that, under the null hypothesis, assignment of points to treatment groups is random, the treatment having no effect. The test, accordingly, randomly assigns each of the replicate points to groups, and recomputes the W/B ratio a large number of times (500 in our tests). If the null hypothesis is false, this randomly derived ratio will (probably) be larger than the W/B ratio obtained from the actual treatment groups. By taking a large number of random reassignments, a valid estimate of the probability under the null hypothesis is obtained as $(n+1)/(500+1)$, where n is the number of times a ratio less than or equal to the actual ratio was obtained (Noreen 1989).

In the clustering association test, the data are first clustered independently of the treatment group, using nonmetric clustering and the computer program RIFFLE (Matthews and Hearne 1991). Because the RIFFLE analysis is naive to treatment group, the clusters may or may not correspond to treatment effects. To evaluate whether the clusters were related to treatment groups, whenever the clustering procedure produced four clusters for the sample points, the association between clusters and treatment groups was measured in a 4 x 4 contingency table, each point in treatment group i and cluster j being counted as a point in frequency cell ij. Significance of the association in the table was then measured with Pearson's χ^2 test, defined as

$$\chi^2 = \sum_{ij} \frac{\left(N_{ij} - n_{ij}\right)^2}{n_{ij}}$$

where N_{ij} is the actual cell count and n_{ij} is the expected cell frequency, obtained from the row and column marginal totals N_{+j} and N_{i+} as

$$n_{ij} = \frac{N_{+j} N_{i+}}{N}$$

where $N = 24$ is the total cell count, and a standard procedure for computing the significance (probability) of χ^2 taken from Press et al. (1990).

REFERENCES AND SUGGESTED READINGS

Anderson, R. S. 1975. Phagocytosis by invertebrate cells *in vitro*: Biochemical events and other characteristics compared with vertebrate phagocytic systems. In *Invertebrate Immunity: Mechanisms of Invertebrate Vector-parasite Relations*. Academic Press, San Francisco, pp. 153-180.

Anderson, R. S., C. S. Giam, L. E. Ray, and M. R. Tripp. 1981. Effects of environmental pollutants on immunological competency of the clam *Merceneria merceneria*: impaired bacterial clearance. *Aquat. Toxicol.* 1:187-195.

ASTM E 1366-91. 1991. Standard practice for the standardized aquatic microcosm: fresh water, *Annual Book of ASTM Standards*, Vol. 11.04, American Society for Testing and Materials, Philadelphia, PA, pp. 1017-1051.

Barnthouse, L. W., G. W. Suter, II, and A. E. Rosen. 1989. Inferring population-level significance from individual-level effects: an extrapolation from fisheries science to ecotoxicology. In *Aquatic Toxicology and Environmental Fate:* 11th Volume. ASTM STP 1007, G. W. Suter, II and M. A. Lewis, Eds., American Society for Testing and Materials, Philadelphia, PA, pp. 289-300.

Barnthouse, L. W., G. W. Suter, II, and A. E. Rosen. 1990. Risks of toxic contaminants to exploited fish populations: Influence of life history, data uncertainty, and exploitation intensity. *Environ. Toxicol. Chem.* 9:297-311.

Barnthouse, L. W. 1993. Population-level effects. In *Ecological Risk Assessment,* G. W. Suter II, Ed., Lewis Publishers, Boca Raton, pp. 247-274.

Bradley, B. P. 1990. Stress-proteins: Their detection and uses in biomonitoring. In *Aquatic Toxicology and Risk Assessment:* 13th Volume. ASTM STP-1096. W. G. Landis and W. H. van der Schalie, Eds. American Society for Testing and Materials, Philadelphia, PA, pp. 338-347.

Cairns, J., Jr. 1992. Paradigms flossed: the coming of age of environmental toxicology. *Environ. Toxicol. Chem.* 11: 285-287.

Connell, J. H. and W. P. Sousa. 1983. On the evidence needed to judge ecological stability or persistence. *Am. Nat.* 121:789-824.

Conquest, L.L. and F.B. Taub. 1989. Repeatability and reproducibility of the Standard Aquatic Microcosm: Statistical properties. In *Aquatic Toxicology and Hazard Assessment:* 12th Volume. ASTM STP 1027. U.M. Cowgill and L.R. Williams, Eds., American Society for Testing and Materials, Philadelphia, PA, pp. 159-177.

Dickson, K.L., W.T. Waller, J.H. Kennedy, and L.P. Ammann, 1992. Assessing the relationship between ambient toxicity and instream biological response. *Environ. Toxicol. Chem.* 11:1307-1322.

Fairbrother, A., R. S. Bennett, and J. K. Bennett. 1989. Sequential sampling of plasma cholinesterase in mallards (Anas platyrhynchos) as an indicator of exposure to cholinesterase inhibitors. *Environ. Toxicol. Chem.* 8:117-122.

Fienberg, S.E. 1985. *The Analysis of Cross-Classified Categorical Data.* MIT Press, Cambridge, MA.

Giddings, J.M., B.R. Parkhurst, C.W. Hehrs, and R.E. Millemann. 1980. Toxicity of a coal liquefaction product to aquatic organisms. *Bull. Environ. Contam. Toxicol.* 25:1-6.

Gleick, J. 1987. *Chaos: Making a New Science.* Penguin Books, New York, NY.

Good, I.J. 1982. An index of separateness of clusters and a permutation test for its significance. *J. Statist. Comp. Simul.* 15:81-84.

Hassell, M. P., H. N. Cumins, and R. M. May. 1991. Spatial structure and chaos in insect population dynamics. *Nature* 353: 255-258.

Hutchinson, G.E. 1959. Concluding remarks. *Cold Spring Harbor Symposium on Quantitative Biology.* 22:415-427.

Hutchinson, G.E. 1961. The paradox of the plankton. *Am. Nat.* 95:137-143.

Johnson, A.R. 1988a. Evaluating ecosystem response to toxicant stress: a state space approach. In *Aquatic Toxicology and Hazard Assessment:* 10th Volume, ASTM STP 971, W.J. Adams, G.A. Chapman, and W.G. Landis, Eds., American Society for Testing and Materials, Philadelphia, PA, pp. 275-285.

Johnson, A.R. 1988b. Diagnostic variables as predictors of ecological risk. *Environ. Manage.* 12:515-523.

Karr, J. R. 1991. Biological integrity: A long-neglected aspect of water resource management. *Ecol. Appl.* 1:66-84.

Karr, J. R. 1993. Defining and assessing ecological integrity: beyond water quality. *Environ. Toxicol. Chem.* 12:1521-1531.

Katz, T.K., T.M. Frost, and J.J. Magnuson. 1987. Inferences from spatial and temporal variability in ecosystems: Long-term zooplankton data from lakes. *Am. Nat.* 129: 830-846.

Kauffman, S. A. and S. Johnsen. 1991. Coevolution to the edge of chaos: coupled fitness landscapes, poised states, and coevolutionary avalanches. *J. Theor. Biol.* 149:467-505.

Kersting, K. 1984. Development and use of an aquatic micro-ecosystem as a test system for toxic substances. Properties of an aquatic micro-ecosystem IV. *Int. Rev. Hydrobiol.* 69:567-607.

Kersting, K. 1985. Properties of an aquatic micro-ecosystem V. Ten years of observations of the prototype. *Verh. Int. Verein. Limnol.* 22:3040-3045.

Kersting, K. 1988. Normalized ecosystem strain in micro-ecosystems using different sets of state variables. *Verh. Int. Verein. Limnol.* 23:1641-1646.

Kersting, K. and R. van Wijngaarden. 1992. Effects of Chloropyrifos on a microecosystem. *Environ. Toxicol. Chem.* 11:365-372.

Kindig, A.C., L.C. Loveday, and F.B. Taub. 1983. Differential sensitivity of new vs. mature synthetic microcosms to streptomycin sulfate treatment. In *Aquatic Toxicology and Hazard Assessment:* Sixth Symposium, ASTM 802. W.E. Bishop, R.D. Cardwell, and B.B. Heidolph, Eds., American Society for Testing and Materials, Philadelphia, PA, pp. 192-203.

Landis, W.G. 1986. Resource competition modeling of the impacts of xenobiotics on biological communities. In *Aquatic Toxicology and Environmental Fate:* 9th Volume. ASTM 921. T.M. Poston and R. Purdy, Eds., American Society for Testing and Materials, Philadelphia, PA, pp. 55-72.

Landis, W. G. 1991. Biomonitoring, Myth or Miracle? In *Pesticides in Natural Systems: How Can Their Effects Be Monitored?* Proc. Conf., December 11th and 12th, 1990. EPA 910/9-91-011. pp 17-38.

Landis, W. G., M. V. Haley, and N. A. Chester. 1993 The use of the standardized aquatic microcosm in the evaluation of degradative bacteria in reducing impacts to aquatic ecosystems. In *Environmental Toxicology and Risk Assessment.* W. G. Landis, J. Hughes, and M. Lewis, ASTM STP-1167. American Society for Testing and Materials, Philadelphia, PA, pp. 159-177

Landis, W. G., R. A. Matthews, A. J. Markiewicz, and G. B. Matthews. *In press.* Multivariate analysis of the impacts of the turbine fuel JP-4 in a microcosm toxicity test with implications for the evaluation of ecosystem dynamics and risk assessment. *Ecotoxicology.*

Landis, W. G., R. A. Matthews, A. J. Markiewicz, and G. B. Matthews. *In press.* Non-linear oscillations detected by multivariate analysis in microcosm toxicity tests with complex toxicants: Implications for biomonitoring and risk assessment. *Environmental Toxicology and Risk Assessment:* 3rd Volume. ASTM 1218. J. S. Hughes, G. R. Biddinger, and E. Mones, Eds., American Society for Testing and Materials, Philadelphia, 1994.

Landis, W. G., G. B. Matthews, R. A. Matthews, and A. Sergeant. *In press.* Application of multivariate techniques to endpoint determination, selection and evaluation in ecological risk assessment. *Environ. Toxicol. Chem.*

Landis, W.G., R.A. Matthews, A.J. Markiewicz, N.J. Shough, and G.B. Matthews. 1993. Multivariate analysis of the impacts of turbine fuel using a standard aquatic microcosm toxicity test. *J. Environ. Sci.* In Press.

Lower, W. R. and R. J. Kendall. 1990. Sentinel species and sentinel bioassay. In *Biomarkers of Environmental Contamination.* J. F. McCarthy and L. R. Shugart, Eds., Lewis Publishers, Boca Raton, FL, pp. 309-331.

Matthews, G. B. and J. Hearne, 1991. Clustering without a metric. *IEEE Transactions on Pattern Analysis and Machine Intelligence* 13(2): 175-184.

Matthews, G.B. and R.A. Matthews. 1990. A model for describing community change. In *Pesticides in Natural Systems: How Can Their Effects Be Monitored?* Proc. Conf. Environmental Research Laboratory/ORD, Corvallis, OR, EPA 9109/9-91/011.

Matthews, G.B. and R.A. Matthews, 1991. A model for describing community change. In *Pesticides in Natural Systems: How Can Their Effects Be Monitored?* Proc. Conf. Environmental Research Laboratory/ORD, Corvallis, OR, EPA 9109/9-91/011.

Matthews, G.B., R.A. Matthews, and B. Hachmoller. 1991a. Mathematical analysis of temporal and spatial trends in the benthic macroinvertebrate communities of a small stream. *Can. J. Fish. Aquat. Sci.* 48:2184-2190.

Matthews, G. B., R. A. Matthews, and W. G. Landis. *In press..* Nonmetric clustering and association analysis: Implications for the evaluation of multispecies toxicity tests and field monitoring. *Environmental Toxicology and Risk Assessment:* 3rd Volume. ASTM 1218. J. S. Hughes, G. R. Biddinger, and E. Mones, Eds., American Society for Testing and Materials, Philadelphia, 1994.

Matthews, R.A., G.B. Matthews, and W.J. Ehinger. 1991b. Classification and ordination of limnological data: a comparison of analytical tools. *Ecol. Model.* 53:167-187.

Matz, A. C. 1992. Development of a Method for Monitoring Chemical Contamination Using Galliform Chicks. M.S. thesis, Western Washington University.

May, R.M. 1973. *Stability and Complexity in Model Ecosystems.* Second Edition, Princeton University Press, Princeton, New Jersey.

May, R.M. and G.F. Oster, (1976) Bifurcations and dynamical complexity in simple ecological models. *Am. Nat.* 110:573-599.

McCarthy, J. F. and L. R. Shugart. 1990. *Biomarkers of Environmental Contamination.* Lewis Publishers, Boca Raton, FL.

Miller, R.V. and S.B. Levy. 1989. Horizontal gene transfer in relation to environmental release of genetically engineered microorganisms. *Gene Transfer in the Environment*, Chapter 13, pp. 405-420.

Nelson, W.G. 1990. Use of the blue mussel, *Mytilus edulis,* in water quality toxicity testing and in situ marine biological monitoring. In *Aquatic Toxicology and Risk Assessment:* 13th Volume. ASTM STP 1096, W.G. Landis and W.H. van der Schalie, Eds., American Society for Testing and Materials, Philadelphia, PA, pp. 167-175.

Nicolis, G. and I. Prigogine. 1989. *Exploring Complexity.* W. H. Freeman and Company, New York.

Noreen, E.W. 1989. *Computer Intensive Methods for Testing Hypotheses.* Wiley-Interscience, New York, NY.

Powell, D. B. and Kocan, R. M. 1990. The response of diploid and polyploid rainbow trout cells following exposure to genotoxic compounds. In *Aquatic Toxicology and Risk Assessment:* Thirteenth Volume. ASTM STP-1096. W. G. Landis and W. H. van der Schalie, Eds., American Society for Testing and Materials, Philadelphia, PA. pp. 290-308.

Press, W.H., B.P. Flannery, A.A. Teukolsky, and W.T. Vetterline, 1990. *Numerical Recipes in C, the Art of Scientific Computing.* Cambridge University Press, New York, NY.

Richerson, P., R. Armstrong, and C.R. Goldman. 1970. Contemporaneous disequilibrium, a new hypothesis to explain the "paradox of the plankton". *Proc. Natl. Acad. Sci. U.S.A.* 67:1710-1714.

Schaffer, W. M. 1985. Can nonlinear dynamics elucidate mechanisms in ecology and epidemiology? *IMA J. Math. Appl. Med. Biol.* 2:221-252.

Schaffer, W. M. and M. Kot. 1985. Do strange attractors govern ecological systems? *Bioscience* 35: 342-350.

Shugart, L. R. 1990. DNA damage as an indicator of pollutant induced genotoxicity. In *Aquatic Toxicology and Risk Assessment:* 13th Volume. ASTM STP-1096. W. G. Landis and W. H. van der Schalie, Eds., American Society for Testing and Materials, Philadelphia, PA, pp. 348-355.

Sugihara, G., B. Grenfell, and R.M. May. 1990. Distinguishing error from chaos in ecological time series. *Phil. Trans. R. Soc. Lond.* 330:235-251.

Smith, E.P., K.W. Pontasch, and J. Cairns, Jr. 1990. Community similarity and the analysis of multispecies environmental data: a unified statistical approach. *Water Res.* 24(4): 507-514.

Stickle, W.B., Jr., S.D. Rice, C. Villars, and W. Metcalf. 1985. Bioenergetics and survival of the marine mussel, *Mytilus edulis* L., during long-term exposure to the water-soluble fraction of Cook Inlet crude oil. In *Marine Pollution and Physiology: Recent Advances.* F.J. Vernberg, F.P. Thurberg, A. Calabrese, and W.B. Vernberg, Eds., University of South Carolina Press, Columbia, SC, pp. 427-446.

Suter, G. W. 1993. A critique of ecosystem health concepts and indexes. *Environ. Toxicol. Chem.* 12: 1533-1539.

Taub, F.B. 1988. Standardized aquatic microcosm — development and testing. In *Aquatic Ecotoxicology: Fundamental Concepts and Methodologies.* Volume II. A. Boudou and F. Ribeyre, Eds., CRC Press, Boca Raton, pp. 47-92.

Taub, F.B. 1989. Standardized aquatic microcosms. *Environ. Sci. Technol.* 23:1064-1066.

Taub, F.B., A.C. Kindig, and L.L. Conquest. 1987. Interlaboratory testing of a standardized aquatic microcosm. In *Aquatic Toxicology and Hazard Assessment:* 10th Volume. ASTM STP 971. W.J. Adams, G.A. Chapman and W.G. Landis, Eds., American Society for Testing and Materials, Philadelphia, PA, pp. 385-405.

Taub, F.B., A.C. Kindig, L.L. Conquest, and J.P. Meador. 1988. Results of the interlaboratory testing of the Standardized Aquatic Microcosm protocol. In *Aquatic Toxicology and Hazard Assessment:* 11th Volume. ASTM STP 1007. G. Suter and M. Lewis, Eds., American Society for Testing and Materials, Philadelphia, PA, pp. 368-394.

Theodorakis, C. W., S. J. D'Surney, J. W. Bickham, T. B. Lyne, B. P. Bradley, W. E. Hawkins, W. L. Farkas, J. R. McCarthy, and L. R. Shugart. 1992. Sequential expression of biomarkers in Bluegill Sunfish exposed to contaminated sediment. *Ecotoxicology* 1:45-73.

Tilman, D. and S. Kilham. 1976. Phosphate and silicate growth and uptake kinetics of the diatoms *Asterionella formosa* and *Cyclotella meneghiniana* in batch and semicontinuous culture. *F. Phycol.* 12:375-383.

Tilman, D. 1982. *Resource Competition and Community Structure.* Princeton University Press, Princeton, NJ, pp. 11-138.

Tilman, D. and D. Weldin. 1991. Oscillations and chaos in the dynamics of a perennial grass. *Nature* 353:653-655.

van der Schalie, W. H. 1986. Can biological monitoring early warning systems be useful in detecting toxic materials in water? In: *Aquatic Toxicology and Environmental Fate:* 9th Volume. ASTM STP 921. T. M. Poston and R. Purdy, Eds., American Society for Testing and Materials, Philadelphia, PA, pp 107-121.

STUDY QUESTIONS

1. What are the two categories of biomonitoring programs?
2. List the six current organizational levels of biomonitoring and explain.
3. Discuss some examples of means by which past and current exposure to toxic xenobiotics are detected.
4. Of what value are biomarkers as predictors of the effects of toxicants?
5. Discuss the inhibition of specific enzymes, enzyme synthesis induction, stress proteins, DNA and chromosomal damage, immunological endpoints, and nutritional state as biomarkers of exposure to xenobiotics.
6. Describe physiological and behavioral indictors of toxicant impact.
7. What are toxicity identification and reduction evaluations?
8. What are the advantages and disadvantages to the toxicity tests given as examples in the text?
9. Define and discuss sentinel organisms as ecosystem monitors. What are the advantages of using this method?
10. Discuss alterations in genetic structure as a means of measuring xenobiotic effects on a population.
11. How can species diversity indicate stress on an ecosystem? What drawbacks does the structure of biological communities have as an indicator of stress?
12. What questions should biological diversity raise if it is used as an indicator of xenobiotic impacts upon biological communities?
13. What is resource competition? What is a resource consumption vector? What is a ZNGI?
14. Describe a two-species resource space graph.
15. Describe how resource heterogeneity can be incorporated into a two-species resource space graph.
16. How can toxicant input vs. natural variation be evaluated when community structure has altered?
17. Discuss nonlinear systems and their role in modelling xenobiotic impacts to ecological systems.
18. Describe the characteristics necessary in multivariate statistical tests used for evaluating complex data sets.
19. What is normalized ecosystem strain? What did Kersting find occurring with the NES as time increased after a perturbation?
20. Explain Johnson's state space of ecosystems.
21. What is the major difficulty with the Johnson and Kersting methods?
22. What is nonmetric clustering and what statistical importance does it have for ecosystem analysis?
23. What is one of the most difficult analytical challenges in ecology?
24. Discuss an assessment baseline and measurement endpoints as analysis of ecological risk assessment.
25. What would be a critical development in the formulation of risk assessment methodologies?
26. Discuss the benefits evolving from the use of multivariate techniques.

CHAPTER 10

Ecological Risk Assessment and Environmental Toxicology

INTRODUCTION

A great deal of environmental toxicology is performed with the eventual goal of performing a risk assessment. A great deal of the research performed in the field is geared toward the determination of the risk of producing a new product or releasing a pesticide or effluent to the environment. Because of the interaction between environmental toxicology and risk assessment, a basic and clear understanding of ecological risk assessment in necessary. Appendix A is a reprint of the recent U.S. EPA document "A Framework for Ecological Risk Assessment". This document is a relatively clear review of the basics of ecological risk assessment as perceived in the early 1990s. In addition to this document, there are two recent reviews of ecological risk assessment. *Ecological Risk Estimation* (Bartell et al. 1992) is an integrated approach to the assessment of aquatic ecological systems with an emphasis upon simulation modeling. *Ecological Risk Assessment* (Suter 1993) is a broader overview with excellent sections on the application of population biology and ecology to risk assessment. This chapter reviews the structure of ecological risk assessment as presented in the U.S. EPA document with some comments on applicability and future directions.

Two points should be considered carefully as regards the relationship between environmental toxicology and risk assessment. First, environmental toxicology should not be seen as dependent upon risk assessment as its justification. Risk assessment is a management tool used for making decisions, often with a great deal of uncertainty. The science of environmental toxicology, as with any science, attempts to answer specific questions. In the case of environmental toxicology the question is primarily how xenobiotics interact with the components of ecological systems. Second, risk assessment is not necessarily a scientific pursuit. The assessment endpoints of risk assessment are often set by societal perceptions and values. Although the scientific process may be used in the gathering of information in the assignment of risks, unless a testable hypothesis can be formulated, the scientific

Table 10.1   Comparison of Hazard Assessment with Risk Assessment

Characteristic	Hazard assessment	Risk assessment
Probabilistic results	No	Yes
Scales of results	Dichotomous	Continuous
Basis for regulation	Scientific judgment	Risk management
Assessment endpoints	Not explicit	Explicit
Expression of contamination	Concentration	Exposure
Tiered assessment	Necessary	Unnecessary
Decision criteria	Judgment	Formal criteria
Use of models	Deterministic fate models	Probabilistic exposure and effects models

Note: The primary distinguishing characteristic of risk assessment is its emphasis upon probabilistic criteria and explicit assessment endpoints. Both methods of assessing the impact of toxicants are in use.

After Suter, G. W., II. 1990. In *Aquatic Toxicology and Risk Assessment:* 15th Volume. ASTM STP 1096. W. G. Laudis and W. H. van der Schalie, Eds., American Society for Testing and Materials, Philadelphia, PA, pp. 5–15.

method is not being applied. As a management tool risk assessment has certainly demonstrated its worth in the past 15 years.

BASICS OF RISK ASSESSMENT

Perhaps the easiest definition of ecological risk assessment is the probability of an effect occurring to an ecological system. Note that the word "probability" is key here. Important components of a risk assessment are the estimations of hazard and exposure due to a stressor.

A stressor is a substance, circumstance, or energy field that causes impacts, either positive or negative, upon a biological system. Stressors could be as wide ranging as chemical effects, ionizing radiation, or rapid changes in temperature.

Hazard is the potential of a stressor to cause particular effects upon a biological system. The determination of an LD_{50} or the mutagenicity of a material are attempts to estimate the hazard posed by a stressor.

Exposure is a measure of the concentrations or persistence of a stressor within the defined system. Exposure can be expressed as a dose, but in environmental toxicology it is often possible to measure environmental concentration. One of the values of determining tissue concentrations in fish and mammals is that it is possible to estimate the actual dose of a chemical to the organism. Biomarkers may also provide clues to dosage.

A stressor poses no risk to an environment unless there is exposure. This is an extremely crucial point. Virtually all materials have as a characteristic some biological effect. However, unless enough of the stressor interacts with biological systems, no effects can occur. Risk is a combination of exposure and effects expressed as a probability. In contrast, hazard assessment does not deal with concentration and is not probabilistic in nature. Table 10.1 compares the two assessments as outlined by Suter (1990).

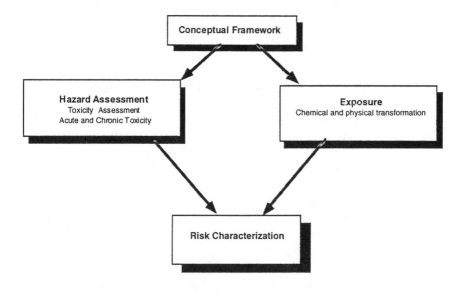

Figure 10.1 Classical Risk Assessment Paradigm. Originally developed for human health risk assessment, this framework does not include the close interaction between effects and exposure in ecosystems.

ECOLOGICAL RISK ASSESSMENT

Two basic frameworks for ecological risk assessment have been proposed over the last 10 years. The first was based upon the National Academy of Sciences report detailing risk assessments for federal agencies. Though simple, this framework forms the basis of human health and ecological risk assessments. Later refinements owe a great deal to this basic description of the risk assessment process. A diagram of the basic format is presented in Figure 10.1. Basically, four boxes contain the critical steps in the risk assessment. First, problem formulation determines the specific questions that are to be asked during the risk assessment process. Second, the hazard assessment details the biological effects of the stressor under examination. Simultaneously, the exposure potential of the material to the critical biological components is calculated as part of an exposure assessment. Last, the probabilistic determination of the likelihood of an effect is formalized as risk characterization.

Recently, the original framework was updated to specifically apply to estimating the risks of stressors to ecological systems. Perhaps of singular importance is the fact that exposure and hazard are not easily separated in ecological systems. When considering effects upon single organisms it is usually easy to separate exposure and effect terms. However, since ecosystems are comprised of many populations, the single-species example is a subset of ecological risk assessment. For instance, once a chemical comes out of the pipe it has already entered the ecosystem. As the material is incorporated into the ecosystem, biological and abiotic components transport or alter the structure of the original material. Even as the ecosystem is affected by the

chemical, the ecosystem is altering the material. In light of this and other considerations a revised framework was presented in 1992.

ECOLOGICAL RISK ASSESSMENT FRAMEWORK

The ecological risk assessment framework attempts to incorporate refinements to
the original ideas of risk assessment and apply them to the general case of ecological
risk assessment. The overall structure is delineated in Figure 10.2.

As before, the ecological risk assessment itself is characterized by a problem
formulation process, analysis containing characterizations of exposure and effects,
and a risk characterization process. Several outlying boxes serve to emphasize the
importance of discussions during the problem formulation process between the risk
assessor and the risk manager, and the critical nature of the acquisition of new data
and verification of the risk assessment and monitoring. The next few sections detail
each aspect of this framework.

PROBLEM FORMULATION

The problem formulation component of the risk assessment process is the beginning of a hopefully iterative process. This critical step defines the question under
consideration and directly affects the scientific validity and policy-making usefulness of the risk assessment. Initiation of the process can begin due to numerous
causes; for example, a request to introduce a new material into the environment,
examination of cleanup options for a previously contaminated site, or as a component
of examining land-use options. The process of formulation is itself comprised of
several subunits (Figure 10.3), discussion between the risk assessor and risk manager, stressor characteristics, identification of the ecosystems potentially at risk,
ecological effects, endpoint selection, conceptual modeling, and input from data
acquisition, verification, and monitoring.

The discussion between the risk assessor and risk manager is crucial in helping
to set the boundaries created by societal goals and scientific reality for the scope of
the risk assessment. Often societal goals are presented in ambiguous terms such as
protection of endangered species, protection of a fishery, or the even vaguer, preserve
the structure and function of an ecosystem. The interaction between the risk assessor
and the risk manager can aid in consolidating these goals into definable components
of a risk assessment.

Stressor characteristics form an important aspect of the risk assessment process.
Stressors can be biological, physical, or chemical in nature. Biological stressors
could include the introduction of a new species or the application of degradative
microorganisms. Physical stressors are generally thought of as a change in temperature, ionizing or nonionizing radiation, or geological processes. Chemical stressors
generally comprise such materials as pesticides, industrial effluents, or waste streams
from manufacturing processes. In the following discussion chemical stressors are
used as the typical example, but often different classes of stressors occur together.

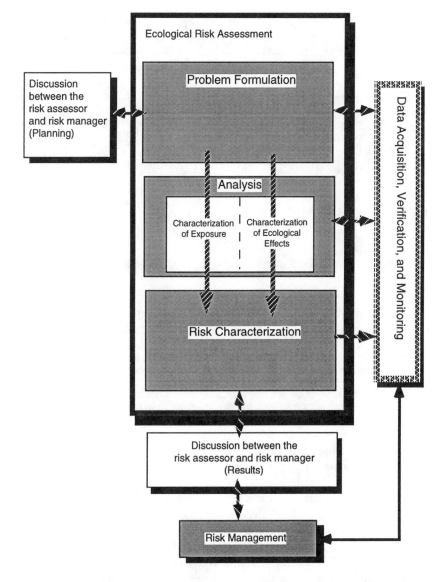

Figure 10.2 Schematic of the framework for ecological risk assessment (U.S. EPA 1992). Especially important is the interaction between exposure and hazard and the inclusion of a data acquisition, verification, and monitoring component. Multivariate analyses will have a major impact upon the selection of assessment and measurement endpoints and will play a major role in the data acquisition, verification, and monitoring phase.

Radionucleotides often produce ionizing radiation and can also produce toxic effects. Plutonium is not only radioactive, but is also highly toxic.

Stressors vary not only in their composition, but also in other characteristics derived in part from their use patterns. These characteristics are usually listed as

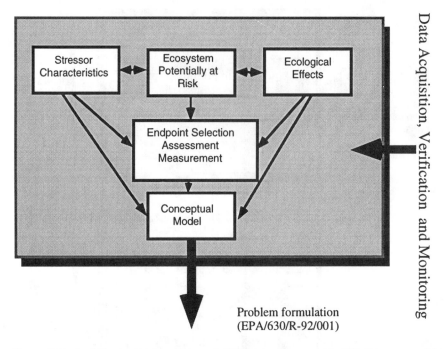

Figure 10.3   Problem formulation. This part of the risk assessment is critical because of the selection of assessment and measurement endpoints. The ability to choose these endpoints generally relies upon professional judgment and the evaluation of the current state of the art. However, *a priori* selection of assessment and measurement endpoints may lock the risk assessor from consideration of unexpected impacts.

intensity (concentration or dose), duration, frequency, timing, and scale. Duration, frequency, and timing address the temporal characteristics of the contamination while the characteristic scale addresses the spatial aspects.

Ecosystems potentially at risk can be one of the more difficult characteristics of problem formulation to address. Even if the risk assessment was initiated by the discovery of a problem in a particular system, the range of potential effects cannot be confined to that locality because atmospheric and water-borne transport materials can impact a range of aquatic and terrestrial ecosystems. Pesticides, although applied to crops, can find their way into ponds and streams adjacent to the agricultural fields. Increased UV intensity may be more damaging to certain systems, such as those at higher altitudes or elevations, but the ramifications are global. For instance, the microlayer interface between an aquatic ecosystem and the atmosphere receives a higher exposure to chemical contamination or UV radiation due to the characteristics of this zone. However, alterations in the microlayer affect the remainder of the system since many eggs and larval forms of aquatic organisms congregate in this microlayer.

Ecosystems have a great number of abiotic and biotic characteristics to be considered during this process. Sediments have both biotic and abiotic components

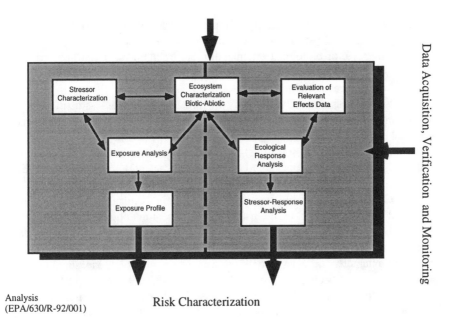

Figure 10.4 Analysis. Although separated into different sides of the analysis box, exposure and ecological responses are intimately connected. Often the biological response to a toxicant alters the exposure for a different compartment of the ecosystem.

that can dramatically affect contaminant availability or half-life. History is an often overlooked characteristic of an ecosystem, but it is one that directly affects species composition and the system's ability to degrade toxic materials. Geographic relationship to nearby systems is another key characteristic influencing species migration and, therefore, recovery rates from stressor impacts. The size of the ecosystem is also an important variable influencing species number and system complexity. All of these characteristics are crucial in accurately describing the ecosystem in relationship to the stressor.

Ecological effects are broadly defined as any impact upon a level of ecosystem organization. Figure 10.4 lists many of the potential interactions between a xenobiotic and a biotic system. Information is typically derived as part of a hazard assessment process but is not limited to detrimental effects of the toxicant. Numerous interactions between the stressor and the ecological system exist and each should be considered as part of the potential ecological effects. Such interactions, which include biotransformation, biodegradation, bioaccumulation, acute and chronic toxicity, reproductive effects, predator-prey interactions, production, community metabolism, biomass generation, community resilience and connectivity, evolutionary impacts, genetics of degradation, and many other factors that represent a direct impact upon the biological aspects of the ecosystem as well as the effects of the ecosystem upon the toxicant, are crucial if an accurate understanding of ecological effects is to be reached.

Endpoint selection is perhaps the most critical aspect of this stage of risk assessment as it sets the stage for the remainder of the process. Any component from virtually any level of biological organization or structural form can be used as an endpoint. Over the last several years two types of endpoints have emerged: assessment and measurement endpoints.

Assessment endpoints serve to focus the thrust of the risk assessment. Selection of appropriate and relevant assessment endpoints can ultimately decide the success or failure of a risk assessment. Assessment endpoints should accurately describe the characteristic of the ecosystem that is to be protected as set by policy. Several characteristics of assessments should be used in the selection of relevant variables. These include ecological relevance, policy goals as defined by societal values, and susceptibility to the stressor. Often, assessment endpoints cannot be directly measured and must be inferred by the use of measurement endpoints.

Measurement endpoints are measurable factors that respond to the stressor and describe or measure characteristics that are essential for the maintenance of the ecosystem characteristic classified as the assessment endpoint. Measurement endpoints can be virtually any aspect of the ecosystem that can be used to provide a more complete picture of the status of the assessment endpoint. Measurement endpoints can range from biochemical responses to changes in community structure and function. The more complete the description of the assessment endpoint that can be provided by the measurement endpoints, the more accurate will be the prediction of impacts.

The design and selection of measurement endpoints should be based on the following criteria:

- Relevance to assessment endpoint
- Measurement of indirect effects
- Sensitivity and response time
- Signal-to-noise ratio
- Consistency with assessment endpoint exposure scenarios
- Diagnostic ability
- Practicality

Some of these aspects are discussed below.

The relevance of a measurement endpoint is the degree to which the measurement can be associated to the assessment endpoint under consideration. Perhaps the most direct measurement endpoints are those that reflect the mechanism of action, such as inhibition of a protein, or mortality of members of the species under protection. Although correlated functions can be and are used as measurement endpoints, correlations do not necessarily imply cause and effect.

Consistency with assessment endpoint scenarios simply means that the measurement endpoint is exposed to the stressor in a manner similar to that of the assessment endpoint. Consistency is important when an organism is used as a surrogate for the assessment endpoint or if a laboratory test is being used to examine residual toxicity.

However, this is not consistent with the approach that secondary effects are important. Other components of the ecosystem essential to the survivorship of the assessment endpoint may be exposed by different means.

Diagnostic ability is related to the relevance issue. Mechanistic scenarios are perhaps the most relevant and diagnostic.

Finally, the practicality of the measurement is essential. The gross physical and chemical parameters of the system are perhaps the easiest to measure. Data on population dynamics, genetic history, and species interactions tend to be more difficult to obtain although they often are the more important parameters. Trade-offs must also be considered in the methods to be used. In many cases in ecological systems the absolute precision and accuracy of only a few of the measurement endpoints may not be as important as obtaining many measurements that are only ranked high, medium, or low. Judgment calls such as this require the input from the data acquisition, verification, and monitoring segment of the risk assessment process.

The conceptual model of the risk assessment is the framework into which the data are placed. Like the selection of endpoints, the selection of a useful conceptual model is crucial to the success or failure of the risk assessment process. In some cases a simple single species model may be appropriate. Typically, models in ecological risk assessment are comprised of many parts and attempt to deal with the variability and plasticity of natural systems. Exposure to the system may come from many different sources. The consideration of organisms at risk depends upon the migratory and breeding habits of numerous organisms, many rare and specialized.

As crucial as the above steps are, they are all subject to revision based upon the acquisition of additional data, verification that the endpoints selected do in fact perform as expected, and that the process has proven successful in predicting ecosystem risks. The data acquisition, verification, and monitoring segment of risk assessment is what makes this a scientific process as opposed to a religious or philosophical debate. Analysis of the response of the measurement endpoints and their power in predicting and corroborating assessment endpoints is essential to the development of better methodologies.

ANALYSIS

As the problem formulation aspect of the risk assessment is completed an analysis of the various factors detailed above comes into play. Central to this process is the characterization of the ecosystem of concern.

Characterization of the ecosystem of concern is often a most difficult process. In many cases involving restoration of damaged ecosystems, there may not be a functional ecosystem and a surrogate must be used to understand the interactions and processes of the system. Often the delineation of the ecosystem is difficult. If the protection of a marine hatchery is considered the assessment endpoint, large areas of the coastal shelf, tidewater, and marine marsh systems have to be included in the process. Even many predominately terrestrial systems have aquatic components that play a major role in nutrient and toxicant input. Ecosystems are also not stagnant systems. They undergo succession and respond to the heterogeneity of climatic inputs in ways that are difficult to predict.

In addition to the gross extent and composition of the system, the resource undergoing protection and its role in the ecosystem needs to be understood. Behavioral changes due to the stressor may preclude successful reproduction or alter migratory patterns. Certain materials with antimicrobial and antifungal properties can alter nutrient cycling. It is also not clear what part ecosystem stability places in dampening deviations due to stressors or if such a property as stability at the ecosystem level exists.

In the traditional risk assessment, exposure and biological response have been separated. In the new framework for ecological risk assessment each of these components has been incorporated into the analysis component. However, as detailed in preceding chapters, organisms degrade, detoxify, sequester, and even use xenobiotics as resources. Conversely, the nature and mixture of the pollutants and the resources of the ecosystem affect the ability of organisms to modify or destroy chemical stressors. Although treated separately, this is as much for convenience and the reality of the intimate interaction between the chemical and the physical and biological components of the ecosystem should not be forgotten.

Exposure Analysis

Characterization of exposure is a straightforward determination of the environmental concentration range or, if available, the actual dose received by the biota of a particular stressor. Although simple in concept, determining or predicting the environmental exposure has proven to be difficult.

First there is the end-of-pipe or deposition exposure. This component is determined more by the use patterns of the material or the waste stream and effluent discharges from manufacturing. In some cases the overall statistics as to production and types of usage, such as the fluorohydorcarbons, is well documented. Manufacturers often can document processes and waste-stream components. Effluents are often regulated as to toxicity and composition. Problem areas often occur due to past practices, illegal dumping of toxic materials, or accidents. In these instances the types of materials, rate of release, and total quantities may not be known.

However, as the material leaves the pipe and enters the ecosystem it is almost immediately affected by both the biotic and abiotic components of the receiving system. All of the substrate and medium heterogeneity as well as the inherent temporal and spatial characteristics of the biota affect the incoming material. In addition to the state of the system at the time of pollution, the history of the environment as contained in the genetic make up of the populations plus the presence in the past or present of additional stressors all impact the chemical-ecosystem interaction. The goal of the exposure analyses is to quantify the occurrence and availability of the stressor within the ecosystem.

Perhaps the most common way of determining exposure is by the use of analytical chemistry to determine concentrations in the substrates and media as well as the biological components of the ecosystem. Analytical procedures have been developed for a number chemicals and the detection ranges are often in the microgram per liter range. Analytical procedures, however, have difficulty in determining degradation products due to microbial activity and do not quantify the exposure of a material to the various biological components. The analysis of tissue samples of representative biota gives a more accurate picture of exposure to materials that are not rapidly detoxified or eliminated. Molecular markers such as DNA damage or enzyme induction or inhibition can also provide useful clues as to actual exposure. Since exposure can occur through different modes and at varying rates through those modes, the total burden upon the organism is difficult to estimate.

It should not be forgotten that a great deal of biotransformation does occur, especially for metals such as mercury and for many organics. In many cases the result is a less toxic form of the original input, but occasionally more toxic materials are created.

Lastly, models attempting to predict the fate and resultant exposure to a stressor can be used and often they are applied in a variety of scenarios. Models, however, are simplifications of our imperfect understanding of exposure and should be tested whenever possible against comparable data sets.

As the temporal and spatial distribution of the stressor has been quantified in the exposure analysis step, it should prove possible to provide the distribution curve for exposure of the biotic components of interest to the stressor. Dose and concentration probabilities are the typical units used in environmental toxicology.

Characterization of Ecological Effects

The characterization of ecological effects is perhaps the most critical aspect of the risk assessment process. Several levels of confidence exist in our ability to measure the relationship between dose and effect. Toxicity measured under set conditions in a laboratory can be made with a great deal of accuracy. Unfortunately, as the system becomes more realistic and includes multiple species and additional routes of exposure, the ability to even measure effects is decreased.

Evaluation of relevant effects data has long been left to professional judgment. Criteria typically used to judge the importance of the data usually include the quality of the data, number of replicates and repeatability, relevance to the selected endpoints, and realism of the study compared to the ecosystem for which the risk assessment is being prepared.

Toxicity data from several sources is usually compiled and compared. Generally, there are acute and chronic data for the stressor on one or several species. Toxicity data are usually limited as to species and the species of interest as an assessment endpoint may not have appropriate data available. This situation often occurs with threatened or endangered species since even a small-scale toxicity test involves relatively large numbers of animals to acquire data of sufficient quality.

Field observations and controlled microcosm and large-scale tests can provide additional data on which to base the risk assessment. Only in these systems can an indication of the importance of indirect effects become apparent. Field research also has limitations. No two fields are alike, requiring extrapolation. If, as discussed in Chapter 9, ecological systems are sensitive to initial conditions, then no two fields will act exactly alike.

Ecological Response Analyses

The combining of the exposure analysis with the ecological effects data results in the stressor-response profile. This profile is an attempt to match ecosystem impacts at the levels of stressor concentration under study. Relationships between the xenobiotic and the measurement endpoint are evaluated with a consideration of how this interaction affects the assessment endpoint. Rarely is this process straightforward. Often, some model is used to specifically state the relationship between the measurement and assessment endpoint; when this relationship is not specifically stated it is then left to professional judgment.

The EPA framework lists the relationships between assessment and measurement endpoints:

1. Phylogenetic extrapolation — relationship of toxicity data from one species to another or perhaps more often, class to class. Often only a 96-h green algal toxicity test is available to use as a representative of all photosynthetic eucaryotes.
2. Response extrapolation — relationship between two toxicity endpoints such as the NOAEL and the EC_{50}.
3. Laboratory-to-field extrapolation — relationship of the estimate of toxicity gathered in the laboratory to the effects expected in the field situation. Laboratory situations are purposefully kept simple compared to the reality of the field and are designed to rank toxicity rather than to mimic the field situation. Laboratory tests have limited the route of exposure and behavior. In the field these restrictions are not in place often leading to unexpected results.
4. Field-to-field (or habitat-to-habitat) extrapolation — relationship of one field or habitat to another. It may be highly unlikely that any two habitats can be identical. Streams on one side of a continental divide tend to have different flora and fauna than a comparable stream on the other side. Even controlled field studies exhibited differences in the replicates. The effect of a toxicant in the streams may be the same in a qualitative fashion but quantification may not be possible.
5. Indirect effects — the toxicant impacts due to the disruption of the ecosystem apart from direct impacts upon the ecosystem components. The elimination of photosynthetic organisms in a pond by a herbicide will eventually eliminate the invertebrate herbivores and the fish that rely upon them as a food source.
6. Organizational levels — examines the transmission of effects up and down levels of biological organization. An alteration in fecundity at the organismal level will generally decrease the rate of growth of a population. Conversely, the decrease elimination of a herbivore population, eliminating much of the top-down control at the community level, will allow the plant populations to grow in an exponential fashion even if the toxicant has some effect upon maximum rate of growth.

7. Spatial and temporal scales — exist in a variety of dimensions relating to the life span and size of the organisms and systems under investigation. One day and 10 m represent several generations and the entire world of many microorganisms, but this level of temporal and spatial scaling is relatively insignificant to a redwood of the Northwest. Not only is the size of the scale important, but so is the heterogeneity. Heterogeneity of both of these variables apparently contributes to the diversity of species and genotypes found in a variety of systems. Maintaining heterogeneity of these scalars may be as important as any other environmental variable in a consideration of impacts to the assessment endpoints.

8. Recovery — the rate at which a system can be restored to its original state. Recovery in the sense of a stable system returning to its original state is what is generally meant and this may be difficult, if not impossible, to accomplish. If recovery does occur it generally depends upon the ability of colonizing organisms to become established upon the impacted site and therefore the isolation of the damaged ecosystem is important. Nonlinear and complexity theory also states that initial conditions are extremely important and that several new stable points may be reached given similar initial conditions. Recovery to the initial state may in fact be of a low likelihood and a more realistic goal may be a new stable point that involves the factors selected as assessment endpoints.

In the evaluation of the ecological response, consideration must often be given to the strength of the cause-effect relationship. Such relationships are relatively straightforward in single species.

Stressor-Response Profile

The stressor-response profile is in some ways analogous to a dose-response curve in the sense of a single species toxicity test expanded to the community and ecosystem level. Since many of the responses are extrapolations and based on models from the molecular to ecosystem level it is important to delineate the uncertainties, qualifications, and assumptions made at each step.

One of the difficulties in the quantification of the stressor-response profile is that many of the extrapolations are qualitative in nature. Phylogenetic extrapolations are rarely quantified or assisted with structure activity relationships. Quantification of population level effects is likewise difficult and in some cases probabilities of extinction have been used as the quantified variable, not a subtle population endpoint.

Perhaps the greatest difficulty is evaluating the stressor-response relationship for an ecological risk assessment and the fact that systems are under the influence of many other stressors. Laboratory organisms are generally healthy, but laboratory conditions do not mimic the ration of micronutrients, behavioral opportunities, and many other factors contained within an ecosystem. Field studies include many climatalogical and structural stressors that are separate from the introduced toxicant. Additionally, there is unlikely to be an ecosystem within range of a laboratory that

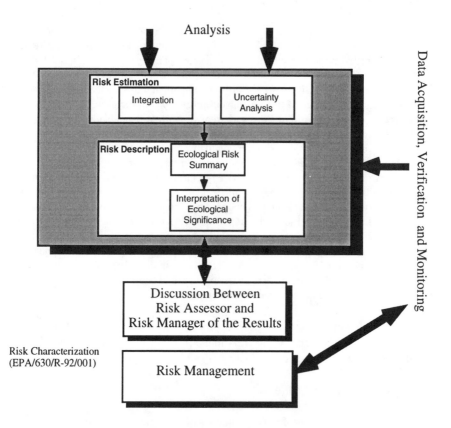

Figure 10.5   Risk characterization. This compartment is comprised of the risk estimation and
risk description boxes. The integration of the exposure and effects data from the
analysis compartment is reconciled in the risk estimation process.

has not been subjected to an anthropogenic stressor, again confounding even the best
designed study.

Data Acquisition, Verification, and Monitoring

Input from this block is most critical at this stage. Basic research on the effects
of stressors to ecosystems, improvement in test methods, molecular mechanisms, and
improvements in modeling provide critical input to this stage of the risk assessment.
An understanding of the phylogenetic relationships of receptors should lead to better
interspecies extrapolations.

RISK CHARACTERIZATION

Risk characterization is the final stage of the risk assessment process (Figure
10.5). This aspect of a risk assessment is comprised of a risk estimation and a risk

description compartment. The overall process is a combining of the ecological effect with the environmental concentration to provide a likelihood of effects given the distribution of the stressor within the system. This process has proven to be difficult to accomplish in a straightforward manner. The probability of toxic impacts is analogous to the weather forecaster's prediction of rain. For instance, say that today there is a 50% chance of rain in the local area. This means that given the conditions observed that a prediction is made, generally from experience, that the chance of rain is 50 out of 100 trials. Notice that this is not a prediction that it will only rain over half of the forecast area. Similarly, toxicology attempts to make predictions regarding the probability of an effect given the conditions of chemical type, concentration, and ecosystem type. This predictive process is still as much an art as weather forecasting.

Integration

The integration of exposure with toxicity has been problematical. As we have previously discussed, environmental toxicology deals with a variety of effects at various levels of biological organization. A fish LD_{50} value is difficult to compare with loss of nitrogen fixation from an ecosystem. Perhaps the most widely used method of estimating risk is the quotient method.

The quotient method is simple and straightforward. The method simply divides the expected environmental concentration by the hazard

$$\text{Quotient} = \frac{\text{Expected environmental concentration}}{\text{Concentration producing an unacceptable environmental effect}}$$

Of course the equation produces a ratio which is generally judged by the criteria below.

Quotient	Risk
> 1	Potential or high risk
1	Potential risk
<<1	Low risk

The difficulty with such an analysis is that it is a qualitative expression of risk without regard to the probability distributions of the chemical concentrations or the effects. Distributions of each can be plotted and the distribution of expected effects calculated. In this example, although the probability of a high concentration is low, the probability of the effect is high, and at low concentrations, the probability of the effect is significantly reduced but the likelihood of the concentration is much higher. Analyses such as these may prove more accurate although more difficult to calculate and perhaps interpret. Time and spatial factors should also be included, complicating the functions but better modeling the interaction between xenobiotic and biota.

Uncertainty analysis goes hand-in-hand with the integration process and has many points of origin. In some instances the conceptual model and the assessment and measurement endpoints associated with it may be inaccurate descriptions of the system under investigation. Only with rigorous monitoring and follow-up validation of the risk assessment is it likely that these types of errors will be eliminated. Fundamental misunderstandings or ignorance of ecosystem processes and interactions may be corrected in this manner.

The quality and source of the data incorporated into the risk assessment again contributes to the uncertainty associated with the risk assessment. Toxicological data routinely vary according to the strain or test organism used. QSARs have an associated uncertainty although this is not routinely quantified. Field studies are noteworthy for the difficulty of interpretation. One of the most perplexing areas of uncertainty is the necessity of using data from studies that were not originally designed to address the question specific to the risk assessment.

Many multispecies tests and field studies are designed to look at only certain populations or other attributes of the ecosystem. This is not the fault of the study per se, since the funding, personnel, and physical resources are usually limited. The danger lies in the picking and choosing of secondary results from these studies. For example, the standardized aquatic microcosm contains 16 species that are initially inoculated into the system. However, in the reporting of the results for publication the dynamics and interactions of all species and the combinations are not reported. To do so would be cumbersome and expensive. Only the dynamics of the organisms and interactions that are the apparently critical components are reported. Assuming that the other components are not affected because of their omission or lack of space in the article could be erroneous. Anecdotal data from field or multispecies tests are similarly difficult to interpret. Omission or inclusion of a report may reflect more the nature of the researcher than the presence or absence of the effect.

Risk Description

The next step in this framework is the risk description. The two aspects of this segment include an ecological risk summary and the interpretation of ecological significance. Although this division is somewhat artificial, it can be paraphrased as, "What are the potential effects and do I believe them?" and "How big a problem is this really going to be?"

The ecological risk summary summarizes the risk estimation results and the uncertainties. The crucial aspect to this section is the decision making regarding the accuracy of the risk estimation. These decisions revolve around three general aspects of the analysis:

- Sufficiency of the data
- Corroborative information
- Evidence of causality

Sufficiency of the data relates to the quality of the data and its completeness. Much of the discussion revolves around the quality and appropriateness of the research conducted or cited in the formation of the risk assessment.

Corroborative information is data derived from similar studies with similar stressors that tend to support the conclusions of the risk assessment. Science is inherently conservative and similarity to data and conclusions of related studies enhance the credibility of the current risk assessment. However, nonsimilarity to previous conclusions or ecological theory does not mean that the current study is flawed; it may mean that the previous work is not as similar as originally thought or that the overall paradigm is incorrect.

Evidence of causality is perhaps the most concrete and also the most elusive aspect of the data assessment process. At the single organism or species level it is often possible to assign specific mechanistic connections for mortality or other impacts. Unfortunately, it is not well understood how prevalent and pervasive these impacts are at the community level. Correlational data may be all that are available for impacts at the level of interspecies interactions. Correlations are difficult to assess because correlation does not denote cause and effect. In a system as complex as an ecosystem multiple correlations due to chance may occur. It may also be difficult to separate cause and effect without firmly established criteria.

Perhaps the most critical aspect of the above analysis is the realization that additional data or even a reformulation of the conceptual model is required. In this case the assessment process is rerouted to the data acquisition, verification, and monitoring stage. An integrative process can then occur to obtain a usable and perhaps accurate risk assessment.

Interpretation of Ecological Significance

Finally, an interpretation of ecological significance is produced that details probable magnitudes, temporal and spatial heterogeneity, and the probability of each of these events and characteristics. One of the judgments that is usually called for is the recovery potential of the affected ecosystem. Given that recovery to the initial state may not be probable or even biologically possible, the question is perhaps dubious. Perhaps a better question is whether the system can exhibit at some future time the properties that initially made it valuable in the terms of the assessment endpoints.

Discussion Between the Risk Assessor and Risk Manager

Finally a report is made to the risk manager detailing the important aspects of the risk assessment. Of crucial importance are the range of impacts, uncertainties in the data and the probabilities, and the stressor-response function. These factors are then taken into consideration with social, economic and political realities in the management of risks. An approach to risk assessment as outlined above, however, does not include a risk/benefit type of analysis. Such considerations are in the purview of the risk manager.

Data Acquisition, Verification, and Monitoring

In the above outline the importance of the data acquisition, verification, and monitoring process in the development of accurate risk assessments has been emphasized. The importance of this aspect, often overlooked, is crucial to the development of risk assessments that reflect ecological reality. Models, no matter how sophisticated, are simply attempts to understand processes and codify relationships in a very specific language. Ptolmeic (Earth-centered) astronomy accurately predicted many aspects of the stars and planets and served to make accurate predictions of celestial events. However, the reversing of direction in the celestial sphere of the planets was difficult to account for given the Earth-centered model. Eventually, the Copernican (Sun-centered) model replaced the Ptolmeic model as the descriptions of solar system dynamics and the insights from the new framework led to other discoveries about the nature of gravity and the motion of the planets. How many of our current models are "Earth-centered"? Only the reiteration of the predictive (risk assessment) and experimental (data acquisition, verification, and monitoring) process can answer that question.

One of the difficulties of ecosystem level analysis has been our inability to accurately present the dynamics of these multidimensional relationships. Conventional univariate statistics are still prevalent although the shortcomings of these methods are well known. Recently, several researchers have proposed different methods of visualizing ecosystems and the risks associated with xenobiotic inputs.

REFERENCES AND SUGGESTED READINGS

Bartell, S.M., R.H. Gardner, and R.V. O'Neill. 1992. *Ecological Risk Estimation*. Lewis Publishers, Chelsea, MI, p. 252.

Suter, G.W., II. 1990. Environmental risk assessment/environmental hazard assessment: Similarities and differences. In *Aquatic Toxicology and Risk Assessment:* 13th Volume. ASTM STP 1096. W.G. Landis and W.H. van der Schalie, Eds., American Society for Testing and Materials, Philadelphia, PA, pp. 5-15.

Suter, G.W., II. 1993. *Ecological Risk Assessment*. Lewis Publishers, Chelsea, MI, p. 538.

STUDY QUESTIONS

1. What are ecological risk assessment, stressor, hazard, and exposure?
2. Define problem formulation, hazard assessment, exposure assessment, and risk characterization as in Figure 10.1.
3. Which aspect of the ecological risk assessment framework defines the question under consideration? What are subunits to this formulation?

4. Stresses can be of what three categories? What five characteristics can stressors have that are derived in part from use patterns?
5. What are some interactions between the stressor and the ecological system?
6. What is an endpoint? An assessment endpoint? A measurement endpoint?
7. How can the variance-to-mean relationship classify the type of sampling distribution?
8. What scenario is the most relevant and diagnostic?
9. What factors make risk assessment a "scientific process"?
10. What two components have been incorporated into the analysis component in the new framework for ecological risk assessment (as opposed to their separation in traditional risk assessment)?
11. What is the goal of the exposure analysis?
12. What are several ways to determine exposure?
13. What is the most critical aspect of the risk assessment process?
14. What are the criteria used to judge the importance of data when characterizing ecological effects?
15. Describe the stressor-response profile.
16. Describe the eight EPA framework-listed relationships between assessment and measurement endpoints.
17. What is one of the difficulties in evaluating the stressor-response relationship?
18. Describe risk characterization.
19. What is the quotient method of estimating risk? Discuss a difficulty with this analysis.
20. Discuss possible erroneous conclusions that may be drawn if secondary results are deduced or extrapolated from multispecies tests and field studies.
21. List the three general aspects of the analysis for the ecological risk summary and describe each.
22. What is a good question to be examined concerning the interpretation of ecological significance?
23. List the factors of crucial importance in the report to the risk manager.

INTRODUCTION

We have decided to include the latest version of the U.S. Environmental Protection Agency's Ecological Risk Assessment Framework. This document is fast becoming the standard conceptual model for ecological risk assessment and the standardized glossary. In order to understand the current literature on ecological risk assessment it is critical to have a working knowledge of the document. In its original form the document has a purple cover, hence it may be referred to as the purple document in transcripts of discussions.

The importance of the document is that it does serve as a framework to organize future research and methods for ecological risk assessment. As with any evolving document, some of the ideas regarding ecosystem resilience and stability mentioned in the framework are dated. However, the inclusion of verification and validation into a risk assessment framework are certainly forward looking. We strongly urge the reader to carefully read this document and to consider the material contained in the preceding chapters. Note inconsistencies and be critical as soon the reader will likely be implementing the framework in a real world situation or be developing improvements.

U.S. EPA Document "A Framework for Ecological Risk Assessment"

EPA/630/R-92/001, February 1992.
Risk Assessment Forum,
U.S. Environmental Protection Agency,
Washington, D.C. 20460

DISCLAIMER

This document has been reviewed in accordance with U.S. Environmental Protection Agency policy and approved for publication. Mention of trade names or commercial products does not constitute endorsement or recommendation for use.

CONTENTS

Acknowledgments 273
Foreword. 273
Preface 275
Contributors and reviewers 275
Executive summary 279
1.   Introduction 281
     1.1. Purpose and scope of the framework report 281
     1.2. Intended audience 282
     1.3. Definition and applications of ecological risk assessment 282
     1.4. Ecological risk assessment framework 283
     1.5. The importance of professional judgment. 286
     1.6. Organization 287
2.   Problem formulation 287
     2.1. Discussion between the risk assessor and risk manager
           (planning) 288
     2.2. Stressor characteristics, ecosystem potentially at risk, and
           ecological effects 289
           2.2.1. Stressor characteristics 289
           2.2.2. Ecosystem potentially at risk 289
           2.2.3. Ecological effects 290
     2.3. Endpoint selection 290
     2.4. The conceptual model 291
3.   Analysis phase 294
     3.1. Characterization of exposure 295
           3.1.1. Stressor characterization: distribution or pattern of change 295
           3.1.2. Ecosystem characterization 296
           3.1.3. Exposure analyses 297
           3.1.4. Exposure profile 298
     3.2. Characterization of ecological effects 299
           3.2.1. Evaluation of relevant effects data. 299
           3.2.2. Ecological response analyses 300
           3.2.3. Stressor-response profile 302
4.   Risk characterization 304
     4.1. Risk estimation 304
           4.1.1. Integration of stressor-response and exposure profiles 304
           4.1.2. Uncertainty 306
     4.2. Risk description 308
           4.2.1. Ecological risk summary 308
           4.2.2. Interpretation of ecological significance 309
     4.3. Discussion between the risk assessor and risk manager
           (results). 311
5.   Key terms. 312
References 313

LIST OF BOXES

Physical and chemical stressors as a focus of the framework
Relationship of the framework to a paradigm for human health risk assessment
Use of the term "exposure"
Characterization of ecological effects used instead of hazard assessment
Additional issues related to the framework
Example stressor characteristics
Endpoint terminology
Considerations in selecting assessment endpoints
Considerations in selecting measurement endpoints
Additional issues in problem formulation
Extrapolations and other analyses relating measurement and assessment endpoints
Hill's Criteria for evaluating causal associations (Hill, 1965)
Additional issues related to the analysis phase
Additional issues related to the risk characterization phase

LIST OF FIGURES

Figure 1. Framework for ecological risk assessment
Figure 2. Problem formulation
Figure 3. Analysis
Figure 4. Risk characterization

ACKNOWLEDGMENTS

This U.S. Environmental Protection Agency (EPA) report has been developed under the auspices of EPA's Risk Assessment Forum, a standing committee of EPA scientists charged with developing risk assessment guidance for Agency-wide use. An interoffice work group chaired by Susan Norton (Office of Health and Environmental Assessment), Donald Rodier (Office of Toxic Substances), and Suzanne Marcy (Office of Water) led this effort. Other members of the work group are Michael Brody, David Mauriello, Anne Sergeant, and Molly Whitworth. William van der Schalie and William Wood of the Risk Assessment Forum staff coordinated the project, which included peer review by scientists from EPA, other federal agencies, and the private sector.

FOREWORD

Publication of this report, "Framework for Ecological Risk Assessment" (Framework Report), is a first step in a long-term program to develop risk assessment guidelines for ecological effects. EPA has been developing risk assessment guidelines primarily for human health effects for several years. In 1986, EPA issued five such guidelines, including cancer, developmental toxicity, and exposure assessment (51 *Federal Register* 33992–34054, 24 September 1986). Although EPA had issued guidance for cancer risk assessment 10 years earlier (41 *Federal Register* 21402, 1976), the 1986 guidelines substantially enlarged the scope of EPA's formal guidance by covering additional health topics and by covering all areas in much greater depth. Each of the guidelines was a product of several years of discussion and review involving scientists and policymakers from EPA, other federal agencies, universities, industry, public interest groups, and the general public.

Preliminary work on comparable guidelines for ecological effects began in 1988. As part of this work, EPA studied existing assessments and identified issues to help develop a basis for articulating guiding principles for the assessment of ecological risks (U.S. EPA, 1991). At the same time, EPA's Science Advisory Board urged EPA to expand its consideration of ecological risk issues to include the broad array of chemical and nonchemical stressors for which research and regulation are authorized in the environmental laws administered by EPA (U.S. EPA, 1990b). As a result, EPA has embarked on a new program to develop guidelines for ecological risk assessment. Like the program for health effects guidance, this activity depends on the expertise of scientists and policymakers from a broad spectrum and draws principles, information, and methods from many sources.

In May 1991, EPA invited experts in ecotoxicology and ecological effects to Rockville, Maryland, to attend a peer review workshop on the draft Framework Report (56 *Federal Register* 20223, 2 May 1991). The workshop draft proposed a framework for ecological risk assessment complemented by preliminary guidance on some of the ecological issues identified in the draft. On the basis of the Rockville workshop recommendations (U.S. EPA, 1991), the revised Framework Report is now limited to discussion of the basic framework (see Figure 1), complemented by second-order diagrams that give structure and content to each of the major elements in the Framework Report (see Figures 2 through 4). Consistent with peer review recommendations, substantive risk assessment guidance is being reserved for study and development in future guidelines.

The Framework Report is the product of a variety of activities that culminated in the Rockville workshop. Beginning early in 1990, EPA work groups and the National Academy of Sciences' (NAS) Committee on Risk Assessment Methodology began to study the 1983 NAS risk assessment paradigm (NRC, 1983), which provides the organizing principles for EPA's health risk guidelines, as a possible foundation for ecological risk assessment. Early drafts of EPA's Framework Report received preliminary peer comment late in 1990.

In February 1991, NAS sponsored a workshop in Warrenton, Virginia, to discuss whether any single paradigm could accommodate all the diverse kinds of ecological risk assessments. There was a consensus that a single paradigm is feasible but that the 1983 paradigm would require modification to fulfill this role. In April 1991, EPA sponsored a strategic planning workshop in Miami, Florida. The structure and elements of ecological risk assessment were further discussed. Some participants in both of these earlier meetings also attended the Rockville workshop. EPA then integrated information, concepts, and diagrams from these workshop reviews with EPA practices and needs to propose a working framework for interim use in EPA programs and for continued discussion as a basis for future risk assessment guidelines.

Use of the framework described in this report is not a requirement within EPA, nor is it a regulation of any kind. Rather, it is an interim product that is expected to evolve with use and discussion. EPA is publishing the Framework Report before proposing risk assessment guidelines for public comment to generate discussion within EPA, among government agencies, and with the public to develop concepts, principles, and methods for use in future guidelines. To facilitate such discussion, EPA is presenting the framework at scientific meetings and inviting the public to submit information relevant to use and development of the approaches outlined for ecological risk assessment in the report.

Dorothy E. Patton, Ph.D.
Chair
Risk Assessment Forum

PREFACE

Increased interest in ecological issues such as global climate change, habitat loss, acid deposition, reduced biological diversity, and the ecological impacts of pesticides and toxic chemicals prompts this Framework Report. This report describes basic elements, or a framework, for evaluating scientific information on the adverse effects of physical and chemical stressors on the environment. The framework offers starting principles and a simple structure as guidance for current ecological risk assessments and as a foundation for future EPA proposals for risk assessment guidelines.

The Framework Report is intended primarily for EPA risk assessors, EPA risk managers, and persons who perform work under EPA contract or sponsorship. The terminology and concepts described in the report may also assist other regulatory agencies, as well as members of the public who are interested in ecological issues.

CONTRIBUTORS AND REVIEWERS

This report was prepared by members of the EPA technical panel listed below, with assistance from the staff of EPA's Risk Assessment Forum. Technical review was provided by numerous individuals, including EPA scientists and participants in the May 1991 peer review workshop. Editorial assistance was provided by R.O.W. Sciences, Inc.

TECHNICAL PANEL
CO-CHAIRS

Suzanne Macy Marcy
(through December 1990)
Office of Water

Susan Braen Norton
Office of Research and
Development

Donald J. Rodier
Office of Toxic Substances

MEMBERS

Michael S. Brody
Office of Policy, Planning and
Evaluation

David A. Mauriello
Office of Toxic Substances

Anne Sergeant
Office of Research and Development

Molly R. Whitworth
Office of Policy, Planning and
Evaluation

RISK ASSESSMENT FORUM STAFF

William H. van der Schalie
Office of Research and Development

William P. Wood
Office of Research and Development

REVIEWERS

M. Craig Barber
U.S. Environmental Protection Agency
Environmental Research Laboratory
Athens, GA

Janet Burris
U.S. Environmental Protection Agency
Office of Emergency and Remedial
Response
Washington, D.C.

Richard S. Bennett, Jr.
U.S. Environmental Protection Agency
Environmental Research Laboratory
Corvallis, OR

David W. Charters
U.S. Environmental Protection Agency
Office of Solid Waste and Emergency
Response
Edison, NJ

Steven Bradbury
U.S. Environmental Protection Agency
Environmental Research Laboratory
Duluth, MN

Patricia A. Cirone
U.S. Environmental Protection Agency
Region 10
Seattle, WA

James R. Clark
U.S. Environmental Protection Agency
Environmental Research Laboratory
Gulf Breeze, FL

Robert Davis
U.S. Environmental Protection Agency
Region 3
Philadelphia, PA

Anne Fairbrother
U.S. Environmental Protection Agency
Environmental Research Laboratory
Corvallis, OR

Jay Garner
U.S. Environmental Protection Agency
Environmental Monitoring Systems
  Laboratory
Las Vegas, NV

Jack H. Gentile
U.S. Environmental Protection Agency
Environmental Research Laboratory
Narragansett, RI

Sarah Gerould
U.S. Geological Survey
Reston, VA

George R. Gibson, Jr.
U.S. Environmental Protection Agency
Office of Water
Washington, D.C.

Alden D. Hinckley
U.S. Environmental Protection Agency
Office of Policy, Planning and
  Evaluation
Washington, D.C.

Erich Hyatt
U.S. Environmental Protection Agency
Office of Research and Development
Washington, D.C.

Norman E. Kowal
U.S. Environmental Protection Agency
Systems Laboratory
Cincinnati, OH

Ronald B. Landy
U.S. Environmental Protection Agency
Office of Technology Transfer and
  Regulatory Support
Washington, D.C.

Foster L. Mayer
U.S. Environmental Protection Agency
Environmental Research Laboratory
Gulf Breeze, FL

Melissa McCullough
U.S. Environmental Protection Agency
Office of Air Quality Planning and
  Standards
Washington, D.C.

J. Gareth Pearson
U.S. Environmental Protection Agency
Environmental Monitoring Systems
  Laboratory
Las Vegas, NV

Ronald Preston
U.S. Environmental Protection Agency
Region 3
Philadelphia, PA

John Schneider
U.S. Environmental Protection Agency
Region 5
Chicago, IL

Harvey Simon
U.S. Environmental Protection Agency
Region 2
New York, NY

Michael W. Slimak
U.S. Environmental Protection Agency
Office of Environmental Processes and
  Effects Research
Washington, D.C.

Q. Jerry Stober
U.S. Environmental Protection Agency
Region 4
Atlanta, GA

Greg R. Susanke
U.S. Environmental Protection Agency
Office of Pesticide Programs
Washington, D.C.

Leslie W. Touart
U.S. Environmental Protection Agency
Office of Pesticide Programs
Washington, D.C.

Michael E. Troyer
U.S. Environmental Protection Agency
Office of Technology Transfer and
  Regulatory Support
Washington, D.C.

Douglas J. Urban
U.S. Environmental Protection Agency
Office of Pesticide Programs
Washington, D.C.

Maurice G. Zeeman
U.S. Environmental Protection Agency
Office of Toxic Substances
Washington, D.C.

PEER REVIEW WORKSHOP PARTICIPANTS

William J. Adams
ABC Laboratories
Columbia, MO

Lawrence W. Barnthouse
Oak Ridge National Laboratory
Oak Ridge, TN

John Bascietto
U.S. Department of Energy
Washington, D.C.

Raymond Beaumier
Ohio Environmental Protection Agency
Columbus, OH

Harold Bergman
University of Wyoming
Laramie, WY

Nigel Blakeley
Washington Department of Ecology
Olympia, WA

James Falco
Battelle Pacific Northwest Laboratory
Richland, WA

James A. Fava
Roy F. Weston, Inc.
West Chester, PA

Alyce Fritz
National Oceanic and Atmospheric
  Administration
Seattle, WA

James W. Gillett
Cornell University
Ithaca, NY

Michael C. Harrass
Food and Drug Administration
Washington, D.C.

Mark Harwell
University of Miami
Miami, FL

Ronald J. Kendall
Clemson University
Pendleton, SC

Wayne G. Landis
Western Washington University
Bellingham, WA

Ralph Portier
Louisiana State University
Baton Rouge, LA

Kenneth Reckhow
Duke University
Durham, NC

John H. Rodgers
University of Mississippi
University, MS

Peter Van Voris
Battelle Pacific Northwest Laboratory
Richland, WA

James Weinberg
Woods Hole Oceanographic Institution
Woods Hole, MA

Randall S. Wentsel
U.S. Army Chemical Research,
    Development and Engineering Center
Aberdeen Proving Ground, MD

EXECUTIVE SUMMARY

This report, "Framework for Ecological Risk Assessment", is the first step in a long-term effort to develop risk assessment guidelines for ecological effects. Its primary purpose is to offer a simple, flexible structure for conducting and evaluating ecological risk assessment within EPA. Although the Framework Report will serve as a foundation for development of future subject-specific guidelines, it is neither a procedural guide nor a regulatory requirement within EPA and is expected to evolve with experience. The Framework Report is intended to foster consistent approaches to ecological risk assessment within EPA, identify key issues, and define terms used in these assessments.

Ecological risk assessments evaluate ecological effects caused by human activities such as draining of wetlands or release of chemicals. The term "stressor" is used here to describe any chemical, physical, or biological entity that can induce adverse effects on individuals, populations, communities, or ecosystems. Thus, the ecological risk assessment process must be flexible while providing a logical and scientific structure to accommodate a broad array of stressors.

The framework is conceptually similar to the approach used for human health risk assessment, but it is distinctive in its emphasis in three areas. First, ecological risk assessment can consider effects beyond those on individuals of a single species and may examine a population, community, or ecosystem. Second, there is no single set of ecological values to be protected that can be generally applied. Rather, these values are selected from a number of possibilities based on both scientific and policy considerations. Finally, there is an increasing awareness of the need for ecological risk assessments to consider nonchemical as well as chemical stressors.

The framework consists of three major phases: (1) problem formulation, (2) analysis, and (3) risk characterization. Problem formulation is a planning and scoping process that establishes the goals, breadth, and focus of the risk assessment. Its end product is a conceptual model that identifies the environmental values to be protected (the assessment endpoints), the data needed, and the analyses to be used.

The analysis phase develops profiles of environmental exposure and the effects of the stressor. The exposure profile characterizes the ecosystems in which the stressor may occur as well as the biota that may be exposed. It also describes the magnitude and spatial and temporal patterns of exposure. The ecological effects profile summarizes data on the effects of the stressor and relates them to the assessment endpoints.

Risk characterization integrates the exposure and effects profiles. Risks can be estimated using a variety of techniques including comparing individual exposure and effects values, comparing the distributions of exposure and effects, or using simulation models. Risk can be expressed as a qualitative or quantitative estimate, depending on available data. In this step, the assessor also:

- describes the risks in terms of the assessment endpoint;
- discusses the ecological significance of the effects;
- summarizes overall confidence in the assessment; and
- discusses the results with the risk manager.

The framework also recognizes several activities that are integral to, but separate from, the risk assessment process as defined in this report. For example, discussions between the risk assessor and risk manager are important. At the initiation of the risk assessment, the risk manager can help ensure that the risk assessment will ultimately provide information that is relevant to making decisions on the issues under consideration, while the risk assessor can ensure that the risk assessment addresses all relevant ecological concerns. Similar discussions of the results of the risk assessment are important to provide the risk manager with a full and complete understanding of the assessment's conclusions, assumptions, and limitations.

Other important companion activities to ecological risk assessment include data acquisition and verification and monitoring studies. New data are frequently required to conduct analyses that are performed during the risk assessment. Data from verification studies can be used to validate the predictions of a specific risk assessment as well as to evaluate the usefulness of the principles set forth in the Framework. Ecological effects or exposure monitoring can aid in the verification process and suggest additional data, methods, or analyses that could improve future risk assessments.

1. INTRODUCTION

Public, private, and government sectors of society are increasingly aware of ecological issues including global climate change, habitat loss, acid deposition, a decrease in biological diversity, and the ecological impacts of xenobiotic compounds such as pesticides and toxic chemicals. To help assess these and other ecological problems, the U.S. Environmental Protection Agency (EPA) has developed this report, "Framework for Ecological Risk Assessment", which describes the basic elements, or framework, of a process for evaluating scientific information on the adverse effects of stressors on the environment. The term "stressor" is defined here as any physical, chemical, or biological entity that can induce an adverse effect (see box[1]). Adverse ecological effects encompass a wide range of disturbances ranging from mortality in an individual organism to a loss in ecosystem function.

This introductory section describes the purpose, scope, and intended audience for this report; discusses the definition and application of ecological risk assessment; outlines the basic elements of the proposed framework; and describes the organization of this report.

1.1. PURPOSE AND SCOPE OF THE FRAMEWORK REPORT

An understanding of the finite purpose and scope of this Framework Report is important. EPA, other regulatory agencies, and other organizations need detailed, comprehensive guidance on methods for evaluating ecological risk. However, in discussing tentative plans for developing such guidance with expert consultants (U.S. EPA, 1991; U.S. EPA, in press-a), EPA was advised to first develop a simple framework as a foundation or blueprint for later comprehensive guidance on ecological risk assessment.

> **Physical and Chemical Stressors as a Focus of the Framework**
>
> This report does not discuss accidentally or deliberately introduced species, genetically engineered organisms, or organisms used to control horticultural or agricultural pests. While the general principles described in the framework may be helpful in addressing risks associated with these organisms, the capacity of such organisms for reproduction and biological interaction introduces additional considerations that are not addressed in this document.

With this background, the framework (see Section 1.4) has two simple purposes, one immediate and one longer term. As a broad outline of the assessment process, the framework offers a basic structure and starting principles for EPA's ecological

[1] The boxes used throughout this document serve several purposes. Some boxes provide additional background and rationale for terms, whereas other boxes expand on concepts described in the text. The boxes at the end of each chapter highlight issues that are integral components of the risk assessment process but require more research, analysis, and debate. Further discussion of these issues is reserved for later guidelines.

risk assessments. The process described by the framework provides wide latitude for planning and conducting individual risk assessments in many diverse situations, each based on the common principles discussed in the framework. The process also will help foster a consistent EPA approach for conducting and evaluating ecological risk assessments, identify key issues, and provide operational definitions for terms used in ecological risk assessments.

In addition, the framework offers basic principles around which long-term guidelines for ecological risk assessment can be organized. With this in mind, this report does not provide substantive guidance on factors that are integral to the risk assessment process such as analytical methods, techniques for analyzing and interpreting data, or guidance on factors influencing policy. Rather, on the basis of EPA experience and the recommendations of peer reviewers, EPA has reserved discussion of these important aspects of any risk assessment for future guidelines, which will be based on the process described in this report.

1.2. INTENDED AUDIENCE

The framework is primarily intended for EPA risk assessors, EPA risk managers, and other persons who either perform work under EPA contract or sponsorship or are subject to EPA regulations. The terminology and concepts described here also may be of assistance to other federal, state, and local agencies as well as to members of the general public who are interested in ecological issues.

1.3. DEFINITION AND APPLICATIONS OF ECOLOGICAL RISK ASSESSMENT

Ecological risk assessment is defined as a process that evaluates the likelihood that adverse ecological effects may occur or are occurring as a result of exposure to one or more stressors. A risk does not exist unless (1) the stressor has the inherent ability to cause one or more adverse effects and (2) it co-occurs with or contacts an ecological component (i.e., organisms, populations, communities, or ecosystems) long enough and at a sufficient intensity to elicit the identified adverse effect. Ecological risk assessment may evaluate one or many stressors and ecological components.

Ecological risk may be expressed in a variety of ways. While some ecological risk assessments may provide true probabilistic estimates of both the adverse effect and exposure elements, others may be deterministic or even qualitative in nature. In these cases, the likelihood of adverse effects is expressed through a semiquantitative or qualitative comparison of effects and exposure.

Ecological risk assessments can help identify environmental problems, establish priorities, and provide a scientific basis for regulatory actions. The process can identify existing risks or forecast the risks of stressors not yet present in the environment. However, while ecological risk assessments can play an important role in identifying and resolving environmental problems, risk assessments are not a solution for addressing all environmental problems, nor are they always a prerequisite for environmental management. Many environmental matters such as the protection of habitats and endangered species are compelling enough that there may not be enough time or data to do a risk assessment. In such cases, professional judgment and the mandates of a particular statute will be the driving forces in making decisions.

1.4. ECOLOGICAL RISK ASSESSMENT FRAMEWORK

The distinctive nature of the framework results primarily from three differences in emphasis relative to previous risk assessment approaches. First, ecological risk assessment can consider effects beyond those on individuals of a single species and may examine population, community, or ecosystem impacts. Second, there is no one set of assessment endpoints (environmental values to be protected) that can be generally applied. Rather, assessment endpoints are selected from a very large number of possibilities based on both scientific and policy considerations. Finally, a comprehensive approach to ecological risk assessment may go beyond the traditional emphasis on chemical effects to consider the possible effects of nonchemical stressors.

The ecological risk assessment framework is shown in Figure 1. The risk assessment process is based on two major elements: characterization of exposure and characterization of ecological effects. Although these two elements are most prominent during the analysis phase, aspects of both exposure and effects also are considered during problem formulation, as illustrated by the arrows in the diagram. The arrows also flow to risk characterization, where the exposure and effects elements are integrated to estimate risk. The framework is conceptually similar to the National Research Council (NRC) paradigm for human health risk assessments (NRC, 1983).

Relationship of the Framework to a Paradigm for Human Health Risk Assessment

In 1983, NRC published a paradigm that has been used in the development of EPA's human health risk assessment guidelines. The paradigm has four phases: hazard identification, dose-response assessment, exposure assessment, and risk characterization (NRC, 1983). Although the framework's problem formulation phase is not explicitly identified in the NRC paradigm, comparable planning issues are addressed in practice at the beginning of all EPA risk assessments. In the framework's analysis phase, characterization of exposure is analogous to exposure assessment, while characterization of ecological effects includes aspects of both hazard identification and dose-response assessment. (The framework uses the term "stressor response" rather than "dose response" because many Agency programs must address stressors other than chemicals, and dose has been used only for chemicals.) Risk characterization is a similar process in both the framework and the NRC paradigm.

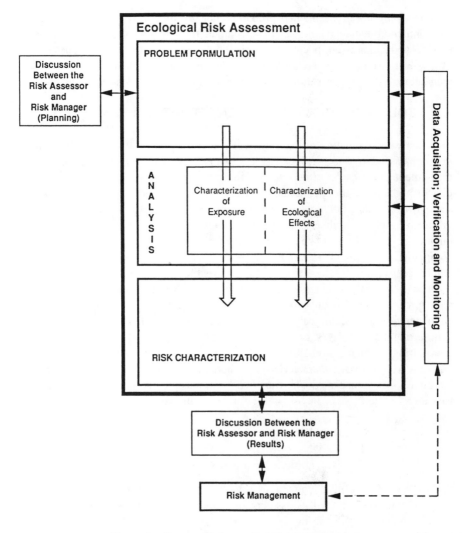

Figure 1. Framework for ecological risk assessment

The first phase of the framework is problem formulation. Problem formulation includes a preliminary characterization of exposure and effects, as well as examination of scientific data and data needs, policy and regulatory issues, and site-specific factors to define the feasibility, scope, and objectives for the ecological risk assessment. The level of detail and the information that will be needed to complete the assessment also are determined. This systematic planning phase is proposed because ecological risk assessments often address the risks of stressors to many species as well as risks to communities and ecosystems. In addition, there may be many ways a stressor can elicit adverse effects (e.g., direct effects on mortality and growth and indirect effects such as decreased food supply). Problem formulation provides an early identification of key factors to be considered, which in turn will produce a more scientifically sound risk assessment.

The second phase of the framework is termed analysis and consists of two activities: characterization of exposure is to predict or measure the spatial and temporal distribution of a stressor and its co-occurrence or contact with the ecological components of concern, while the purpose of characterization of ecological effects is to identify and quantify the adverse effects elicited by a stressor and, to the extent possible, to evaluate cause-and-effect relationships.

The third phase of the framework is risk characterization. Risk characterization uses the results of the exposure and ecological effects analyses to evaluate the likelihood of adverse ecological effects associated with exposure to a stressor. It includes a summary of the assumptions used, the scientific uncertainties, and the strengths and weaknesses of analyses. In addition, the ecological significance of the risks is discussed with consideration of the types and magnitudes of the effects, their spatial and temporal patterns, and the likelihood of recovery. The purpose is to provide a complete picture of the analysis and results.

In addition to showing the three phases of the framework, Figure 1 illustrates the need for discussions between the risk assessor and risk manager. At the initiation of the risk assessment, the risk manager can help ensure that the risk assessment will ultimately provide information that is relevant to making decisions on the issues under consideration, while the risk assessor can ensure that the risk assessment addresses all relevant ecological concerns. Similar discussions of the results of the risk assessment are important to provide the risk manager with a full and complete understanding of the assessment's conclusions, assumptions, and limitations.

Use of the Term "Exposure"

Some reviewers of earlier drafts of this interim framework proposed that the term "exposure" — which, as used in human health risk assessment, generally refers to chemical stressors — not be used for the nonchemical stressors that can affect a variety of ecological components. Other terms, including "characterization of stress", have been suggested. At this time, EPA prefers exposure, partly because characterization of stress does not convey the important concept of the co-occurrence and interaction of the stressor with an ecological component as well as exposure does.

Characterization of Ecological Effects Used Instead of Hazard Assessment

The framework uses characterization of ecological effects rather than hazard assessment for two reasons. First, the term "hazard" can be ambiguous, because it has been used in the past to mean either evaluating the intrinsic effects of a stressor (U.S. EPA, 1979) or defining a margin of safety or quotient by comparing a toxicological endpoint of interest with an estimate of exposure concentration (SETAC, 1987). Second, many reviewers believed that hazard is more relevant to chemical than to nonchemical stressors.

Figure 1 also indicates a role for verification and monitoring in the framework. Verification can include validation of the ecological risk assessment process as well as confirmation of specific predictions made during a risk assessment. Monitoring can aid in the verification process and may identify additional topics for risk assessment. Verification and monitoring can help determine the overall effectiveness of the framework approach, provide necessary feedback concerning the need for future modifications of the framework, help evaluate the effectiveness and practicality of policy decisions, and point out the need for new or improved scientific techniques (U.S. EPA, in press-a).

The interaction between data acquisition and ecological risk assessment is also shown in Figure 1. In this report, a distinction is made between data acquisition (which is outside of the risk assessment process) and data analysis (which is an integral part of an ecological risk assessment). In the problem formulation and analysis phases, the risk assessor may identify the need for additional data to complete an analysis. At this point, the risk assessment stops until the necessary data are acquired. When a need for additional data is recognized in risk characterization, new information generally is used in the analysis or problem formulation phases. The distinction between data acquisition and analysis generally is maintained in all of EPA's risk assessment guidelines; guidance on data acquisition procedures are provided in documents prepared for specific EPA programs.

The interactions between data acquisition and ecological risk assessment often result in an iterative process. For example, data used during the analysis phase may be collected in tiers of increasing complexity and cost. A decision to advance from one tier to the next is based on decision triggers set at certain levels of effect or exposure. Interactions of the entire risk assessment process also may occur. For example, a screening-level risk assessment may be performed using readily available data and conservative assumptions; depending on the results, more data then may be collected to support a more rigorous assessment.

1.5. THE IMPORTANCE OF PROFESSIONAL JUDGMENT

Ecological risk assessments, like human health risk assessments, are based on scientific data that are frequently difficult and complex, conflicting or ambiguous, or incomplete. Analyses of such data for risk assessment purposes depends on professional judgment based on scientific expertise. Professional judgment is necessary to:

- design and conceptualize the risk assessment;
- evaluate and select methods and models;
- determine the relevance of available data to the risk assessment;
- develop assumptions based on logic and scientific principles to fill data gaps; and
- interpret the ecological significance of predicted or observed effects.

Because professional judgment is so important, specialized knowledge and experience in the various phases of ecological risk assessment is required. Thus, an interactive multidisciplinary team that includes biologists and ecologists is a prerequisite for a successful ecological risk assessment.

1.6. ORGANIZATION

The next three sections of this report are arranged to follow the framework sequentially. Section 2 describes problem formulation; this section is particularly important for assessors to consider when specific assessment endpoints are not determined *a priori* by statute or other authority. Sections 3 and 4 discuss analysis and risk characterization, respectively. Section 5 defines the terms used in this report. The lists of ecological risk assessment issues at the end of Sections 1 through 4 highlight areas for further discussion and research. EPA believes that these issues will require special attention in developing ecological risk assessment guidelines.

Additional Issues Related to the Framework

- Use of the framework for evaluating risks associated with biological stressors.
- Use of the term exposure (versus characterization of stress) for both chemical and nonchemical stressors.
- Use of the term characterization of ecological effects rather than hazard assessment.

---

2. PROBLEM FORMULATION

Problem formulation is the first phase of ecological risk assessment and establishes the goals, breadth, and focus of the assessment. It is a systematic planning step that identifies the major factors to be considered in a particular assessment, and it is linked to the regulatory and policy context of the assessment.

Entry into the ecological risk assessment process may be triggered by either an observed ecological effect, such as visible damage to trees in a forest, or by the identification of a stressor or activity of concern, such as the planned filling of a marsh or the manufacture of a new chemical. The problem formulation process (Figure 2) then begins with the initial stages of characterizing exposure and ecological effects, including evaluating the stressor characteristics, the ecosystem potentially at risk, and the ecological effects expected or observed. Next, the assessment and measurement endpoints are identified. (Measurement endpoints are ecological characteristics that can be related to the assessment endpoint.) The outcome of problem formulation is a conceptual model that describes how a given stressor might affect the ecological components in the environment. The conceptual model also describes the relationships among the assessment and measurement endpoints, the data required, and the methodologies that will be used to analyze the data. The conceptual model serves as input to the analysis phase of the assessment.

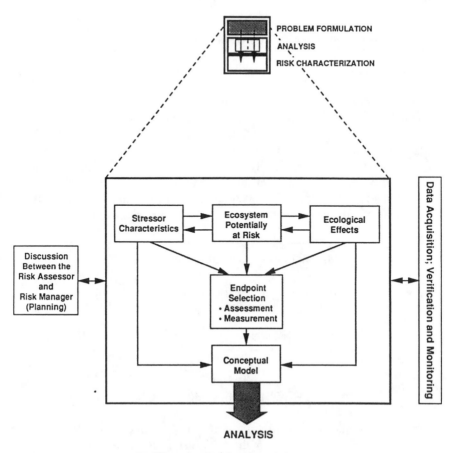

Figure 2. Problem formulation

2.1. DISCUSSION BETWEEN THE RISK ASSESSOR AND RISK MANAGER (PLANNING)

To be meaningful and effective, ecological risk assessments must be relevant to regulatory needs and public concerns as well as scientifically valid. Although risk assessment and risk management are distinct process, establishing a two-way dialogue between risk assessors and risk managers during the problem formulation phase can be a constructive means of achieving both societal and scientific goals. By bringing the management perspective to the discussion, risk managers charged with protecting societal values can ensure that the risk assessment will provide relevant information to making decisions on the issue under consideration. By bringing scientific knowledge to the discussion, the ecological risk assessor ensures that the assessment addresses all important ecological concerns. Both perspectives are necessary to appropriately utilize resources to produce scientifically sound risk assessments that are relevant to management decisions and public concerns.

2.2. STRESSOR CHARACTERISTICS, ECOSYSTEM POTENTIALLY AT RISK, AND ECOLOGICAL EFFECTS

The initial steps in problem formulation are the identification and preliminary characterization of stressors, the ecosystem potentially at risk, and ecological effects. Performing this analysis is an interactive process that contributes to both the selection of assessment and measurement endpoints and the develop of a conceptual model.

2.2.1. Stressor Characteristics

The determination of stressor characteristics begins with the identification of potential chemical or physical stressors. Chemical stressors include a variety of inorganic and organic substances. Some chemicals may result in secondary stressors, as in the case of stratospheric ozone depletion caused by chlorofluorocarbons that could result in increased exposures to ultraviolet radiation. Physical stressors include extremes of natural conditions (e.g., temperature and hydrologic changes) and habitat alteration or destruction. Stressors that may result from management practices, such as harvesting of fishery or forest resources, also may be considered. Example stressor characteristics are summarized in the box below. Gathering information on the characteristics of a stressor helps define the ecosystems potentially at risk from the stressor as well as the ecological effects that may result.

Example Stressor Characteristics

Type — Chemical or physical
Intensity — Concentration or magnitude
Duration — Short or long term
Frequency — Single event, episodic, or continuous
Timing — Occurrence relative to biological cycles
Scale — Spatial heterogeneity and extent

2.2.2. Ecosystem Potentially at Risk

The ecosystem within which effects occur provides the ecological context for the assessment. Knowledge of the ecosystem potentially at risk can help identify ecological components that may be affected and stressor-ecosystem interactions relevant to developing exposure scenarios. The approach to identifying the ecosystem potentially at risk from a stressor depends in part on how the risk assessment was initiated. If a stressor first was identified, information on the spatial and temporal distribution patterns of the stressor can be helpful in identifying ecosystems potentially at risk. Similarly, if the risk assessment is initiated by observing effects, these effects can directly indicate ecosystems or ecological components that may be considered in the assessment.

A wide range of ecosystem properties may be considered during problem formulation. These properties include aspects of the abiotic environment (such as climatic conditions and soil or sediment properties), ecosystem structure (including the types and abundances of different species and their trophic level relationships), and ecosystem function (such as the ecosystem energy source, pathways of energy utilization, and nutrient processing) (U.S. EPA, in press-b). In addition, knowledge of the types and patterns of historical disturbances may be helpful in predicting ecological responses to stressors.

The need to evaluate spatial and temporal distribution and variation is inherent in many of these example characteristics. Such information is especially useful for determining potential exposure, that is, where there is co-occurrence of or contact between the stressor and ecological components.

2.2.3. Ecological Effects

Ecological effects data may come from a variety of sources. Relevant sources of information include field observations (e.g., fish or bird kills, changes in aquatic community structure), field tests (e.g., microcosm or mesocosm tests), laboratory tests (e.g., single species or microcosm tests), and chemical structure-activity relationships. Available information on ecological effects can help focus the assessment on specific stressors and on ecological components that should be evaluated.

Many factors can influence the utility of available ecological effects data for problem formulation. For example, the applicability of laboratory-based tests may be affected by any extrapolations required to specific field situations, while the interpretation of field observations may be influenced by factors such as natural variability or the possible presence of stressors other than the ones that are the primary focus of the risk assessment.

2.3. ENDPOINT SELECTION

Information compiled in the first stage of problem formulation is used to help select ecologically based endpoints that are relevant to decisions made about protecting the environment. An endpoint is a characteristic of an ecological component (e.g., increased mortality in fish) that may be affected by exposure to a stressor (Suter, 1990a). Two types of endpoints are distinguished in this report. Assessment endpoints are explicit expressions of the actual environmental value that is to be protected. Measurement endpoints are measurable responses to a stressor that are related to the valued characteristics chosen as the assessment endpoints (Suter, 1990a).

Assessment endpoints are the ultimate focus in risk characterization and link the measurement endpoints to the risk management process (e.g., policy goals). When an assessment endpoint can be directly measured, the measurement and assessment endpoints are the same. In most cases, however, the assessment endpoint cannot be directly measured, so a measurement endpoint (or a suite of measurement endpoints) is selected that can be related, either qualitatively or quantitatively, to the assessment endpoint. For example, a decline in a sport fish population (the assessment endpoint) may be evaluated using laboratory studies on the mortality of surrogate species, such as the fathead minnow (the measurement endpoint). Sound professional judgment is necessary for proper assessment and measurement endpoint selection, and it is important that both the selection rationale and the linkages between measurement endpoints, assessment endpoints, and policy goals be clearly stated.

Assessment and measurement endpoints may involve ecological components from any level of biological organization, ranging from individual organisms to the ecosystem itself. In general, the use of a suite of assessment and measurement endpoints at different

> **Endpoint Terminology**
>
> Several reviewers have suggested using the term "indicator" in place of "measurement endpoint". At this time, measurement endpoint is preferred because it has a specific meaning (a characteristic of an ecological system that can be related to an assessment endpoint), whereas indicator can have several different meanings. For example, indicator has been used at EPA to mean (1) measures of administrative accomplishments (e.g., number of permits issued), (2) measures of exposure (e.g., chemical levels in sediments), or (3) measures of ecosystem integrity. These indicators cannot always be related to an assessment endpoint.

organizational levels can build greater confidence in the conclusions of the risk assessment and ensure that all important endpoints are evaluated. In some situations, measurement endpoints at one level of organization may be related to an assessment endpoint at a higher level. For example, measurement endpoints at the individual level (e.g., mortality, reproduction, and growth) could be used in a model to predict effects on an assessment endpoint at the population level (e.g., viability of a trout population in a stream).

General considerations for selecting assessment and measurement endpoints are detailed in the following boxes. More detailed discussions of endpoints and selection criteria can be found in Suter (1989, 1990a), Kelly and Harwell (1990), U.S. Department of the Interior (1987), and U.S. EPA (1990a).

2.4. THE CONCEPTUAL MODEL

The major focus of the conceptual model (Figure 2) is the development of working hypotheses regarding how the stressor might affect ecological components of the natural environment (NRC, 1986). The conceptual model also includes descriptions of the ecosystem potentially at risk and the relationship between measurement and assessment endpoints.

During conceptual model development, a preliminary analysis of the ecosystem, stressor characteristics, and ecological effects is used to define possible exposure

Considerations in Selecting Assessment Endpoints

Ecological Relevance

Ecologically relevant endpoints reflect important characteristics of the system and are functionally related to other endpoints. Selection of ecologically relevant endpoints requires some understanding of the structure and function of the ecosystem potentially at risk. For example, an assessment endpoint could focus on changes in a species known to have a controlling influence on the abundance and distribution of many other species in its community. Changes at higher levels of organization may be significant because of their potential for causing major effects at lower organizational levels.

Policy Goals and Societal Values

Good communication between the risk assessor and risk manager is important to ensure that ecologically relevant assessment endpoints reflect policy goals and societal values. Societal concerns can range from protection of endangered or commercially or recreationally important species to preservation of ecosystem attributes for functional reasons (e.g., flood water retention by wetlands) or aesthetic reasons (e.g., visibility in the Grand Canyon).

Susceptibility to the Stressor

Ideally, an assessment endpoint would be likely to be both affected by exposure to a stressor and sensitive to the specific type of effects caused by the stressor. For example, if a chemical is known to bioaccumulate and is suspected of causing eggshell thinning, an appropriate assessment endpoint might be raptor population viability.

scenarios. Exposure scenarios consist of a qualitative description of how the various ecological components co-occur with or contact the stressor. Each scenario is defined in terms of the stressor, the type of biological system and principal ecological components, how the stressor will contact or interact with the system, and the spatial and temporal scales.

For chemical stressors, the exposure scenario usually involves consideration of sources, environmental transport, partitioning of the chemical among various environmental media, chemical/biological transformation or speciation processes, and identification of potential routes of exposure (e.g., ingestion). For nonchemical stressors such as water level or temperature changes or physical disturbance, the exposure scenario describes the ecological components exposed and the general temporal and spatial patterns of their co-occurrence with the stressor. For example, for habitat alterations, the exposure scenario may describe the extent and distributional pattern of disturbance, the populations residing within or using the disturbed areas, and the spatial relationship of the disturbed area to undisturbed areas.

Considerations in Selecting Measurement Endpoints

Relevance to an Assessment Endpoint
When an assessment endpoint cannot be directly measured, measurement endpoints are identified that are correlated with or can be used to infer or predict changes in the assessment endpoint.

Consideration of Indirect Effects
Indirect effects occur when a stressor acts on elements of the ecosystem that are required by the ecological component of concern. For example, if the assessment endpoint is the population viability of trout, measurement endpoints could evaluate possible stressor effects on trout prey species or habitat requirements.

Sensitivity and Response Time
Rapidly responding measurement endpoints may be useful in providing early warnings of ecological effects, and measurement endpoints also may be selected because they are sensitive surrogates of the assessment endpoint. In many cases, measurement endpoints at lower levels of biological organization may be more sensitive than those at higher levels. However, because of compensatory mechanisms and other factors, a change in a measurement endpoint at a lower organizational level (e.g., a biochemical alteration) may not necessarily be reflected in changes at a higher level (e.g., population effects).

Signal-to-Noise Ratio
If a measurement endpoint is highly variable, the possibility of detecting stressor-related effects may be greatly reduced even if the endpoint is sensitive to the stressor.

Consistency with Assessment Endpoint Exposure Scenarios
The ecological component of the measurement endpoint should be exposed by similar routes and at similar or greater stressor levels as the ecological component of the assessment endpoint.

Diagnostic Ability
Measurement endpoints that are unique or specific responses to a stressor may be very useful in diagnosing the presence or effects of a stressor. For example, measurement of acetylcholinesterase inhibition may be useful for demonstrating responses to certain types of pesticides.

Practicality Issues
Ideal measurement endpoints are cost effective and easily measured. The availability of a large database for a measurement endpoint is desirable to facilitate comparisons and develop models.

Although many hypotheses may be generated during problem formulation, only those that are considered most likely to contribute to risk are selected for further evaluation in the analysis phase. For these hypotheses, the conceptual model describes the approach that will be used for the analysis phase and the types of data and analytical tools that will be needed. It is important that hypotheses that are not carried forward in the assessment because of data gaps be acknowledged when uncertainty is addressed in risk characterization. Professional judgment is needed to select the most appropriate risk hypotheses, and it is important to document the selection rationale.

Additional Issues in Problem Formulation

- Role of risk management concerns in establishing assessment endpoints. Although it is important to consider risk management concerns when assessment endpoints are selected, there is still uncertainty as to how these inputs should influence the goals of the risk assessment, the ecological components to be protected, and the level of protection required.
- Identifying specific assessment and measurement endpoints for different stressors and ecosystems.

---

3. ANALYSIS PHASE

The analysis phase of ecological risk assessment (Figure 3) consists of the technical evaluation of data on the potential effects and exposure of the stressor. The analysis phase is based on the conceptual model developed during problem formulation. Although this phase consists of characterization of ecological effects and characterization of exposure, the dotted line in Figure 3 illustrates that the two are performed interactively. An interaction between the two elements will ensure that the ecological effects characterized are compatible with the biota and exposure pathways identified in the exposure characterization. The output of ecological effects characterization and exposure characterization are summary profiles that are used in the risk characterization phase (Section 4). Discussion of uncertainty analysis, which is an important part of the analysis phase, may be found in Section 4.1.2.

Characterization of exposure and ecological effects often requires the application of statistical methods. While the discussion of specific statistical methods is beyond the scope of this document, selection of an appropriate statistical method involves both method assumptions (e.g., independence of errors, normality, equality of variances) and data set characteristics (e.g., distribution, presence of outliers or influential data). It should be noted that statistical significance does not always reflect biological significance, and profound biological changes may not be detected by statistical tests. Professional judgment often is required to evaluate the relationship between statistical and biological significance.

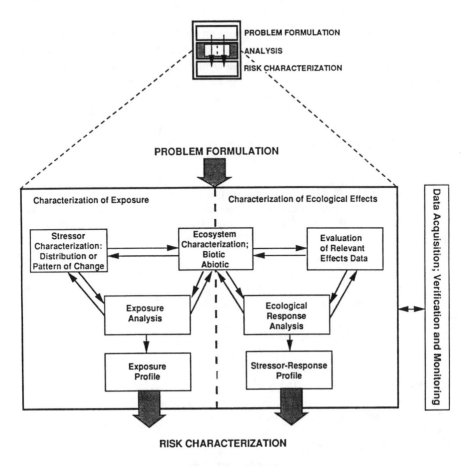

Figure 3. Analysis

3.1. CHARACTERIZATION OF EXPOSURE

Characterization of exposure (half of the analysis phase shown in Figure 3) evaluates the interaction of the stressor with the ecological component. Exposure can be expressed as co-occurrence or contact depending on the stressor and the ecological component involved. An exposure profile is developed that quantifies the magnitude and spatial and temporal distributions of exposure for the scenarios developed during problem formulation and serves as input to the risk characterization.

3.1.1. Stressor Characterization: Distribution or Pattern of Change

Stressor characterization involves determining the stressor's distribution or pattern of change. Many techniques can be applied to assist in this stressor characterization process. For chemical stressors, a combination of modeling and monitoring data often is used. Available monitoring data may include measures of releases into the environment and media concentrations over space and time. Fate and transport models often are used that rely on physical and chemical characteristics of the chemical coupled with the characteristics of the ecosystem. For nonchemical stressors such as physical alterations or harvesting, the pattern of change may depend on resource management or land-use practices. Depending on the scale of the disturbance, the data for stressor characterization can be provided by a variety of techniques, including ground reconnaissance, aerial photographs, or satellite imagery.

During stressor characterization, one considers not only the primary stressor but also secondary stressors that can arise as a result of various processes. For example, removal of riparian (stream-side) vegetation not only alters habitat structure directly, but can have additional ramifications such as increased siltation and temperature rise. For chemicals, secondary stressors can be produced by a range of environmental fate processes.

The timing of the stressor's interaction with the biological system is another important consideration. If the stressor is episodic in nature, different species and life stages may be affected. In addition, the ultimate distribution of a stressor is rarely homogeneous; it is important to quantify such heterogeneity whenever possible.

3.1.2. Ecosystem Characterization

During ecosystem characterization, the ecological context of the assessment is further analyzed. In particular, the spatial and temporal distributions of the ecological component are characterized, and the ecosystem attributes that influence the distribution and nature of the stressor are considered.

Characteristics of the ecosystem can greatly modify the ultimate nature and distribution of the stressor. Chemical stressors can be modified through biotransformation by microbial communities or through other environmental fate processes, such as photolysis, hydrolysis, and sorption. The bioavailability of chemical stressors also can be affected by the environment, which in turn influences the exposure of ecological components.

Physical stressors can be modified by the ecosystem as well. For example, siltation in streams depends not only on sediment volume, but on flow regime and physical stream characteristics. Similarly, nearby wetlands and levees influence water behavior during flood events.

The spatial and temporal distributions of ecological components also are considered in ecosystem characterization. Characteristics of ecological components that influence their exposure to the stressor are evaluated, including habitat needs, food preferences, reproductive cycles, and seasonal activities such as migration and selective use of resources. Spatial and temporal variations in the distribution of the ecological component (e.g., sediment invertebrate distribution) may complicate evaluations of exposure. When available, species-specific information about activity patterns, abundance, and life histories can be very useful in evaluating spatial and temporal distributions.

Another important consideration is how exposure to a stressor may alter natural behavior, thereby affecting further exposure. In some cases, this may lead to enhanced exposure (e.g., increased preening by birds after aerial pesticide spraying), while in other situations initial exposure may lead to avoidance of contaminated locations or food sources (e.g., avoidance of certain waste effluents or physically altered spawning beds by some fish species).

3.1.3. Exposure Analyses

The next step is to combine the spatial and temporal distributions of both the ecological component and the stressor to evaluate exposure. In the case of physical alterations of communities and ecosystems, exposure can be expressed broadly as co-occurrence. Exposure analyses of individuals often focus on actual contact with the stressor, because organisms may not contact all of the stressors present in an area. For chemical stressors, the analyses may focus further on the amount of chemical that is bioavailable, that is, available for uptake by the organism. Some chemical exposure analyses also follow the chemical within the organism's body and estimate the amount that reaches the target organ. The focus of the analyses will depend on the stressors being evaluated and the assessment and measurement endpoints.

The temporal and spatial scales used to evaluate the stressor need to be compatible with the characteristics of the ecological component of interest. A temporal scale may encompass the lifespan of a species, a particular life stage, or a particular cycle, for example, the long-term succession of a forest community. A spatial scale may encompass a forest, a lake, a watershed, or an entire region. Stressor timing relative to organism life stage and activity patterns can greatly influence the occurrence of adverse effects. Even short-term events may be significant if they coincide with critical life stages. Periods of reproductive activity may be especially important, because early life stages often are more sensitive to stressors, and adults also may be more vulnerable at this time.

The most common approach to exposure analysis is to measure concentrations or amounts of a stressor and combine them with assumptions about co-occurrence, contact, or uptake. For example, exposure of aquatic organisms to chemicals often is expressed simply as concentration in the water column; aquatic organisms are assumed to contact the chemical. Similarly, exposures of organisms to habitat alteration often are expressed as hectares of habitat altered; organisms that utilize the habitat are assumed to co-occur with the alteration. Stressor measurements can also be combined with quantitative parameters describing the frequency and magnitude of contact. For example, concentrations of chemicals in food items can be combined with ingestion rates to estimate dietary exposure of organisms.

In some situations, the stressor can be measured at the actual point of contact while exposure occurs. An example is the use of food collected from the mouths of nestling birds to evaluate exposure to pesticides through contaminated food (Kendall, 1991). Although such point-of-contact measurements can be difficult to obtain, they reduce the need for assumptions about the frequency and magnitude of contact.

Patterns of exposure can be described using models that combine abiotic ecosystem attributes, stressor properties, and ecological component characteristics. Model selection is based on the model's suitability for the ecosystem or component of interest, the availability of the requisite data, and the study objectives. Model choices range from simple, screening-level procedures that require a minimum of data to more sophisticated methods that describe processes in more detail but require a considerable amount of data.

Another approach to evaluating exposure uses chemical, biochemical, or physiological evidence (e.g., biomarkers) of a previous exposure. This approach has been used primarily for assessing chemical exposures and is particularly useful when a residue or biomarker is diagnostic of exposure to a particular chemical. These types of measurements are most useful for exposure characterization when they can be quantitatively linked to the amount of stressor originally contacted by the organism. Pharmacokinetic models are sometimes used to provide this linkage.

3.1.4. Exposure Profile

Using information obtained from the exposure analysis, the exposure profile quantifies the magnitude and spatial and temporal patterns of exposure for the scenarios developed during problem formulation and serves as input to risk characterization. The exposure profile is only effective when its results are compatible with the stressor-response profile. For example, appraisals of potential acute effects of chemical exposure may be averaged over short time periods to account for short-term pulsed stressor events. It is important that characterizations for chronic stressors account for both long-term, low-level exposure and possible shorter term, higher level contact that may elicit similar adverse chronic effects.

Exposure profiles can be expressed using a variety of units. For chemical stressors operating at the organism level, the usual metric is expressed in dose units (e.g., mg/kg body weight/day). For higher levels of organization (e.g., an entire ecosystem), exposure may be expressed in units of concentration/unit area/time. For physical disturbance, the exposure profile may be expressed in other terms (e.g., percentage of habitat removed or the extent of flooding/year).

An uncertainty assessment is an integral part of the characterizations of exposure. In the majority of assessments, data will not be available for all aspects of the characterization of exposure, and those data that are available may be of questionable or unknown quality. Typically, the assessor will have to rely on a number of assumptions with varying degrees of uncertainty associated with each. These assumptions will be based on a combination of professional judgment, inferences based on analogy with similar chemicals and conditions and estimation techniques, all of which contribute to the overall uncertainty. It is important that the assessor characterize each of the various sources of uncertainty and carry them forward to the risk characterization so that they may be combined with a similar analysis conducted as part of the characterization of ecological effects.

3.2. CHARACTERIZATION OF ECOLOGICAL EFFECTS

The relationship between the stressor and the assessment and measurement endpoints identified during problem formulation is analyzed in characterization of ecological effects (Figure 3). The evaluation begins with the evaluation of effects data that are relevant to the stressor. During ecological response analysis, the relationship between the stressor and the ecological effects elicited is quantified, and cause-and-effect relationships are evaluated. In addition, extrapolations from measurement endpoints to assessment endpoints are conducted during this phase. The product is a stressor-response profile that quantifies and summarizes the relationship of the stressor to the assessment endpoint. The stressor-response profile is then used as input to risk characterization.

3.2.1. Evaluation of Relevant Effects Data

The type of effects data that are evaluated depends largely on the nature of the stressor and the ecological component under evaluation. Effects elicited by a stressor may range from mortality and reproductive impairment in individuals and populations to disruptions in community and ecosystem function such as primary productivity. The evaluation process relies on professional judgment, especially when few data are available or when choices among several sources of data are required. If available data are inadequate, new data may be needed before the assessment can be completed.

Data are evaluated by considering their relevance to the measurement and assessment endpoints selected during problem formulation. The analysis techniques that will be used also are considered; data that minimize the need for extrapolation are desirable. Data quality (e.g., sufficiency of replications, adherence to good laboratory practices) is another important consideration. Finally, characteristics of the ecosystem potentially at risk will influence what data will be used. Ideally, the test system reflects the physical attributes of the ecosystem and will include the ecological components and life stages examined in the risk assessment.

Data from both field observations and experiments in controlled settings can be used to evaluate ecological effects. In some cases, such as for chemicals that have yet to be manufactured, test data for the specific stressor are not available. Quantitative structure-activity relationships (QSARs) are useful in these situations (Auer et al., 1990; Clements et al., 1988; McKim et al., 1987).

Controlled laboratory and field tests (e.g., mesocosms) can provide strong causal evidence linking a stressor with a response and can also help discriminate between multiple stressors. Data from laboratory studies tend to be less variable than those from field studies, but because environmental factors are controlled, responses may differ from those in the natural environment.

Observational field studies (e.g., comparison with reference sites) provide environmental realism that laboratory studies lack, although the presence of multiple stressors and other confounding factors (e.g., habitat quality) in the natural environment can make it difficult to attribute observed effects to specific stressors. Confidence in causal relationships can be improved by carefully selecting comparable reference sites or by evaluating changes along a stressor gradient where differences in other environmental factors are minimized. It is important to consider potential confounding factors during the analysis.

3.2.2. Ecological Response Analyses

The data used in characterization of ecological effects are analyzed to quantify the stressor-response relationship and to evaluate the evidence for causality. A variety of techniques may be used, including statistical methods and mathematical modeling. In some cases, additional analyses to relate the measurement endpoint to the assessment endpoint may be necessary.

Stressor-Response Analyses

The stressor-response analysis describes the relationship between the magnitude, frequency, or duration of the stressor in an observational or experimental setting and the magnitude of response. The stressor-response analysis may focus on different aspects of the stressor-response relationship, depending on the assessment objectives, the conceptual model, and the type of data used for the analysis. Stressor-response analyses, such as those used for toxicity tests, often portray the magnitude of the stressor with respect to the magnitude of response. Other important aspects to consider include the temporal (e.g., frequency, duration, and timing) and spatial distributions of the stressor in the experimental or observational setting. For physical stressors, specific attributes of the environment after disturbance (e.g., reduce forest stand age) can be related to the response (e.g., decreased use by spotted owls) (Thomas et al., 1990).

Analyses Relating Measurement and Assessment Endpoints

Ideally, the stressor-response evaluation quantifies the relationship between the stressor and the assessment endpoint. When the assessment endpoint can be measured, this analysis is straightforward. When it cannot be measured, the relationship between the stressor and measurement endpoint is established first, then additional extrapolations, analyses, and assumptions are used to predict or infer changes in the assessment endpoint. The need for analyses relating measurement and assessment endpoints also may be identified during risk characterization, after an initial evaluation of risk.

Measurement endpoints are related to assessment endpoints using the logical structure present in the conceptual model. In some cases, quantitative methods and models are available, but often the relationship can be described only qualitatively. Because of the lack of standard methods for many of these analyses, professional

---

Extrapolations and Other Analyses Relating
Measurement and Assessment Endpoints

Extrapolation between Taxa

example: from bluegill sunfish mortality to rainbow trout mortality

Extrapolation between Responses

example: from bobwhite quail LC_{50} to bobwhite quail NOEL (no observed effect level)

Extrapolation from Laboratory to Field

example: from mouse mortality under laboratory conditions to mouse mortality in the field

Extrapolation from Field to Field

example: from reduced invertebrate community diversity in one stream to another stream

Analysis of Indirect Effects

example: relating removal of long-leaf pine to reduced populations of red-cockaded woodpecker

Analysis of Higher Organizational Levels

example: relating reduced individual fecundity to reduced population size

Analysis of Spatial and Temporal Scales

example: evaluation of the loss of a specific wetland used by migratory birds in relation to the larger scale habitat requirements of the species

Analysis of Recovery

example: relating short-term mortality to long-term depauperation

---

judgment is an essential component of the evaluation. It is important to clearly explain the rationale for any analyses and assumptions.

Extrapolations commonly used include those between species, between responses, from laboratory to field, and from field to field. Differences in responses among taxa depend on many factors, including physiology, metabolism, resource utilization, and life history strategy. The relationship between responses also depends on many factors, including the mechanism of action and internal distribution of the stressor within the organism. When extrapolating between different laboratory and field settings, important considerations include differences in the physical environment and organism behavior that will alter exposure, interactions with other stressors, and interactions with other ecological components.

In addition to these extrapolations, an evaluation of indirect effects, other levels of organization, other temporal and spatial scales, and recovery potential may be necessary. Whether these analyses are required in a particular risk assessment will depend on the assessment endpoints identified during problem formulation.

Important factors to consider when evaluating indirect effects include interspecies interactions (e.g., competition, disease), trophic-level relationships (e.g., predation), and resource utilization. Effects on higher (or lower) organizational levels depend on the severity of the effect, the number and life stage of organisms affected, the role of those organisms in the community or ecosystem, and ecological compensatory mechanisms.

The implications of adverse effects at spatial scales beyond the immediate area of concern may be evaluated by considering ecological characteristics such as community structure and energy and nutrient dynamics. In addition, information from the characterization of exposure on the stressor's spatial distribution may be useful. Extrapolations between different temporal scales (e.g., from short-term impacts to long-term effects) may consider the stressors' distribution through time (intensity, duration, and frequency) relative to ecological dynamics (e.g., seasonal cycles, life cycle patterns).

In some cases, evaluation of long-term impacts will require consideration of ecological recovery. Ecological recovery is difficult to predict and depends on the existence of a nearby source of organisms, life history and dispersal strategies of the ecological components, and the chemical-physical environmental quality following exposure to the stressor (Cairns, 1990; Poff and Ward, 1990; Kelly and Harwell, 1990). In addition, there is some evidence to suggest that the types and frequency of natural disturbances can influence the ability of communities to recover (Schlosser, 1990).

Evaluation of Causal Evidence

Another important aspect of the ecological response analysis is to evaluate the strength of the causal association between the stressor and the measurement and assessment endpoints. This information supports and complements the stressor-response assessment and is of particular importance when the stressor-response relationship is based on field observations. Although proof of causality is not a requirement for risk assessment, an evaluation of causal evidence augments the risk assessment. Many of the concepts applied in human epidemiology can be useful for evaluating causality in observational field studies. For example, Hill (1965) suggested nine evaluation criteria for causal associations. An example of ecological causality analysis was provided by Woodman and Cowling (1987), who evaluated the causal association between air pollutants and injury to forests.

3.2.3. Stressor-Response Profile

The results of the characterization of ecological effects are summarized in a stressor-response profile that describes the stressor-response relationship, any extrapolations and additional analyses conducted, and evidence of causality (e.g., field effects data).

Ideally, the stressor-response relationship will relate the magnitude, duration, frequency, and timing of exposure in the study setting to the magnitude of effects.

Hill's Criteria for Evaluating Causal Associations (Hill, 1965)

1. **Strength:** A high magnitude of effect is associated with exposure to the stressor.
2. **Consistency:** The association is repeatedly observed under different circumstances.
3. **Specificity:** The effect is diagnostic of a stressor.
4. **Temporality:** The stressor precedes the effect in time.
5. **Presence of a biological gradient:** A positive correlation between the stressor and response.
6. **A plausible mechanism of action.**
7. **Coherence:** The hypothesis does not conflict with knowledge of natural history and biology.
8. **Experimental evidence.**
9. **Analogy:** Similar stressors cause similar responses.

Not all of these criteria must be satisfied, but each incrementally reinforces the argument for causality. Negative evidence does not rule out a causal association but may indicate incomplete knowledge of the relationship (Rothman, 1986).

For practical reasons, the results of stressor-response curves are often summarized as one reference point, for instance, a 48-h LC_{50}. Although useful, such values provide no information about the slope or shape of the stressor-response curve. When the entire curve is used, or when points on the curve are identified, the difference in magnitude of effect at different exposure levels can be reflected in risk characterization.

It is important to clearly describe and quantitatively estimate the assumptions and uncertainties involved in the evaluation, where possible. Examples include natural variability in ecological characteristics and responses and uncertainties in the test system and extrapolations. The description and analysis of uncertainty in characterization of ecological effects are combined with uncertainty analyses for the other ecological risk assessment elements during risk characterization.

Additional Issues Related to the Analysis Phase

- Quantifying cumulative impacts and stress-response relationships for multiple stressors.
- Improving the prediction of ecosystem recovery.
- Improving the quantification of indirect effects.
- Describing stressor-response relationships for physical perturbations.
- Distinguishing ecosystem changes due to natural processes from those caused by man.

4. RISK CHARACTERIZATION

Risk characterization (Figure 4) is the final phase of risk assessment. During this phase, the likelihood of adverse effects occurring as a result of exposure to a stressor are evaluated. Risk characterization contains two major steps: risk estimation and risk description. The stressor-response profile and the exposure profile from the analysis phase serve as input to risk estimation. The uncertainties identified during all phases of the risk assessment also are analyzed and summarized. The estimated risks are discussed by considering the types and magnitude of effects anticipated, the spatial and temporal extent of the effects, and recovery potential. Supporting information in the form of a weight-of-evidence discussion also is presented during this step. The results of the risk assessment, including the relevance of the identified risks to the original goals of the risk assessment, then are discussed with the risk manager.

4.1. RISK ESTIMATION

Risk estimation consists of comparing the exposure and stressor-response profiles as well as estimating and summarizing the associated uncertainties.

4.1.1. Integration of Stressor-Response and Exposure Profiles

Three general approaches are discussed to illustrate the integration of the stressor-response and exposure profiles: (1) comparing single effect and exposure values; (2) comparing distributions of effects and exposure; and (3) conducting simulation modeling. Because these are areas of active research, particularly in the assessment of community- and landscape-level perturbations, additional integration approaches are likely to be available in the future. The final choice as to which approach will be selected depends on the original purpose of the assessment as well as time and data constraints.

Comparing Single Effect and Exposure Values

Many risk assessments compare single effect values with predicted or measured levels of the stressor. The effect values from the stressor-response profile may be used as is, or more commonly, uncertainty or safety factors may be used to adjust the value. The ratio or quotient of the exposure value to the effect value provides the risk estimate. If the quotient is one or more, an adverse effect is considered likely to occur. This approach, known as the quotient method (Barnthouse et al., 1986), has been used extensively to evaluate the risks of chemical stressors (Nabholz 1991; Urban and Cook, 1986). Although the quotient method is commonly used and accepted, it is the least probabilistic of the approaches described here. Also, correct usage of the quotient method is highly dependent on professional judgment, particularly in instances when the quotient approaches one. Greater insight into the magnitude

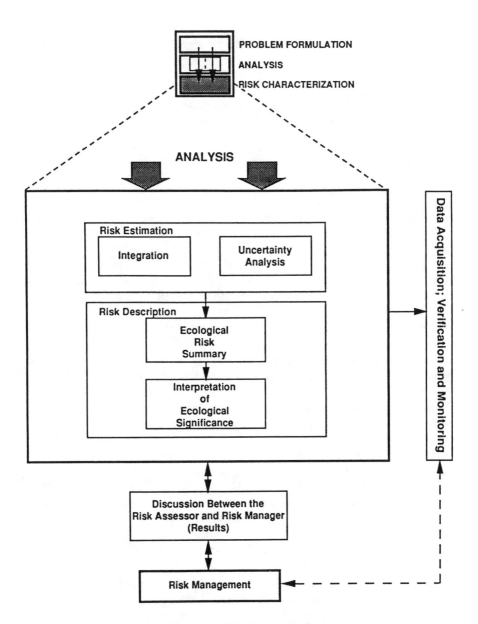

Figure 4. Risk characterization

of the effects expected at various levels of exposure can be obtained by evaluating the full stressor-response curve instead of a single point and by considering the frequency, timing, and duration of the exposure.

Comparing Distributions of Effects and Exposure

This approach uses distributions of effects and exposure (as opposed to single values) and thus makes probabilistic risk estimates easier to develop. Risk is quantified by the degree of overlap between the two distributions; the more overlap, the greater the risk. An example of this approach, analysis of extrapolation error, is given in Barnthouse et al. (1986). To construct valid distributions, it is important that sufficient data amenable to statistical treatment are available.

Conducting Simulation Modeling

Simulation models that can integrate both the stressor-response profile and exposure profile are useful for obtaining probabilistic estimates of risk. Two categories of simulation models are used for ecological risk assessment: single-species population models are used to predict direct effects on a single population of concern using measurement endpoints at the individual level, while multispecies models include aquatic food web models and terrestrial plant succession models and are useful for evaluating both direct and indirect effects.

When selecting a model, it is important to determine the appropriateness of the model for a particular application. For example, if indirect effects are of concern, a model of community-level interactions will be needed. Direct effects to a particular population of concern may be better addressed with population models. The validation status and use history of a model also are important considerations in model selection. Although simulation models are not commonly used for ecological risk assessment at the present time, this is an area of active research, and the use of simulation models is likely to increase.

In addition to providing estimates of risks, simulation models also can be useful in discussing the results of the risk characterization to the risk manager. This dialogue is particularly effective when the relationship between risks to certain measurement endpoints and the assessment endpoint are not readily apparent (e.g., certain indirect effects and large-scale ecosystem-level disturbances).

4.1.2. Uncertainty

The uncertainty analysis identifies, and, to the extent possible, quantifies the uncertainty in problem formulation, analysis, and risk characterization. The uncertainties from each of these phases of the process are carried through as part of the total uncertainty in the risk assessment. The output from the uncertainty analysis is an evaluation of the impact of the uncertainties on the overall assessment and, when feasible, a description of the ways in which uncertainty could be reduced.

A complete discussion of uncertainty is beyond the scope of this report, and the reader is referred to the works of Finkel (1990), Holling (1978), and Suter (1990b). However, a brief discussion of the major sources of uncertainty in ecological risk assessment is appropriate. For illustrative purposes, four major areas of uncertainty are presented below. These are not discrete categories, and overlap does exist among them. Any specific risk assessment may have uncertainties in one or all of these categories.

Conceptual Model Formulation

As noted earlier, the conceptual model is the product of the problem formulation phase, which, in turn, provides the foundation for the analysis phase and the development of the exposure and stressor-response profiles. If incorrect assumptions are made during conceptual model development regarding the potential effects of a stressor, the environments impacted, or the species residing within those systems, then the final risk assessment will be flawed. These types of uncertainties are perhaps the most difficult to identify, quantify, and reduce.

Information and Data

Another important contributor of uncertainty is the incompleteness of the data or information upon which the risk assessment is based. In some instances, the risk assessment may be halted temporarily until additional information is obtained. In other cases, certain basic information such as life history data may be unobtainable with the resources available to the risk assessment. In yet other cases, fundamental understanding of some natural processes with an ecosystem may be lacking. In instances where additional information cannot be obtained, the role of professional judgment and judicial use of assumptions are critical for the completion of the assessment.

Stochasticity (Natural Variability)

Natural variability is a basic characteristic of stressors and ecological components as well as the factors that influence their distribution (e.g., weather patterns, nutrient availability). As noted by Suter (1990b), of all the contributions to uncertainty, stochasticity is the only one that can be acknowledged and described but not reduced. Natural variability is amenable to quantitative analyses, including Monte Carlo simulation and statistical uncertainty analysis (O'Neill and Gardner, 1979; O'Neill et al., 1982).

Error

Errors can be introduced through experimental design or the procedures used for measurement and sampling. Such errors can be reduced by adherence to good laboratory practices and adherence to established experimental protocols. Errors also can be introduced during simulation model development. Uncertainty in the development and use of models can be reduced through sensitivity analyses, comparison with similar models, and field validation.

In summary, uncertainty analyses provide the risk manager with an insight into the strengths and weaknesses of an assessment. The uncertainty analysis also can serve as a basis for making rational decisions regarding alternative actions as well as for obtaining additional information to reduce uncertainty in the risk estimates.

4.2. RISK DESCRIPTION

Risk description has two primary elements. The first is the ecological risk summary, which summarizes the results of the risk estimation and uncertainty analysis and assesses confidence in the risk estimates through a discussion of the weight of evidence. The second element is interpretation of ecological significance, which describes the magnitude of the identified risks to the assessment endpoint.

4.2.1.   Ecological Risk Summary

The ecological risk summary summarizes the results of the risk estimation and discusses the uncertainties associated with problem formulation, analysis, and risk characterization. Next, the confidence in the risk estimates is expressed through a weight-of-evidence discussion. The ecological risk summary may conclude with an identification of additional analyses or data that might reduce the uncertainty in the risk estimates. These three aspects of the ecological risk summary are discussed in the following sections.

Summary of Risk Estimation and Uncertainty

Ideally, the conclusions of the risk estimation are described as some type of quantitative statement (e.g., there is a 20% change of 50% mortality). However, in most instances, likelihood is expressed in a qualitative statement (e.g., there is a high likelihood of mortality occurring). The uncertainties identified during the risk assessment are summarized either quantitatively or qualitatively, and the relative contribution of the various uncertainties to the risk estimates are discussed whenever possible.

Weight of Evidence

The weight-of-evidence discussion provides the risk manager with insight about the confidence of the conclusions reached in the risk assessment by comparing the positive and negative aspects of the data, including uncertainties identified throughout the process. The considerations listed below are useful in a weight-of-evidence discussion:

- The sufficiency and quality of the data. A risk assessment conducted with studies that completely characterize both the effects and exposure of the stressor has more credibility and support than an assessment that contains data gaps. It is important to state if the data at hand were sufficient to support the findings of the assessment. In addition, data validity (e.g., adherence to protocols, having sufficient replications) is an important facet of the weight-of-evidence analysis.
- Corroborative information. Here the assessor incorporates supplementary information that is relevant to the conclusions reached in the assessment. Examples include reported incidences of effects elicited by the stressor (or similar stressor) and studies demonstrating agreement between model predictions and observed effects.
- Evidence of causality. The degree of correlation between the presence of a stressor and some adverse effect is an important consideration for many ecological risk assessments. This correlation is particularly true when an assessor is attempting to establish a link between certain observed field effects and the cause of those effects. Further discussions of the evaluation of causal relationships may be found in the section on characterization of ecological effects (Section 3.2.2.).

Identification of Additional Analyses

The need for certain analyses may not be identified until after the risk estimation step. For example, the need to analyze the risks to a fish population (an assessment endpoint) due to an indirect effect such as zooplankton mortality (a measurement endpoint) may not be established until after the risk to zooplankton has been characterized. In such cases, another iteration through analysis or even problem formulation may be necessary.

4.2.2. Interpretation of Ecological Significance

The interpretation of ecological significance places risk estimates in the context of the types and extent of anticipated effects. It provides a critical link between the estimation of risks and the communication of assessment results. The interpretation step relies on professional judgment and may emphasize different aspects depending on the assessment. Several aspects of ecological significance that may be considered include the nature and magnitude of the effects, the spatial and temporal patterns of the effects, and the potential for recovery once a stressor is removed.

Nature and Magnitude of the Effects

The relative significance of different effects may require further interpretation, especially when changes in several assessment or measurement endpoints are observed or predicted. For example, if a risk assessment is concerned with the effects of stressors on several ecosystems in an area (such as a forest, stream, and wetland), it is important to discuss the types of effects associated with each ecosystem and where the greatest impact is likely to occur.

The magnitude of an effect will depend on its ecological context. For example, a reduction in the reproductive rate may have little effect on a population that reproduces rapidly, but it may dramatically reduce the numbers of a population that reproduces slowly. Population-dependent and -independent factors in the ecosystem also may influence the expression of the effect.

Finally, it is important to consider the effects in the context of both magnitude and the likelihood of the effect occurring. In some cases, the likelihood of exposure to a stressor may be low, but the effect resulting from the exposure would be devastating. For example, large oil spills may not be common, but they can cause severe and extensive effects in ecologically sensitive areas.

Spatial and Temporal Patterns of the Effects

The spatial and temporal distributions of the effect provide another perspective important to interpreting ecological significance. The extent of the area where the stressor is likely to occur is a primary consideration when evaluating the spatial pattern of effects. Clearly, a stressor distributed over a larger area has a greater potential to affect more organisms than one confined to a small area. However, a stressor that adversely affects small areas can have devastating effects if those areas provide critical resources for certain species. In addition, adverse effects to a resource that is small in scale (e.g., acidic bogs) may have a small spatial effect but may represent a significant degradation of the resource because of its overall scarcity.

The duration of any effect is dependent on the persistence of the stressor as well as how often the stressor is likely to occur in the environment. It is important to remember that even short-term effects can be devastating if such exposure occurs during critical life stages of organisms.

Recovery Potential

A discussion of the recovery potential may be an integral part of risk description, although the need for such an evaluation will depend on the objective of the assessment and the assessment endpoints. An evaluation of the recovery potential may require additional analyses, as discussed in Section 3.1., and will depend on the nature, duration, and extent of the stressor.

Depending on the assessment objectives, all of the above factors may be used to place the risks into the broader ecological context. This discussion may consider the ramifications of the effects on other ecological components that were not specifically addressed in the assessment. For example, an assessment that focused on the decline of alligator populations may include a discussion of the broader ecological role of the alligator, such as the construction of wallows that act as water reservoirs during droughts. In this way, the potential effects on the community that depends on the alligator wallows can be brought out in risk characterization.

4.3. DISCUSSION BETWEEN THE RISK ASSESSOR AND RISK MANAGER (RESULTS)

Risk characterization concludes the risk assessment process and provides the basis for discussions between the risk assessor and risk manager that pave the way for regulatory decision-making. The purpose of these discussions is to ensure that the results of the risk assessment are clearly and fully presented and to provide an opportunity for the risk manager to ask for any necessary clarification. Proper presentation of the risk assessment is essential to reduce the chance of over- or under-interpretation of the results. To permit the risk manager to evaluate the full range of possibilities contained in the risk assessment, it is important that the risk assessor provide the following types of information:

- the goal of the risk assessment;
- the connection between the measurement and assessment endpoints;
- the magnitude and extent of the effect, including spatial and temporal considerations and, if possible, recovery potential;
- the assumptions used and the uncertainties encountered during the risk assessment;
- a summary profile of the degrees of risk as well as a weight-of-evidence analysis; and
- the incremental risk from stressors other than those already under consideration (if possible).

The results of the risk assessment serve as input to the risk management process, where they are used along with other inputs defined in EPA statutes, such as social and economic concerns, to evaluate risk management options.

In addition, based on the discussions between the risk assessor and risk manager, follow-up activities to the risk assessment may be identified, including monitoring, studies to verify the predictions of the risk assessment, or the collection of additional data to reduce the uncertainties in the risk assessment. While a detailed discussion of the risk management process is beyond the scope of this report, consideration of the basic principles of ecological risk assessment described here will contribute to a final product that is both credible and germane to the needs of the risk manager.

Additional Issues Related to the Risk Characterization Phase

- Predicting the time required for an ecological component to recover from a stressor.
- Combining chemical and nonchemical stressors in risk characterization.
- Incorporating critical effect levels into risk characterization.
- Better quantification of uncertainty.
- Developing alternative techniques for expressing uncertainty in risk characterization.

5. KEY TERMS

Assessment endpoint — An explicit expression of the environmental value that is to be protected.

Characterization of ecological effects — A portion of the analysis phase of ecological risk assessment that evaluates the ability of a stressor to cause adverse effects under a particular set of circumstances.

Characterization of exposure — A portion of the analysis phase of ecological risk assessment that evaluates the interaction of the stressor with one or more ecological components. Exposure can be expressed as co-occurrence, or contact depending on the stressor and ecological component involved.

Community — An assemblage of populations of different species within a specified location in space and time.

Conceptual model — The conceptual model describes a series of working hypotheses of how the stressor might affect ecological components. The conceptual model also describes the ecosystem potentially at risk, the relationship between measurement and assessment endpoints, and exposure scenarios.

Direct effect — An effect where the stressor acts on the ecological component of interest itself, not through effects on other components of the ecosystem (compare with definition for indirect effect).

Ecological component — Any part of an ecological system, including individuals, populations, communities, and the ecosystem itself.

Ecological risk assessment — The process that evaluates the likelihood that adverse ecological effects may occur or are occurring as a result of exposure to more or more stressors.

Ecosystem — The biotic community and abiotic environment within a specified location in space and time.

Exposure — Co-occurrence of or contact between a stressor and an ecological component.

Exposure profile — The product of characterization of exposure in the analysis phase of ecological risk assessment. The exposure profile summarizes the magnitude and spatial and temporal patterns of exposure for the scenarios described in the conceptual model.

Exposure scenario — A set of assumptions concerning how an exposure may take place, including assumptions about the exposure setting, stressor characteristics, and activities that may lead to exposure.

Indirect effect — An effect where the stressor acts on supporting components of the ecosystem, which in turn have an effect on the ecological component of interest.

Measurement endpoint — A measurable ecological characteristic that is related to the valued characteristic chosen as the assessment endpoint. Measurement endpoints are often expressed as the statistical or arithmetic summaries of the observations that comprise the measurement.

Median lethal concentration (LC_{50}) — A statistically or graphically estimated concentration that is expected to be lethal to 50% of a group of organisms under specified conditions (ASTM, 1990).

No observed effect level (NOEL) — The highest level of a stressor evaluated in a test that does not cause statistically significant differences from the controls.

Population — An aggregate of individuals of a species within a specified location in space and time.

Recovery — The partial or full return of a population or community to a condition that existed before the introduction of the stressor.

Risk characterization — A phase of ecological risk assessment that integrates the results of the exposure and ecological effects analyses to evaluate the likelihood of adverse ecological effects associated with exposure to a stressor. The ecological significance of the adverse effects is discussed, including consideration of the types and magnitudes of the effects, their spatial and temporal patterns, and the likelihood of recovery.

Stressor — Any physical, chemical, or biological entity that can induce an adverse response.

Stressor-response profile — The product of characterization of ecological effects in the analysis phase of ecological risk assessment. The stressor-response profile summarizes the data on the effects of a stressor and the relationship of the data to the assessment endpoint.

Trophic levels — A functional classification of taxa within a community that is based on feeding relationships (e.g., aquatic and terrestrial green plants comprise the first trophic level and herbivores comprise the second).

Xenobiotic — A chemical or other stressor that does not occur naturally in the environment. Xenobiotics occur as a result of anthropogenic activities such as the application of pesticides and the discharge of industrial chemicals to air, land, or water.

REFERENCES

American Society for Testing and Materials. (1990). Standard terminology relating to biological effects and environmental fate. E943–90. In: *ASTM; 1990 Annual Book of ASTM Standards,* Section 11, Water and Environmental Technology, ASTM, Philadelphia, PA.

ASTM See American Society for Testing and Materials.

Auer, C. M.; Nabholz, J. V.; Baetcke, K. P. (1990). Mode of action and the assessment of chemical hazards in the presence of limiting data: use of structure-activity relationships (SAR) under TSCA, Section 5. *Environmental Health Perspectives* (87):183–197.

Barnthouse, L. W.; Suter, G. W., II; Bartell, S. M.; Beauchamp, J. J.; Gardner, R. H.; Linder, E.; O'Neill, R. V.; Rosen, A. E. (1986). *User's Manual for Ecological Risk Assessment.* Publication No. 2679, ORNL-6251. Environmental Sciences Division, Oak Ridge National Laboratory, Oak Ridge, TN.

Cairns, J., Jr. (1990). Lack of theoretical basis for predicting rate and pathways of recovery. In: Yount, J. D.; Niemi, G. J., Eds. Recovery of Lotic Communities and Ecosystems Following Disturbance: Theory and Application. *Environmental Management* 14(5):517–526.

Clements, R. G.; Johnson, D. W.; Lipnick, R. L.; Nabholz, J. V.; Newsome, L. D. (1988). Estimating toxicity of industrial chemicals to aquatic organisms using structure activity relationships. EPA-560-6-88-001. U.S. Environmental Protection Agency, Washington, D.C. (available from NTIS, Springfield, VA, PB89–117592).

Finkel, A. M. (1990). *Confronting Uncertainty in Risk Management: A Guide for Decision-Makers.* Center for Risk Management, Resources for the Future, Washington, D.C.

Hill, A. B. (1965). The environment and disease: association or causation? *Proceedings of the Royal Society of Medicine.* 58:295–300.

Holling, C. S. (1978). *Adaptive Environmental Assessment and Management.* John Wiley and Sons, New York, NY.

Kelly, J. R.; Harwell, M. A. (1990). Indicators of ecosystem recovery. In: Yount, J. D.; Niemi, G. J., Eds. Recovery of Lotic Communities and Ecosystems Following Disturbance: Theory and Application. *Environmental Management* 14(5):527–546.

Kendall, R. J. (1991). Ecological risk assessment for terrestrial wildlife exposed to agrochemicals: a state-of-the-art review and recommendations for the future. Presented at the Ecological Risk Assessment Workshop sponsored by the National Academy of Sciences Committee on Risk Assessment Methodology, 26 February–1 March 1991.

McKim, J. B.; Bradbury, S. P.; Niemi, G. J. (1987). Fish acute toxicity syndromes and their use in the QSAR approach to hazard assessment. *Environmental Health Perspectives* 71:171–186.

Nabholz, J. V. (1991). Environmental hazard and risk assessment under the United States Toxic Substances Control Act. *Science of the Total Environment* 109/110:649–665.

National Research Council. (1983). *Risk Assessment in the Federal Government: Managing the Process*. National Research Council, National Academy Press, Washington, D.C.

National Research Council. (1986). *Ecological Knowledge and Environmental Problem-Solving: Concepts and Case Studies*. National Research Council, National Academy Press, Washington, D.C.

NRC. See National Research Council.

O'Neill, R. V. (1979). Natural variability as a source of error in model predictions. In: *Systems Analysis of Ecosystems*. Innis, G. S. and O'Neill, R. V., Eds. International Cooperative Publishing House, Burtonsville, Maryland. pp. 23–32.

O'Neill, R. V.; Gardner, R. H. (1979). Sources of uncertainty in ecological models. In: *Methodology in Systems Modeling and Simulation*. Zeigler, B. P., Elzas, M. S., Klir, G. J., and Orens, T. I., Eds. North Holland Publishing Company. pp. 447–463.

Poff, N. L.; Ward, J. V. (1990). Physical habitat template of lotic systems: recovery in the context of historical pattern of spatiotemporal heterogeneity. In: Young, J. D.; Niemi, G. J., Eds. Recovery of Lotic Communities and Ecosystems Following Disturbance: Theory and Application. *Environmental Management* 14(5):629–646.

Rothman, K. J. (1986). *Modern Epidemiology*. 1st ed. Little, Brown and Company, Boston, MA.

Schlosser, I. J. (1990). Environmental variation, life history attributes, and community structure in stream fishes: implications for environmental management and assessment. *Environmental Management* 14(5):621–628.

SETAC. See Society of Environmental Toxicology and Chemistry.

Society of Environmental Toxicology and Chemistry. (1987). *Research Priorities in Environmental Risk Assessment*. Report of a workshop held in Breckenridge, CO, August 16–21, 1987. Society of Environmental Toxicology and Chemistry, Washington, D.C.

Suter, G. W., II. (1989). Ecological endpoints. In: *U.S. EPA. Ecological Assessments of Hazardous Waste Sites:* A Field and laboratory reference document. Warren-Hicks, W.; Parkhurst, B. R.; Baker, S. S., Jr., Eds. EPA 600/3–89/013. March 1989.

Suter, G. W., II. (1990a). Endpoints for regional ecological risk assessments. *Environmental Management* 14(1):19–23.

Suter, G. W., II. (1990b). Uncertainty in environmental risk assessment. In: von Furstenberg, G. M., Ed. *Acting Under Uncertainty: Multidisciplinary Conceptions*. Kluwer Academic Publishers, Boston, MA. pp. 203–230.

Thomas, J. W.; Forsman, E. D.; Lint, J. B.; Meslow, E. C.; Noon, B. R.; J. Verner. (1990). *A Conservation Strategy for the Spotted Owl*. Interagency Scientific Committee to Address the Conservation of the Northern Spotted Owl. 1990–791/20026. U.S. Government Printing Office, Washington, D.C.

Urban, D. J.; Cook, N. J. (1986). *Standard Evaluation Procedure for Ecological Risk Assessment*. EPA/540/09–86/167, Hazard Evaluation Division, Office of Pesticide Programs, U.S. Environmental Protection Agency, Washington, D.C.

U.S. Department of the Interior. (1987). *Injury to Fish and Wildlife Species.* Type B technical information document. CERCLA 301 Project, Washington, D.C.

U.S. EPA. See U.S. Environmental Protection Agency.

U.S. Environmental Protection Agency. (1979). Toxic Substances Control Act. Discussion of premanufacture testing policies and technical issues; Request for comment. 44 *Federal Register* 16240–16292.

U.S. Environmental Protection Agency. (1990a). *Environmental Monitoring and Assessment Program. Ecological Indicators.* EPA/600/3–90/060, Office of Research and Development, Washington, D.C.

U.S. Environmental Protection Agency. (1990b). *Reducing Risk: Setting Priorities and Strategies for Environmental Protection.* Science Advisory Board SAB-EC-90–021, Washington, D.C.

U.S. Environmental Protection Agency. (1991). *Summary Report on Issues in Ecological Risk Assessment.* EPA/625/3–91/018, Risk Assessment Forum, Washington, D.C.

U.S. Environmental Protection Agency. (in press-a). *Peer Review Workshop Report on a Framework for Ecological Risk Assessment.* EPA/625/3–91/022, Risk Assessment Forum, Washington, D.C.

U.S. Environmental Protection Agency. (in press-b). *Ecological Risk Assessment Guidelines Strategic Planning Workshop.* EPA/630/R-92/002, Risk Assessment Forum, Washington, D.C.

Woodman, J. N.; Cowling, E. B. (1987). Airborne chemicals and forest health. *Environmental Science and Technology* 21(2):120–126.

INDEX

A

Acartia sp., 52, 54, 78
Acetylcholine, organophosphates, 97–101
Acetylcholinesterase, 100, 101
Acetylcholinesterase inhibitors, 106, 201
Achromobacter sp., toxicant biodegradation, 172, 173
Acinetobacter calcoaceticus, toxicant biodegradation, 172
Aconitase, 92, 93, 102
Acute toxicity tests, 27
  aquatic vertebrates and invertebrates, 51–55
  Daphnia, 46–48
Adaptation, 147
Additive model, 119–120
AEP, 189
Aeromonas sp., toxicant biodegradation, 173
Aflatoxin, toxicity, 125, 127
Aging, 123
  ozone, effect, 146
Air pollution, see Atmospheric pollutants; Smog
Alcaligenes denitrificans, 233–234
Alcaligenes eutrophus, toxicant biodegradaion, 172, 173
Algae, 52–54, 78–79
  growth toxicity test, 48–51
  halometabolites, 190
Allergic response, mechanism, 92
Aluminum, metal shift, 95
American kestrel, 58
American Society for Testing and Materials, see ASTM
American Type Culture Collection, 32
Amino acids, ozone, 147
4-Amino-3,5-dichlorobenzoic acid, microbial degradation, 173
2-Aminoethylphosphonic acid, 189

δ-Aminolevulinate dehydratase, inhibition, 91
Amnicola limosa, 52, 53, 77–78
Amphibians, frog embryo, teratogenesis assay, 61–63
Amphipods, 52–54, 75–76
Anabaena flos-aquae, 79
Analysis of variance (ANOVA), 37–39, 41, 227–228
Analytical chemistry, 1, 260–261
Anaphylaxis, 92
Anas platyrhynchos, 57–59, 79
Anas rubripes, 58
Anemia, 203
Aniline, microbial degradation, 172
Animals, see also Birds; Fish; Invertebrates; Mammals; Vertebrates
  fluoride, 154–155
  indicators of ecological impact, 202, 203, 206
  nitrogen oxides, 142–143
  ozone, 146–147
  sulfur dioxide, 139
  toxicity, 123–131
ANOVA, 37–39, 41, 227–228
Antagonism, toxicants, 118, 130
Antergan, action, 92
Anthracene, microbial degradation, 172
Antibodies, as biomarkers, 202
Antihistamines, action, 92
Antioxidants, 128
  biotransformation, 167
  cancer, 131
  vitamin C, 129
  vitamin E, 128, 130
Aplysia, OPA anhydrase, 186
Apoenzymes, 91
Aquatic toxicology
  algae, 48–54, 78–79, 190
  chemical defense, 190

Daphnia, 32, 33, 46–48, 75, 104–106, 110–111
FIFRA microcosm, 68–70
halometabolites, 190
invertebrates, 51–55, 75–78, 186
microcosms, 64
mixed flask culture, 67–68
multispecies toxicity tests, 63, 65–67
OPA anhydrolases, 181, 182
standardized aquatic microcosm, 65–67
testing, 29, 31, 33–34
vertebrates, 51–55, 72–75
Aromatic compounds, microbial degradation,
    169, 171–175
Arthrobacter sp., toxicant biodegradation, 172,
    173
Ascorbate, plants, 87
Ascorbic acid, 128–129
Aspergillus sp., toxicant biodegradation, 172, 173
Assessment endpoint, 258, 262, 290–292, 300–
    301
Asthmatics, 140, 143
ASTM, 2, 24–27, 46, 49, 57, 59–60
ASTM-PROBIT, 36, 37
Atmospheric pollutants
    carbon monoxide, 90, 148–151
    excretion, 89
    exposure routes, 31
    fluorides, 95, 127, 152–156
    metabolism, 89
    nitrogen oxides, 118, 140–143
    ozone, 87, 90, 95, 116, 118, 122, 128, 143–148
    plants, 85–87, 90, 122
    sulfur oxides, 135–140
    vertebrates, 87–89
Avrainvillea longicalulis, 190
Azotobacter sp., toxicant biodegradation, 172

B

Bacillus sp., toxicant biodegradation, 172
Bacillus stearothermophilus, 183
Bacteria
    genetics, 179–180
    OPA anhydrolases, 182, 190
    toxicant degradation, 167–175
    isolation and engineering, 179–180
Baetis sp., 52, 53, 76–77
Bay shrimp, 52, 54
Beijerinckia sp., toxicant biodegradation, 172
Benzene, microbial degradation, 172
Benzoic acid, microbial degradation, 172
Beryllium, 92

Bioaccumulation, 9, 200
Biodegradation, 9, 167–175, 200
    genetics, 179–180
    isolation and engineering, organism, 177–178
    OPA anhydrolases, 180–190
Biological integrity, index, 209–210
Biological stressors, 254, 281
Biomarkers, 200–204
Biometrics, 1
Biomonitoring, 197–204
Bioremediation, 175–190
    microbial degradation, 167–175
Biosensors, 200
Biotransformation, 9, 161–167, 200, 203, 261
Biphenyl, microbial degradation, 172
Birds, 52–54, 79–80
    biomarkers, 202
    pesticide toxicity, dietary lipids, 127
    as sentinel organisms, 206–207
    toxicity tests, 55–57, 60
Black duck, 58
Blue crab, 52, 54
Bluegill, 52–53, 74, 186–187, 204
Blue-green bacteria, 49, 52–54, 79
    halometabolites, 190
Bobwhite, 57–59, 79, 207
Body fat, pollutant toxicity and, 127
Brevibacterium sp., toxicant biodegradation, 172
Bromoacetate, 99, 102
Bromoacetic acid, 99, 102, 103
Brook trout, 52–53, 73
Bufo sp., 52, 53
Butylcholinesterase, 99

C

Cadmium, 87
    antagonism of zinc and selenium, 118, 130
    human body, effect, 90–92
    picolinic acid complexes, 130
Calcium, fluoride toxicity, effect on, 155
Calcium binding protein, 127
Calcium metabolism, vitamin D, 128
Caligoida, 78
Callinectes sapidus, 52, 54
Canadian soldier, 77
Cancer, see Carcinogenesis
Capitella capitata, 52, 54
Carassius auratus, 52, 53, 73–74
Carbohydrate metabolism, ozone and, 145
Carbohydrates, pollutant toxicity and, 125–126
Carbon disulfide, 54, 95

Carbon monoxide, 90, 148–151
Carbon tetrachloride, 92–93, 125, 167
Carboxyhemoglobin, 90, 150
Carcinogenesis
  antioxidants, 131
  chelation, 94
  selenium, 131
  vitamin A, 127
  vitamin C, 129
  vitamin E, 128
β-Carotene, 128
Carotenoids, 128
Carrier molecule, 88
Catechol, microbial degradation, 173
Catfish, 52–53, 74, 186–187
Cellular structure, disruption by toxicant, 90
Central nervous system, organophosphates, 99
Cephalopods, OPA anhydrases, 181
CERCLA (1980), 4
Ceriodaphnia dubia, 48, 49, 75
Channel catfish, 52–53, 74, 186–187
Chaotic dynamics, 221–227, 241
Chelation, 93–94
Chemical defense, marine animals, 190
Chemical stressors, 254, 281, 289, 292, 295–296
Chemical warfare agents, organophosphates, 96–99
Chemostat, 178
Chironomus sp., 52, 53, 77
Chlamydomonas reinhardi, 52, 54, 78–79
Chlamydomonas sp., toxicant biodegradation, 172
Chloramphenicol, 190
Chlorinated hydrocarbon pesticides, 127
Chloroacetate, 99, 102
Chloroacetic acid, 99, 102, 103
Chlorobenzene, microbial degradation, 173
Chlorobenzoic acid, microbial degradation, 172
4-Chlorocatechol, microbial degradation, 173
4-Chloro-3,5-dinitrobenzoic acid, microbial degradation, 172
Chloroform, sex variation in toxicity, 124, 125
Chlorophenol, microbial degradation, 173
Chlorotetracycline, 190
Chlorotoluene, microbial degradation, 173
Cholecalciferol, 128
Chronic toxicity tests, 27, 37–40
Chronomus sp., 52, 53, 77
Citharichthys stigmaeus, 52, 53
Clam, OPA anhydrase, 186
Clean Air and Clean Water standards, 2
Clean Water Act, (1972), 3
CLOGP3 computer program, 108

Clupea harengus, 52, 53
Coal, sulfur oxides, 135
Coenzymes, 91
Cofactors, 91–92
Coho salmon, 52–53, 72–73
Colinus virginianus, 57–59, 79
Combarus sp., 52, 53, 76
Combustion, carbon monoxide, 149
Community structure, 12, 199, 208–210
Comprehensive Environmental Response
    Compensation and Liability Act (1980), 4
Computer software
  nonmetric clustering, 244
  probit data analysis, 36–37
  QSAR, 107–108, 110
Concentration, of toxicant, 115
Conceptual statistical analysis, 42, 231–232
Coniophora pueana, toxicant biodegradation, 173
Conjugation, biotransformation, 162, 164
Contaminated sites, toxicity testing, 205
Continuous endpoints, QSAR, 104
Continuous exposure, 116
Continuous-flow test, 29
Copepods, 52, 54, 78
Copper, 94, 95
Coturnix japonica, 57–59
Crab, 52, 54
Crangon sp., 52, 54
Crassostrea sp., 52, 54
Crayfish, 52–54, 76
Crowned guinea fowl, 58
Cunninghamella elegans, toxicant biodegradation, 172–175
Cuttlefish, OPA anhydrase, 186
Cyclopoida, 78
Cymatogaster aggregata, 52, 53
Cyprinodon variegatus, 52, 53
Cytochrome P-450, 126, 130, 164–165

D

Daphnia, 75
  EC_{50} by QSAR, 104–106
  isopropylamine toxicity, QSAR, 110–111
Daphnia magna, 32, 33
  acute toxicity test, 46–48
  test species, 75
Data analysis, 34–42
  ANOVA, 37–39, 41, 227–228
  conceptual statistical analysis, 42, 231–232
  hypothesis testing, 39–40, 240
  intervals of nonsignificant difference, 41

logit methods, 36–37
multivariate techniques, 41–42, 228–244
nonmetric clustering, 42, 231–237, 240, 243–
    244
normalized ecosystem strain, 228–229
probit methods, 35–37
state space analysis, 230–231
DDT, 125–127, 161
Deakylation, 163
Deamination, 163
Degradation, microbial, see Microbial degrada-
    tion
Deletion, degradative gene, 179–180
Dental fluorosis, 154–155, 161
Dermal exposure, 31
Desulfuration, 164
Detoxification, 125–127
    biotransformation, 166–167, 180–190, 203
    OPA anhydrolases, 180–190
DFP, 180, 181, 184–190
Diagnostic variables, 230–231
Diaulula sandiegensis, 190
Diazinon, 98
Dichapetalum toxicarium, 190
Dichlorobenzoic acid, microbial degradation, 172
3,5-Dichlorocatechol, microbial degradation, 173
Dieldrin, vitamin A, 127
Diet, pollutant toxicity and, 124–131
Diisopropylfluorophosphate, 98
Diphenylhydramine, action, 92
Discriminate analysis, 104
Diseases, toxicity, 123
Dissolved oxygen, 116
Diversity index, 12
DNA, alterations, 10
Dose-response curve, 18–24
Dove, 58
Drake, 77
Duck, 52–54, 79
Dugesia tigrina, 52, 53, 78
DULUTH-TOX, 36, 37
Dun, 77
Dunnett's procedure, 38
Dynamics, ecosystems, 237–243

E

EC_{50}, 21
    by QSAR, 104–106
Ecological response analyses, 262–263, 300–302
Ecological risk assessment, 251–313
Ecological risk summary, 308–309

Ecosystems, 1, 12–15, 296
    ANOVA, 227–228
    biomonitoring, 197–204
    chaotic dynamics, 221–227
    field studies, 28, 34
    indicators of toxicant impact, 204–211
    multivariate techniques, 228–244
    non-equilibrium dynamics, 237–243
    resource competition model, 211–219
    risk assessment, 1–2, 251–268
        EPA framework, 271–313
    stability, 237–243
Ecosystem strain, 41
Efficacy, biomonitoring, 199
Effluents, toxicity testing, 204–205
Electrophorus, OPA anhydrase, 187
Endoplasmic reticulum, 126, 164
Endpoint
    assessment endpoint, 258, 262, 290–292, 300–
        301
    biomonitoring, 201–202
    calculation, 35–36
    continuous endpoint, 104
    measurement endpoint, 258, 262, 290–291, 300
    population parameters, 11, 199, 207–208
    selection, 258, 290–291, 293
English sole, 52, 53
Enolase, inhibition, 91
Environmental Protection Agency, see EPA
Enzyme inhibition
    acetylcholinesterase inhibitors, 106, 201
    enolase, 91
    fluorides, 156
    monohaloacetic acids, 102
    organophosphates, 97–99
    photosynthesis, 142
    transaminases, 91
Enzymes
    antioxidant defense system, 167, 168
    biomonitoring, 201, 203
    biotransformation, 167, 168
    fluoride effect, 154
    OPA anhydrolases, 180–190
    ozone, 147
    toxicants, effect, 90–92
EPA, 2
    avian toxicity tests, 57, 59–60
    ecological risk assessment framework, 271–313
Ephemerella sp, 52, 53, 76–77
Epinephrine, 92–93
Eukaryotes, 167, 168, 185
Excretion, atmospheric pollutants, 89
Exposure, 252, 285, 295, 297–298, 312

risk assessment analysis, 260–261
Exposure routes, 31, 85–87, 116, 292
Exposure time, 116
Extrapolation, 262, 301

F

Fasting, pollutant toxicity and, 124–125, 127
Fathead minnow, 52–53, 74
Fatty acids
  fluorinated, 190
  nitrogen dioxide reaction with, 143
  ozone reaction with, 147
Federal Insecticide, Fungicide, and Rodenticide
    Act (FIFRA) (1972), 3
Federal Pesticide Act (1978), 3
Federal Water Pollution Control Act (1972), 3
FETAX, 61–63
Fiddler crab, 52, 54
Field studies, 28, 34, 226, 266
FIFRA (1972), 3
FIFRA microcosm, 68–70
Fish, 10, 11
  cough response, 204–205
  effluent toxicity testing, 204–205
  index of biological integrity, 209
  mortality, 207
  OPA anhydrase, 186–187
  pesticide toxicity, dietary lipids, 127
  test species, 52–53, 72–75
  thermal pollution, 116
  tumors as indicator of ecological impact, 203
  ventilatory, rate, 204–205
Fishfly, 77
Flatworm, 78
Flavobacterium sp., OPA anhydrolase, 183, 188
Flounder, 52, 53
Flow-through test, 29
Fluorides, 95, 127, 152–156
Fluorine, 152
Fluoroacetate, 99, 102
Fluoroacetic acid, 99, 102, 103
Fluorocitrate, 92, 93
Fluorometabolites, 190
Fluorosis, 154–155, 161
Frateuria sp., toxicant biodegradation, 172
Free radicals, biotransformation, 167
Freshwater algae, 49, 50, 190
Freshwater fish, 52–54
Freshwater invertebrates, 52–54, 75–78
Frog embryo, teratogenesis assay, 61–63
Fundulus sp., 52, 53

Fungi, 33
  biodegradation, 174, 175
  halometabolites, 190

G

Gammarus sp., 52, 53, 75–76
GAPD, 102
Gaseous pollutants, 135–156
  carbon monoxide, 90, 148–151
  fluorides, 95, 127, 152–156
  nitrogen oxides, 118, 140–143
  ozone, 87, 90, 95, 116, 118, 122, 128, 143–148
  sulfur oxides, 135–140
Gasterosteus aculeatus, 52, 53
Gastropods, 52–53, 77–78
Gel electrophoresis, 11
Genetic markers, 11–12, 208
Genetics
  biodegradative organisms, 179–190
  resource competition model, 218
Genetic variation, toxicity, 122, 123
GENETOX, 107
Gentisic acid, microbial degradation, 173
Gills, 10, 31, 203
Gladiolus, 122
Glutathione, biotransformation, 167
Glyceraldehyde-3-phosphate dehydrogenase,
    145–146
Glycolysis, 102, 145–146
Goldfish, 52–53, 73–74
Graphical interpretation, 35
Grass shrimp, 52, 54
Gray partridge, 58
Green algae, 49, 52, 54, 78–79
  halometabolites, 190
Green crab, 52, 54
Green sunfish, 52–53, 74–75
Growth assay, 27
  algae, 48–51
Guaiacols, microbial degradation, 173
Guinea fowl, 58

H

Halometabolites, 190
Hazard, 252
Heat shock proteins, 201
Heavy metals, 91
Hemigrapsus sp., 52, 54
Hemodynamics
  metal shift, 94–95

tumors as indicator of ecological impact, 203
Hemoglobin, carbon monoxide and, 90, 150
Heptachlor, toxicity, effect of protein, 125
Herring, 52, 53
Hexagenia sp., 52, 53, 76–77
Histamine, 92, 93
Histidine, 93
Holoenzyme, 91
Homarus, OPA anhydrase, 186
Homoprotocatechuic acid, microbial degradation, 173
Humans
  cadmium, 90–92
  fluoride, 155–156
  nitrogen oxides, 142–143
  ozone, effect, 146–147
  sulfur dioxide, 139–140
  zinc, 130
Humidity, toxicity, 117
Hyalella azteca, 52, 53, 75–76
25-Hydroxy-D$_3$, 128
Hydroxylation, aromatic, 163
Hydroxyl radical, photochemistry, 136–138
Hypothesis testing, 39–40, 240

I

IBI, 209–210
IC$_{50}$, 21
Ictalurus punctatus, 52, 53, 74
Immunological suppression, 10
IND, 41
Index of biological integrity, 209–210
Indicator, 202–203, 206, 291
Indoor pollution, 143
Inhalation exposure, 31
Inhibitory concentration, 21
Insecticides, see Pesticides
Intermittent exposure, 116
Intermittent-flow test, 29
Intervals of nonsignificant difference, 41
Invertebrates
  acute toxicity tests, 51–55
  aquatic, 52–54, 75–78, 186
  chemical defense, 190
Iodoacetate, 99, 102
Iodoacetic acid, 99, 102, 103
Iron, use by body, 130
Iron disulfide, 135
Isopropylamine, toxicity in *Daphnia*, QSAR, 110–111
Isozymes, biotransformation, 164–165

J

Japanese quail, 57–59

K

Kestrel, 58
3-Ketoadipic acid pathway, microbial toxicant degradation, 169, 170
Killifish, test species, 52, 53

L

Lagodon rhomboides, 52, 53
LC$_{50}$, 19, 21, 312
  moving average method, 36
  QSAR, 108
LD$_{50}$, 19
  probit methods, 37
  QSAR, 107
Lead, 91, 127
  picolinic acid complexes, 130
  plants, 87
Leaf maturity, toxicity, 122
Legislation, 2–4
Leistomus xanthurus, 52, 53
Lepomis cyanellus, 52, 53, 74–78
Lepomis macrochirus, 52, 53, 74
Lernaeopodoida, 78
Lesions, 10, 203
Lethal synthesis, 92, 93
Lifestyle, pollutant toxicity, 123–124
Light intensity, toxicity, 117
Linear notation, QSAR, 110
Linear regression
  QSAR, 104, 105
  toxocity data, 39
Linoleic acid, 126, 127
Lipids, 126–127, 146
Lipophilicity, 88
Litchfield and Wilcoxin method, 36
Liver, 123, 126
  biotransformation, 162, 164, 166–167
  metal shift, 94–95
  organophosphates, 99
  pollutant metabolism, 89
Lobster, OPA anhydrase, 186
LOEC, 22, 38–39
LOEL, 22
Logit method, 36–37
Log P, QSAR models, 104

Longnose killifish, 52, 53
Lowest observed effects concentration, 22
Lowest observed effects level, 22
LT_{50}, 102

M

Magnesium, 91, 92
Magnesium fluorophosphate, 91
Malabsorption syndrome, 123
Malathion, toxicity, effect of protein, 125, 126
Mallard, 79
Malthion, biodegradation, 183
Mammals
  biomarkers, 202
  pesticide toxicity, dietary lipids, 127
  toxicity tests, 55–56, 60
Manganese, action, 92
Marker alleles, 11–12, 208
MATC, 23, 38–39
Mathematical modeling, 1, 118–121
Mayfly, 52–53, 76–77
Measurement endpoint, 258, 262, 290–291, 300
Menidia sp., 52, 53
Mercury, 91, 130
Mesocosm, 28, 40–42
Metabolism
  atmospheric pollutants, 89
  biotransformation, 9, 161–167
  xenobiotics, 9, 161–167
Metal chelation, 93–94
Metallothionein, 90, 201–202
Metal shift, 94–95
Methemoglobinemia, 142–143
MFO system, 123, 125–127, 164, 201
Microbial degradation, 167–175
  genetics, 179–180
  isolation and engineering, organisms, 177–178
  OPA anhydrolases, 180–190
Microbiology, 1
Microcosm, 28
  data analysis, 40–42
  FIFRA microcosm, 68–70
  mixed flask culture microcosm, 67–68
  soil core microcosm, 70–71
  standardized aquatic microcosm, 65–67, 227,
    233–237, 266
Microcystis aeruginosa, 79
Microorganisms, xenobiotic degradation, 167–
  175
Midge, 52–53, 77
Minerals, pollutant toxicity and, 130–131

Minimum allowable toxicant concentration, 23
Minimum threshold concentration, 23
Minnow, 52–53, 74
Mipafox, 180, 181, 185–190
Mixed flask culture microcosm, 67–68
Mixed-function oxygenase system, see MFO
  system
Mixtures, toxicity, 118–121
Modeling, 1, 261
  ecological risk characterization, 306
  population models, 220–227
  QSAR, 102–111
  resource competition model, 211–219
  toxicant effects on biological systems, 211–
    220
  toxicity of mixtures, 118–121
Molecular biology, 1
Molecular genetics, 1
Mollusks, 77–78
  halometabolites, 190
  mortality, 207
  OPA anhydrase, 186
MOLSTAC, 108
Molybdenum, metal shift, 95
Monitoring, 197–204
  biomarkers, 200–204
  risk assessment, 264, 268, 280, 286, 295–296
Monoamine oxidase, 102
Monohaloacetic acids, 99, 102, 103
Mortality, 11
  dose-response curve, 18–24
  mollusks, 207
  resource competition model, 218
Mosses, 33
Moving average method, 36
MTC, 23
Multiple regression, QSAR, 108
Multiple toxicity index, 121
Multispecies toxicity tests, 27–28, 33, 34, 63–71,
  266
  conceptual statistical analysis, 42, 231–232
  data analysis, 37–42, 228–244
  FIFRA microcosm, 68–70
  mixed flask culture, 67–68
  nonmetric clustering, 42, 231–237, 240, 243–
    244
  normalized ecosystem strain, 228–229
  soil core microcosm, 70–71
  standardized aquatic microcosm, 65–67, 227,
    233–237
  state space analysis, 230–231
Multivariate analysis, 41–42
  conceptual statistical analysis, 42, 231–232

nonmetric clustering, 42, 231–237, 240, 243–244

normalized ecosystem strain, 228–229

state space analysis, 230–231

toxicant impact on ecosystems, 228–244

Mummichog, 52, 53

Mussel, 206

Mycobacterium sp., toxicant biodegradation, 173, 174

Mysid, 52, 54

Mysidopsis, 52, 54

Mytilus edulis, 206

No observed adverse effects level, 22

No observed effects concentration, 22

No observed effects level, 22

NOR, 228

Normalized ecosystem strain, 228–229

Normal operating range, 228

Northern bobwhite, 57–59, 79–80

Norway rat, 32

NPDES tests, 3, 204

Nucleocidin, 190

Nutrition, pollutant toxicity and, 124–131

Nutritional toxicology, 124

N

NADH, 91

cytochrome P-450 system, 165–166

ozonization, 148

NADPH, 91, 167

cytochrome P-450 system, 165–166

ozonization, 148

Naphthalene, biodegradation, 174, 175

Narcosis, 9, 95–96

National Environmental Policy Act, 2

National Pollution Discharge Elimination System (NPDES) (1972), 3, 204

n-dimensional hypervolume, 227, 230

Necrosis, 10, 203

Nerve impulses, organophosphates, 97–101

NES, 228–229

Neurospora, toxicant biodegradation, 173

Neurotoxins, organophosphates, 96–101

Niacin, 91

Nitric oxide, 140

Nitrogen dioxide, 118, 140–142

Nitrogen oxides, 140–143

Nitrogen tetroxide, 140

Nitrogen trioxide, 140

Nitrosamine, vitamin E, 128

Nitrosation, vitamin C, 129

Nitrous oxide, 140

NOAEC, 22

NOAEL, 22

Nocardia sp., toxicant biodegradation, 172, 173

NOEC, 22, 38–39

NOEL, 22, 39–40, 312

Non-equilibrium dynamics, 237–243

Nonlinear systems, 222, 226, 227

Nonmetric clustering, 42, 231–237, 240, 243–244

No observed adverse effects concentration, 22

O

Obesity, pesticide toxicity, 127

Obligate thermophile organism, 183

Octopus, OPA anhydrase, 186

Oil, sulfur oxides, 135

Oligocottus maculosus, 52, 53

Oncogenesis, 10–11

Oncorhynchus gairdneri, 52, 53, 72–73

Oncorhynchus kisutch, 52, 53, 72–73

OPA anhydrase, 183–190

OPA anhydrolases, 180–190

opd gene, 179, 180

opd gene product, 182–183, 188

Orconectes sp., 52, 53, 76

Organization for Economic Cooperation and Development, 2

Organochlorine pesticides, 127

Organophosphates, 96–101

opd genes, 180

QSAR, 106

Oscillatoria sp., toxicant biodegradation, 173

Osteofluorosis, 156, 161

Otus asio, 58

Outdoor Aquatic Microcosm Tests to Support Pesticide Registrations, 68

Owl, 58

Oxidation, side chain, 163

Oxohemoglobin, 151

Oxygen depletion, thermal pollution, 116–117

Oyster, 52, 54

Ozone, 118, 143–148

human respiratory system, 90, 146–147

metal shift after exposure, 95

plants, 87, 90, 122, 145–146

rats, effect on, 116, 128

sources, 143

Ozonolysis, fatty acids, 147

P

Pachygrapsus sp., 52, 54
Pacifastacus ieniusculus, 52, 53, 76
Paecilomyces sp., toxicant biodegradation, 173
Palaemonetes sp., 52, 54
PAN, 117, 122, 145
Pandalus sp., 52, 54
Paralichthys sp., 52, 53
Paramecium aurelia, 168
Paramecium bursarea, 168
Paraoxon, 98
  biodegradation, 183, 184, 187
Parasitism, 203
Parathion, 98
  biodegradation, 184
  toxicity, effect of protein, 125, 126
Parophrys vetulus, 52, 53
Partridge, 58
PCB, toxicity, dietary lipids, 127
PCP, biodegradation, 173–175
Penaeus sp., 52, 54
Penicillium sp., toxicant biodegradation, 173
Pentachlorophenol, see PCP
Pentose phosphate pathway, 146
Perch, 52, 53
Peroxyacyl nitrate, 117, 122, 145
Pesticides, 3, 17
  biodegradation, 183, 184
  chlorinated hydrocarbons, 127
  FIFRA microcosm, 68–70
  legislation, 3–4
  microbial degradation, 169, 171–175
  organophosphates, 96–101, 106, 180
Pharmacokinetics, 1
Phasianus colchicus, 57–59, 80
Pheasant, 57–59, 80
Phenanthrene, microbial degradation, 173
Phenobarbital, toxicity, effect of protein, 125, 126
Phosphoglucomutase, 145
Phospholipids, 126
Phosphonolipids, 189
Phosphotriesterase, 183
Photochemical smog, 144–145
Photochemistry, hydroxyl radical, 136–138
Photosynthesis, inhibition by nitrogen oxides, 142
Phylogenetic extrapolation, 262, 301
Physa sp., 52, 53, 77–78
Physical stressors, 289, 296

Picolinic acid, 130
Pimephales promelas, 52, 53, 74
Pinfish, 52, 53
Planaria, 52–53, 78
Plants
  atmospheric pollutants, 85–87, 90, 122
  fluoride, 153–154
  halometabolites, 190
  humidity, effect on pollutant toxicity, 117
  indicators of ecological impact, 202–203, 206
  lead, 87
  light intensity, effect on pollutant toxicity, 117
  nitrogen oxides, 141–142
  ozone, 145–146
  sulfur dioxide, 137, 139
  thermal pollution, 116–117
Platichthys stellatus, 52, 54
Platyhelminthes, 52–53, 78
Pollutants, see Xenobiotics
Polychaete, 52, 54
Pontooporeeia hoyi, 75
Population crash, 11
Population models, 220–227
Population parameters, 11, 199, 207–208
Potentiation, toxicants, 117–118
Predictability, ecosystem data, 237–243
Pregnancy
  carbon monoxide effect, 150
  response to toxins during, 123
Probit method, calculation, 35–37
Procambarus sp., 52, 53, 76
Prokaryotes, 167–169
  toxicant degradation, 171, 179–180
Proportional diluter, 29–31
Proteins
  carbon monoxide, effect, 150–151
  fluorides, effect, 156
  heat shock proteins, 201
  ozone, effect, 147
  pollutant toxicity and, 125
  stress proteins, 201
Protists, 168, 169
Protocatechuic acid, microbial degradation, 173
Pseudomonas sp.
  OPA anhydrolase, 182, 188
  toxicant biodegradation, 172–175
Pteronarcys sp., 52, 53, 76
Public policy, 3–4
Pyrocatechase, toxicant biodegradation, 172

Q

QSAR, 8, 102–111, 266
  models, 104–109
  software, 107–108, 110
Quail, 57–59
Quantitative structure activity relationships, see QSAR
Quotient method, 265

R

Rainbow trout, 52–53, 73
Rana sp., 52, 53
Rat
  metal shift, 94–95
  nutritional modulation, 125
  ozone, effect on, 116, 128
  toxicity tests, 55–56
Rat poison, 92
R. cuneata, OPA anhydrase, 182, 187, 188
Receptor, 9
Recirculating test, 29
Recombination, degradative gene, 180
Recovery, 237–238, 263, 310, 312
Reference toxicant, 47
Regression analysis
  QSAR, 108
  toxocity data, 39, 104
Remediation, see Bioremediation
Reproduction, resource competition model, 218
Reproductive studies, 27, 58
Reproductive success, 11, 206
Resource competition model, 211
Response extrapolation, 262, 301
Rhodococcus sp., toxicant biodegradation, 172
Rhodopseudomonas palustris, toxicant biodegradation, 172
Rhodotorula glutinis, toxicant biodegradation, 173
RIFFLE, 244
Ring dove, 58
Ring-necked pheasant, 57–59, 80
Risk, 252
Risk assessment, 1–2, 251, 268
  EPA framework, 271–313
Risk characterization, 264–268, 280, 285, 304–311, 313

Risk description, 308–310

S

Salmon, 52–53, 72–73
Saltwater algae, 49, 50
Saltwater fish, 52–54
Saltwater invertebrates, 52–54, 78
Salvelinus fontinalis, 52, 53, 73
SAM, see Standardized aquatic microcosm
Sanddab, 52, 53
Sand shrimp, 52, 54
SAS-PROBIT, 36, 37
Scavenger reactions, 87
SCM, 70–71
Scopulariopsis sp., toxicant biodegradation, 173
Screech owl, 58
Scuds, see Amphipods
Sculpin, 52, 53
Sea-hare, OPA anhydrase, 186
Selenium, 95, 118, 130
Sepia, OPA anhydrase, 186
Sex variation, pollutant toxicity, 124
Shadfly, 77
Sheepshead minnow, 52, 53
Shiner perch, 52, 53
Shore crab, 52, 54
Shrimp, 52, 54
Silverside, 52, 53
Simulation models, ecological risk characterization, 306
Single species tests, 46–63
  algae, growth toxicity, 48–51
  Daphnia, acute toxicity, 46–48
  resource competion model, 213
  vertebrates, aquatic, 51–55
Skeletal fluorosis, 154–155, 161
Small cosmos toxicity tests, 27–28
Smelting, 135
SMILES, 110
Smog, 144–145
Smoking, 123–124, 151
Snail, 52–53, 77–78
Society for Environmental Toxicology and Chemistry, 2
Sodium fluoroacetate, 92
Sodium pentachlorophenate, 47, 173
Software, see Computer software
Soil core microcosm, 70–71
Sole, 52, 53

Soman, 181–190
Spearmen-Karber method, 36
Species diversity, 12, 122, 208–209
Spinner, 77
Spisula, OPA anhydrase, 186
SPSS-PROBIT, 36, 37
Squid-type OPA anhydrase, 181, 182, 185, 187
Standardized aquatic microcosm (SAM), 65–67, 227, 233–237, 266
Star Culture Collection, 32
Starry flounder, 52, 53
Starvation, pollutant toxicity and, 124–125, 127
State space analysis, 230–231
Static-renewal test, 29
Static test, 29
Statistical analysis, 34–42
    ANOVA, 37–39, 41, 227–228
    conceptual, 42, 231–232
    hypothesis testing, 39–40, 240
    intervals of nonsignificant difference, 41
    logit methods, 36–37
    multivariate techniques, 41–42, 228–244
    nonmetric clustering, 42, 231–237, 240, 243–244
    normal ecosystem strain, 228–229
    probit methods, 35–37
    QSAR model development, 108–109
    state space analysis, 230–231
Stickleback, 52, 53
Stomata, atmospheric pollutants, 31, 85–87, 90
Stonefly, 52–54, 76
Storage, atmospheric pollutants, 88–89
Streptomyces calvus, 190
Streptomyces sp., toxicant biodegradation, 172
Streptopelia risoria, 58
Stressor, 252, 254–256, 279, 289, 295–296, 313
Stressor-response profile, 263–264, 302–304, 313
Stress proteins, 201
Structure activity model, see QSAR
Sublethal toxicity tests, 27
Substituted benzenes, biodegradaion, 171–174
Sulfoxide formation, 163
Sulfur dioxide, 87, 136–139
Sulfuric acid mist, human respiratory system, 90
Sulfur oxides, 135–140
Sunfish, 52–53, 74–75, 186, 204
Superfund legislation, 4
Superoxide dismutase, 167, 168
Surf clam, OPA anhydrase, 186
Synergism, toxicants, 117–118

T

Tabun, 181–190
TCA cycle, 102, 169
Temperature, toxicity, effect on, 116–117
Teratogenesis assay, FETAX, 61–63
Test animals
    aquatic, 51–53
    care and treatment, 58, 60–61
    choice of, 32–34
    exposure scenario, 31
Text organisms, 32–34
    algae, 49
    sentinel organisms, 206–207
Tetrachlorohydroquinone, biodegradation, 173, 174
2,3,4,6-Tetrachlorophenol, microbial degradation, 173
Tetrahymena thermophila, OPA anhydrolases, 182, 185, 187, 188
Thermal pollution, 116
Thiazolidone, 94
Threespine stickleback, 52, 53
Tidepool sculpin, 52, 53
Tiered testing, 3–4
TIE/TRE program, 205
Toad, 52
α-Tocopherol, 128
Tolerance, 146
Toluene, biodegradation, 171, 173, 174
TOPKAT, 110
Toxaphene, toxicity, effect of protein, 125, 126
Toxicant, see Xenobiotics
Toxicity, definition, 17
Toxicity identification evaluation, 205
Toxicity reduction evaluation, 205
Toxicity testing, 3–4, 17–42, 45–80; see also Test animals; Test organisms
    design parameters, 29–31
    dose-response curve, 18–24
    ecological systems, 204–206
    effluents, 204–205
    exposure scenarios, 31, 85–87, 116, 292
    integration with ecological data, 265–266
    mesocosm, 28
    microcosm, 27–28
    multispecies tests, 63–71
    nonmetric clustering, 42
    single species tests, 46–63
    standard methods, 24–27

statistics, 34–42
tiered testing, 3–4
TIE/TRE programs, 205
Toxic Substance Control Act (TSCA) (1976), 4
Toxic unit model, 119
Transaminases, inhibition, 91
Translocation, 89
Transport, atmospheric pollutants, 85
2,3,6-Trichlorobenzoic acid, microbial degradation, 172
3,4,5-Trichloroguaiacol, microbial degradaion, 173
2,4,6-Trichlorophenol, microbial degradation, 173
2,4,5-Trichlorophenoxyacetic acid, microbial degradation, 173
Trichoderma viride, toxicant biodegradation, 173
Trichosporon cutaneum, toxicant biodegradation, 173
Troposphere, nitrogen oxides, 140–141
Trout, 52–53, 73
t-test, 38
Tumors, 10–11, 203
Turbellaria, 52–53, 78

U

Ubiquiten, 201
Uca sp., 52, 54
UG-PROBIT, 37
Ulothrix sp., 79
Ultrastructure, 87
Uncertainty analysis, ecological risk characterization, 306–308
U.S. Environmental Protection Agency, see EPA
Uptake
    atmospheric pollutants, 86–88
    resource competition model, 211

V

Validation, QSAR, 109, 110, 197
Vanadium, metal shift resulting from, 94
Variance, ANOVA, 37–39, 41
Vertebrates
    aquatic, 51–55, 72–75
    atmospheric pollutants, 87–89
    terrestrial, 55–61
Vitamin A, 127–128
Vitamin C, 128–130, 155, 167
Vitamin D, 128
Vitamin E, 128, 130, 147, 167

Vitamin K, 91

W

Washington State, 3
Water flea, see *Daphnia magna; Ceriodaphnia dubis*
Water Pollution Control Act (1972), 3
Water quality, laboratory toxicity tests, 46–47, 51, 54
Water treatment, bioreactor, 176
Weight-of-evidence, 308–309
Whole-body exposure, 31
Willowfly, 77
Within-between raio, 243

X

Xenobiotics, 7–8, 115–131, 313
    action, mechanism, 89–102
    antagonism, 118, 130
    bioaccumulation, 9, 200
    biodegradation, 9, 167–190, 200
    biomarkers, 200–204
    biomonitoring, 197–204
    biotransformation, 9, 161–167, 200
    detoxification, 125–127, 166–167, 180–190, 203
    dose-response curve, 17–24
    ecological risk assessment, 251–313
    environmental factors, 116–117
    exposure mode, 31, 85–87, 116
    exposure time, 116
    interaction, 117–118
    microbial degradation, 167–190
    mixtures, 118–122
    potentiation, 117–118
    QSAR, 8, 102–111
    synergism, 117–118
    toxicity testing, 17–42
Xenopus, teratogenesis assay, 61–63

Z

Zero net growth isocline, 212, 217
Zinc
    absorption by body, 130
    action, 92
    as antagonist to cadmium toxicity, 118, 130
    metal shift, 95
Zinc sulfide, 135–136
ZNGI, 212–214, 217